The Opacity of Spiral Disks

NATO ASI Series

Advanced Science Institutes Series

*A Series presenting the results of activities sponsored by the NATO Science Committe
which aims at the dissemination of advanced scientific and technological knowledge,
with a view to strengthening links between scientific communities.*

The Series is published by an international board of publishers in conjunction with the NATO
Scientific Affairs Division

A	Life Sciences	Plenum Publishing Corporation
B	Physics	London and New York
C	Mathematical and Physical Sciences	Kluwer Academic Publishers
D	Behavioural and Social Sciences	Dordrecht, Boston and London
E	Applied Sciences	
F	Computer and Systems Sciences	Springer-Verlag
G	Ecological Sciences	Berlin, Heidelberg, New York, London,
H	Cell Biology	Paris and Tokyo
I	Global Environmental Change	

PARTNERSHIP SUB-SERIES

1.	Disarmament Technologies	Kluwer Academic Publishers
2.	Environment	Springer-Verlag / Kluwer Academic Publishers
3.	High Technology	Kluwer Academic Publishers
4.	Science and Technology Policy	Kluwer Academic Publishers
5.	Computer Networking	Kluwer Academic Publishers

*The Partnership Sub-Series incorporates activities undertaken in collaboration with NATO's
Cooperation Partners, the countries of the CIS and Central and Eastern Europe, in Priority Areas
concern to those countries.*

NATO-PCO-DATA BASE

The electronic index to the NATO ASI Series provides full bibliographical references (with keywor
and/or abstracts) to more than 50000 contributions from international scientists published in all
sections of the NATO ASI Series.
Access to the NATO-PCO-DATA BASE is possible in two ways:

– via online FILE 128 (NATO-PCO-DATA BASE) hosted by ESRIN,
Via Galileo Galilei, I-00044 Frascati, Italy.

– via CD-ROM "NATO-PCO-DATA BASE" with user-friendly retrieval software in English, Frencl
and German (© WTV GmbH and DATAWARE Technologies Inc. 1989).

The CD-ROM can be ordered through any member of the Board of Publishers or through NATO-
PCO, Overijse, Belgium.

The Opacity of Spiral Disks

edited by

Jonathan I. Davies

Department of Physics and Astronomy,
UWCC, Cardiff, U.K.

and

David Burstein

Department of Physics and Astronomy,
Arizona State University,
Tempe, Arizona, U.S.A.

Springer Science+Business Media, B.V.

Proceedings of the NATO Advanced Research Workshop on
The Opacity of Spiral Disks
Cardiff, Wales
July 25-29, 1994

A C.I.P. Catalogue record for this book is available from the Library of Congress

ISBN 978-94-010-4171-3   ISBN 978-94-011-0381-7 (eBook)
DOI 10.1007/978-94-011-0381-7

Printed on acid-free paper

TABLE OF CONTENTS

Participants of the NATO ARW
The Opacity of Spiral Disks
Cardiff 25-29 July 1994

1. J. Silk
2. H. Jones
3. R. White
4. E. Huizinga
5. E. Valentijn
6. W. van Driel
7. A. Broeils
8. H. Rix
9. R. Giovanelli
10. N. Devereux
11. L. Bottinelli

12. F. Prada
13. R. de Grys
14. N. Neininger
15. B. Wang
16. D. Block
17. L. Gouguenheim
18. M. Greenberg
19. G. de Vaucouleurs
20. M. Trewhella
21. P. James
22. W. Keel

23. S. Jörsäter
24. M. Roberts
25. Rh. Evans
26. B. Cunow
27. S. Phillipps
28. M. Näslund
29. D. Calzetti
30. A. Boselli
31. S. Courteau
32. A. Witt
33. S. Ryder

34. J. Knapen
35. B. Madore
36. Y. Byun
37. P. Grosbøl
38. H. Thronson
39. C. McKeith
40. R. Arendt
41. D. Burstein
42. A. Bosma
43. M. Disney
44. R. Braun

45. ?
46. R. Peletier
47. G. Bruzual
48. N. Kylafis
49. D. Zaritsky
50. K. Freeman
51. J. Davies

The start

FOREWORD

We are well aware of how dust influences our observations of distant stars and how easily dust may mislead us with regard to the way in which stars are distributed within the Galaxy, but how does dust affect our view of other galaxies? This is the question that was posed to those who attended this meeting.

By its very nature dust is illusive: as dust obscures by both scattering and absorption, it can effectively disguise its very own existence. It was not until the mid-1930's that astronomers generally agreed that dust *did* redden and dim stars in our own Galaxy, and it was not until the late 1950's that astronomers began to seriously inquire of its effects in other galaxies

To the best of our knowledge, this is the first international meeting to have been held devoted solely towards understanding the observational effects of dust in other galaxies. Because of this we have been fortunate in attracting many of the major workers in this field, both observers and theorists. Among these pages the reader will find a wide range of opinion about how much dust there is in the disks of galaxies, where that dust is, and how to model the effects of dust.

We tried to structure this meeting so that there was a ready and easy exchange between the speaker and the audience, and so that there was a large amount of time for discussion. We also tried our best to keep track of the multitude of conversations (sometimes failing to keep up with the volume of words, as often three or more people were trying to talk at once!). Unfortunately, we were not always successful in capturing the comments for each paper. Yet, we hope we were able to preserve the flavour of those lively interactions in the comments that we were able to preserve for posterity.

Perhaps what impressed us most of all is that the meeting came away with a consensus that there is still much to do if we are to decide on how dust affects our view of galactic disks. It is a meeting at which the participants realised as a group that simple models of dust effects are, well, simple, and that the true effects of dust in galaxies are, well, complicated. However, we hope the participants of the meeting and you, the reader of this volume, come away with much more than that. The studies detailed in these pages show that we can observationally study the effects of dust in galaxies in several different ways, all of them complimentary to each other. Many of the contributions point to future observations that will help

1

J. I. Davies and D. Burstein (eds.), The Opacity of Spiral Disks, 1–2.
© 1995 *Kluwer Academic Publishers.*

us nail down this problem once and for all (we hope !). If we were to hold a meeting in Cardiff in five years time (and Cardiff is clearly a commodious place to hold a meeting !), what will we have learned ?

Finally it is with sadness that we note the lack of a written contribution from Ken Freeman. He made such a valuable contribution to the workshop, but illness meant he was unable to put in writing the things he so eloquently put in words. Ken, for all the participants of this meeting, we hope the receipt of these proceedings finds you as active as always.

To all the participants thank you, we enjoyed it.

Jon Davies and Dave Burstein

The overhead projector

INTRODUCTORY REMARKS

MIKE. DISNEY
Department of Physics and Astronomy, University of Wales College of Cardiff, PO Box 913, Cardiff CF2 3YB, UK or m.disney@astro.cf.ac.uk

1.   Introduction

The question of how much starlight, and therefore how much baryonic matter, is obscured from our sight by intervening smoke (or 'dust' as it is more often but incorrectly called - seeing that it originates from condensation and not disintegration) is clearly an important one which impinges on many areas of astrophysics.   Until recently it was generally thought that Holmberg [1957] had demonstrated that spiral galaxies, probably the main receptacles of both stars and smoke in the cosmos, are largely transparent, and that any invisible baryonic component must be rather insignificant.

Modern technical developments, in particular the widespread use of computers, allowed for more sophisticated analyses of the data.   Not only has Holmberg's conclusion been thrown into doubt [Disney et al 1989] but some authors have gone to the opposite extreme and claimed that optical observations access only the upper crusts of spirals and that most of the stars, and maybe most of the mass, lies buried underneath.

So widespread are the implications of this question that authors have often been inclined to adopt extreme points of view, motivated perhaps by their interests and opinions in other areas.   For some a significant obscuration would be a considerable nuisance, threatening to hold up progress in their pet field of research for years to come.   Others, on the contrary, may be too ready to welcome extra obscuration as a way out of some otherwise uncomfortable impasse.   As far as I understood the question there are good arguments on both sides, but none, taken on its own, should be regarded as conclusive.   Even in physics it has been proven unwise to throw out a conjecture on the basis of a single experimental refutation.   In astrophysics it would be downright foolish.   Observations can rarely be disentangled from selection effects on the one hand or from interpretations on the other where the interpretations usually contain embedded working hypotheses which should never pass unquestioned.   In this workshop you will hear of observations which apparently contradict both each other and your own pet interpretation of the subject.   Do not assume that they are necessarily wrong because, on deeper investigation, the contradictions may turn out to be more apparent than real.   One needs to approach this subject more like a juror than a logician.

5

J. I. Davies and D. Burstein (eds.), The Opacity of Spiral Disks, 5–17.
© 1995 *Kluwer Academic Publishers.*

The title of our workshop, "The opacity of spiral discs" is actually rather vague and will have different undertones for different people. Some will be more interested in whether you can see through a spiral galaxy to the cosmos on the far side, while others will be more concerned with the amount of starlight missing from the galaxy itself. We need to be careful both in defining our terms and in formulating the separate questions we are individually trying to answer.

The remainder of this talk is divided up under the following headings: (2) Why does spiral opacity matter in the wider context of astrophysics? (3) An uncritical introduction to the wide variety of arguments on both sides; (4) Some of the main pitfalls in interpretation; (5) Future observational opportunities; (6) An attempt to formulate some of the key questions.

2. **Why Does The Subject Matter To Astrophysics In General?**

(a) Correlations like the HR diagram proved to be the Rosetta Stone which unlocked the secrets of stellar astrophysics. We've had very little luck so far in disentangling the corresponding correlations between the global properties of galaxies [e.g. Roberts 1975]. There are intriguing hints here and there, such as the Fisher-Tully correlation, but these may be little more than observational selection. Could it be that the correlations among spirals, particularly among the large ones which contain most of the light and mass of the galaxy population, are disguised because we don't know how to correct their global properties for internal obscuration?

(b) Our sight of the background universe could well be affected by obscuration in foreground spirals. To what extent is foreground smoke responsible for our failure to find primeval galaxies and for the apparently steep fall in the number of QSOs beyond a redshift of 3? [e.g. Hartwick and Schade 1990]. The number of galaxies intercepted by a typical sightline to $\chi \equiv 1+z$ is:

$$f(\chi) \cong n_G A_G \left(\frac{c}{H_0}\right) \frac{1}{\sqrt{\Omega_0}} \left\{\frac{2}{3}\left(\chi^{3/2} - 1\right)\right\} \tag{1}$$

where n_G and A_G are the number-density and average cross-section of the absorbers.. Various authors [Ostriker and Heisler 1984, Wright 1990, Fall and Pei 1993] have used different combinations of n_G, A_G, extinction-curve and dust-evolution to argue that $f(\chi) \to 1$ for z's between 2 and 10 - see Fig 1.

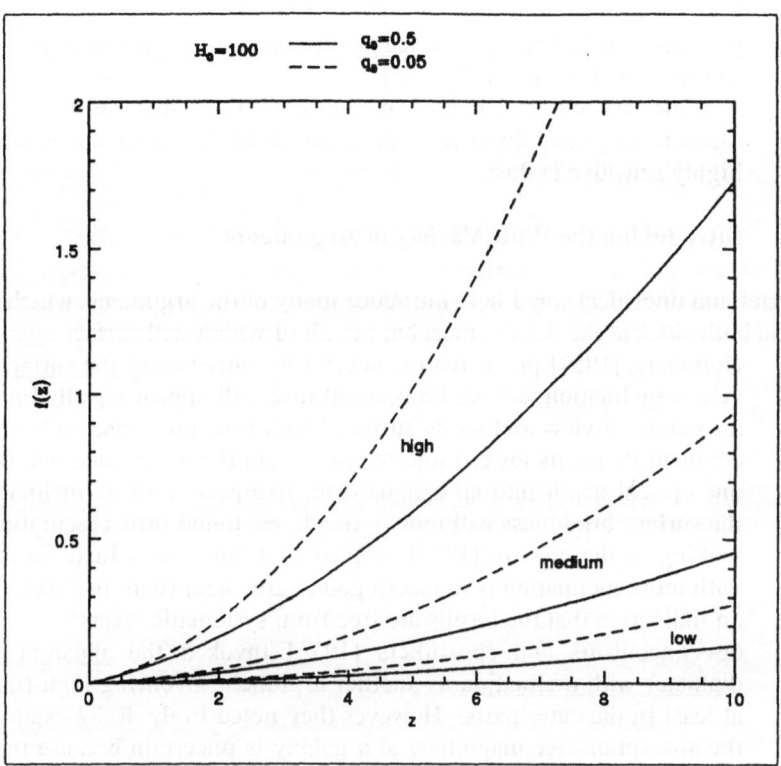

Fig 1: *f(x) for different n_G's and A_G's; $n_G = .04, .02, .01 Mpc^{-3}$; with A_G's of 4, 2 and 1 scalengths.*

(c) The dynamically inferred <u>hidden mass in spiral galaxies</u> may be, to some extent though not entirely, due to shrouded stars [e.g. Davies, J. I., 1991]. In particular the decomposition into bulge and exponential components may be severely affected because decomposition procedures tend to steal light from the spheroid to compensate for light in the central disc that has been lost to dust.

(d) What is <u>the main source of the large Far Infra-Red (FIR) outputs</u> found by IRAS in some disc galaxies? Depending on the precise definition of the ratio L_{FIR}/L_{OPT} (which can differ widely between authors) the FIR output can vary between 10^{-2} and 10^2 times the optical. Shrouded AGN's and 'hidden starbursts' may occasionally be responsible but there is a good chance that much of the FIR is simply starlight reprocessed on large quantities of dust-grains. [Disney et al 1989]

(e) Arguments about <u>the dynamical evolution of galaxies</u> could be wrong-footed by dust. For instance Dressler [1980] has argued that the bulges in late-type spirals are too underluminous for them to be the precursors of SO's. If significant bulge-light is obscured, this may not be true.

(f)   The K-corrections needed to infer the luminosity and spectral evolution of galaxies will be heavily affected by the amount, distribution and evolution of smoke. [e.g. Bruzual et al, 1988].

(g)   Various cosmological measurements, e.g. the Fisher-Tully correlation as a measure of galaxy distance, depend on inclination corrections which may be highly sensitive to dust.

3.   **Introducing the Wide Variety of Arguments**

In a brief and uncritical way I here introduce many of the arguments which have been used on both sides of the debate, most but not all of which will surface again later.

(a)   Holmberg [1957] pioneered the subject by introducing the surface-brightness versus inclination test. A transparent disc will appear equally luminous from all points of view and so its surface-brightness must rise with inclination to compensate for its lower projected area. On the other hand one will see only one optical depth into an opaque disc, irrespective of its inclination, and so the surface brightness will remain fixed. He found little obscuration and later, looking at the colours [1975], argued that "the small increase of reddening with more inclination is in fact in good agreement (with the SB/Inclin. test) ... an indication that the results are free from systematic errors".

(b)   de-Vaucouleurs and co-workers [1976] invoked the apparent increase of diameter with inclination as another argument favouring high transparency - at least in the outer parts. However they noted in the RC-2 catalogue "But ... the absorption-free magnitude of a galaxy is uncertain because the correction varies greatly with the (unknown) details of the dust distribution". However others went ahead (e.g. the R.S.A. catalogue) ignoring such caveats.

(c)   In a prescient paper Jura [1980] showed that the constancy of surface-brightness among discs, discovered by both Holmberg [1957] and Freeman [1970] could be explained by increasing their apparent opacity.

(d)   The advent of IRAS led to the discovery that some spirals at least have FIR outputs as large, and in some cases much larger than their optical outputs [de Jong et al 1984, Soifer et al 1984]. Since galaxies were 'known' to be optically thin alternative mechanisms to general absorption had to be sought to manufacture FIR: e.g. highly obscured starbursts.

(e)   Bruzual et al [1988] modelled the radiative transfer in dusty discs, including scattering, in order to study K-corrections. They "found that the I(r) profiles in discs which are optically thin should change with inclination. Observationally they do not, therefore $1 \leq \tau_r \leq 2$".

(f)   Disney et al [1989], demonstrated that Holmberg's conclusions were based on the wholly unrealistic supposition that all of the smoke lies as a screen in front of all of the stars. With more realistic configurations [for example with the dust sandwiched as a thinner layer within the stellar disc] Holmberg's own observations, both of colour and surface-brightness, could be used to argue for almost infinite optical depth - see Fig 2. after reviewing all the observations,

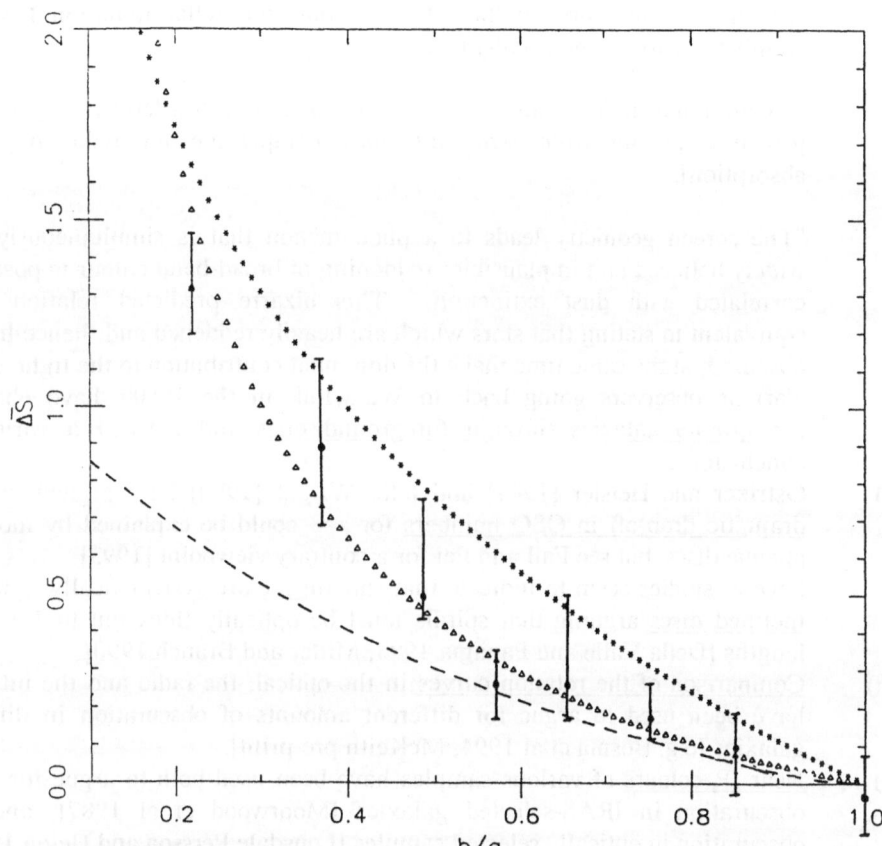

Fig 2: *Holmberg's data on surface-brightness v. inclination. Diamonds show his optically thin screen model. Stars show a slab of infinite optical depth, dashes an optically thick sandwich. Interpretation is obviously ambiguous.*

including those from IRAS they concluded that on the balance of the current evidence spirals were probably opaque.

(g)   Valentijn [1990] set the controversy alight by re-applying Holmberg's test to thousands of spirals in the ESO catalogue, but reaching the entirely opposite conclusion, namely that galaxies are opaque out to their visible extremities. He remarked that optical astronomers may be seeing "only the upper crusts of spiral galaxies".

(h)   Saunders et al [1990] compared the FIR luminosity function of galaxies, as judged from IRAS, with the better known optical luminosity function finding that $\langle L_{FIR} \rangle = 5.6 \pm 0.6 \times 10^7 h \; L_0 / M_{pc}^3$ which is "down by a factor of at least 2, probably 4, and possibly 6 compared to $\langle L_{OPT} \rangle$. Therefore galaxies are optically thin $\left(\tau_B \sim 0.25 \pm 0.1 \right)$".

(i)   Witt et al [1992] modelled radiative transfer including scattering, in a range of spherical models and reached some quotably trenchant conclusions:

"Much previous work on the effects of dust on stellar radiation has been simplistic, wrong, and usually both."

"...common techniques for estimating the amount of dust fail by large factors" [because blueing from scattering may compensate for reddening from absorption].

"The screen geometry leads to a phenomenon that is simultaneously both widely believed and implausible: reddening of broad-band colour is positively correlated with dust extinction.   This bizarre predicted relation ... is equivalent to stating that stars which are heavily reddened and, hence heavily obscured, at the same time make the dominant contribution to the light..."

(j)   Various observers going back to Wesselink in the 1950's have observed <u>background galaxies through foreground ones</u> and reached a variety of conclusions.

(k)   Ostriker and Heisler [1984] and later Wright [1990] have argued that <u>the dramatic drop-off in QSO numbers</u> for $z>3$ could be explained by modestly opaque discs, but see Fall and Pei for a contrary viewpoint [1993].

(l)   Several studies seem to indicate that <u>supernovae are systematically fainter in inclined discs</u> arguing that spirals must be optically thick out to 1.7 scale-lengths [Della Valle and Panagia 1993, Miller and Branch 1990].

(m)   <u>Comparison of the rotation curves</u> in the optical, the radio and the infra-red have been used to argue for different amounts of obscuration in different galaxies [e.g. Bosma et al 1994, McKeith pre-print].

(n)   <u>Near IR colours</u> of various samples have been used both to argue for heavy obscuration in IRAS-selected galaxies [Moorwood et al 1987], and low obscuration in optically selected samples [Lonsdale Persson and Helou 1987].

(o)   <u>Sub-millimetre observations</u> should show up the cool dust {<30°K} and, depending on aperture corrections, have led to the inference of very little [Eales et al 1989] or rather more [Cox and Metzger 1987] extinction.

4.   Pitfalls in Interpretation

It is manifest that all of the above contradictory arguments cannot be right. What can be going wrong? Here we introduce some of the more obvious pitfalls and tangles.

(a)   <u>Historical tangles can arise</u> when an earlier incorrect assumption can spawn a secondary inference which can then take on an independent life of its own. Thus the early assumption of low opacity combined with the high FIR outputs occasionally found by IRAS led to the idea of widespread and optically hidden [in totally shrouded GMCs] starbursts in galaxies.   Once spawned the starburst phenomenon can then be used to discount large amounts of generalised absorption as the explanation of a large FIR output in any galaxy.

(b)   <u>Arguments in the radiative transfer game tend to be more circular than most.</u> For instance if, in a particular galaxy, you assume low absorption, then you correct the photometry with a small A_B. Therefore you conclude that the galaxy is not very luminous. Thus it probably doesn't contain much ISM (dust) - thus confirming your original assumption.

(c)   There are <u>large and uncertain Bolometric corrections</u>, particularly so since much of the dust may be at temperatures almost inaccessible from the ground. If one takes proper account of the band-width variation with frequency then the peak in the energy curve of a black body will occur at [Disney and Sparks 1982]:

$$\lambda_{max}(T) \text{ in } \mu \approx \frac{3600}{T(^\circ K)}$$

Thus for dust at $T_d = 40^\circ K$, $\lambda_{max} \sim 90\mu(\text{IRAS})$; whereas if $T_d = 20^\circ K$, $\lambda_{max} \sim 180_\mu(\text{KAO})$ and if $T_d \sim 12^\circ K$, $\lambda_{max} \sim 300\mu$ a wavelength almost inaccessible from the ground, and, apart from COBE observations of our galaxy, still lacking from our repertoire in space. Apart from a handful of galaxies one is forced to interpolate between 100μ and 1000 μ to find the major source of dust emission [see Fig 3]. It is not surprising that estimates of bolometric corrections can vary between different authors by as much as a factor 10. For instance Trewhella et al in a poster-paper show that the Saunders et al [1990] IRAS L.F. data can be easily re-interpreted to yield a $1 < \tau_B < 4$ instead of the $\tau_B \sim 0.26 \pm 0.1$ quoted originally.

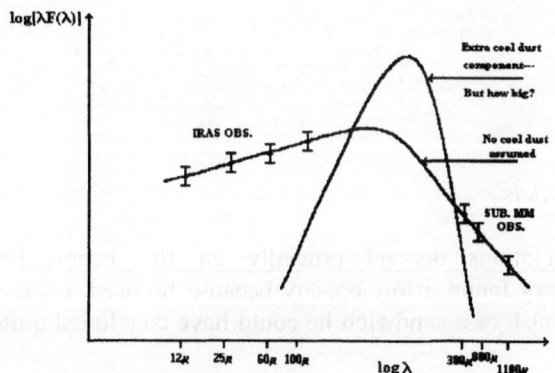

Fig 3:   *How much cold dust radiation lies between IRAS and mm?*

Screen model.

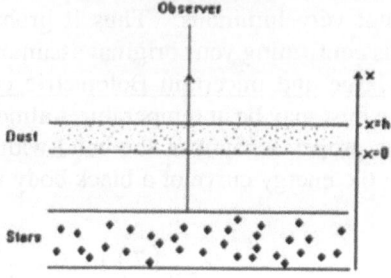

Slab model.

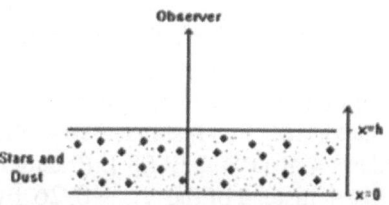

Sandwich model.

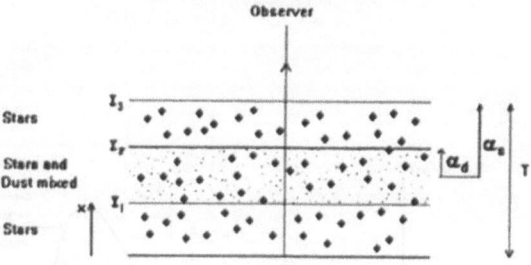

Fig 4:   Toy models.

(d)   <u>Interpretations depend crucially on the geometrical models assumed</u>. Holmberg found a low opacity because he used a screen geometry. Had he used a slab or a sandwich he could have concluded quite differently - see Fig 4.

(e)   As Huizinga [1994] has recently emphasised the <u>methods of measurement</u> may differ from author to author, leading to systematic effects. He illustrates this graphically by comparing diameters measured by eye, by isophote (machine) and by machine algorithms.

(f)   Then <u>there is scattering</u> which, until the advent of powerful computers has been generally hard to treat. Do you leave it in your interpretations or out?

Depending on the observations you intend to analyse the answer may be 'yes' or 'no'. As I understand it scattering is less important at the highest and lowest optical depths, most important when $\tau \sim 1$.

(g)   There is the <u>largely unknown clumpiness</u> of the smoke. If it is highly clumped into GMC's with a low covering factor then there will be little general extinction or background absorption. Different galaxies appear to have different amounts of clumpiness [see Frontpiece] which complicates the issue.

(h)   <u>What you see depends crucially on where you look</u>. That may seem obvious but is not always acknowledged. Analyses based on optical diameters may well lead to low τ's because the outer isophotes are indeed optically thin. At the same time investigations of the FIR output are probing chiefly the inner regions which may well be obscured. Thus two different tests applied to the same sample can reach opposite conclusions. And don't forget that the more luminous a galaxy is, the more ISM it is likely to contain and therefore the more obscuration it will probably suffer from.

(i)   This brings us to the huge subject of the <u>selection effects</u> which may be clouding our judgement. Here are a few in this area, ranging from the obvious to the more subtle:

     (i)   We live in a local hole in the ISM enabling us to see out of our galaxy better than most.

     (ii)   When we see one galaxy through another, as we occasionally do, we necessarily find foreground galaxies with low absorption. This is likewise true for QSO's and supernovae.

     (iii)   How do we choose a sample of galaxies for extinction studies? If we choose a sample either explicitly or implicitly based on an optical magnitude limit then we will tend to pick a class with lower extinction. On the other hand if we choose on the basis of IRAS flux then we will be selecting out galaxies of higher extinction. So how should we choose?

     (iv)   Finally we come to the pernicious business of surface-brightness selection. It has long been known that most galaxies have a surface-brightness close to the sky and close to the optimum required to see them as galaxies. It has been speculated that this is a selection effect [Disney 1976, Disney and Phillipps 1983] and it can be shown that this selection effect operates independently of inclination [Davies 1990]. The "Visibility" of a galaxy, i.e. the volume within which it can be detected in some survey based on limiting isophotal size and/or isophotal magnitude, can be an extremely sharply peaked function of surface brightness. Therefore we should not be surprised to find a large proportion of the galaxies in a catalogue with this same optimal surface-brightness, and to find that their surface-brightnesses are independent of inclination. But this independence is precisely the Holmberg-Valentijn test for high opacity!

14

I cannot resist showing just how powerful this effect can be. Fig (5) [from Davies et al 1994] shows the entire sample of late-tape spirals in the ESO catalogue whose velocities are known. Because we know the velocities one can compute the median velocity (and hence presumably distance) of the galaxies in each surface-brightness bin. Cubing that distance provides an estimate of the 'Visibility' of the galaxies in each surface-brightness class. There is no 'theory' in all this, and yet note how the numbers fall away dramatically on either side of the peak in perfect sympathy with the volumes searched. It is hard to explain this as anything but a selection-effect. The simplest interpretation of the diagram is that, at a given luminosity, there are in truth equal numbers of galaxies in each magnitude bin of surface-brightness but we pick up only these few with optimal surface-brightness. If this is correct, and it is difficult to see how it could be wrong, then beware the use of surface-brightnesses to infer obscuration.

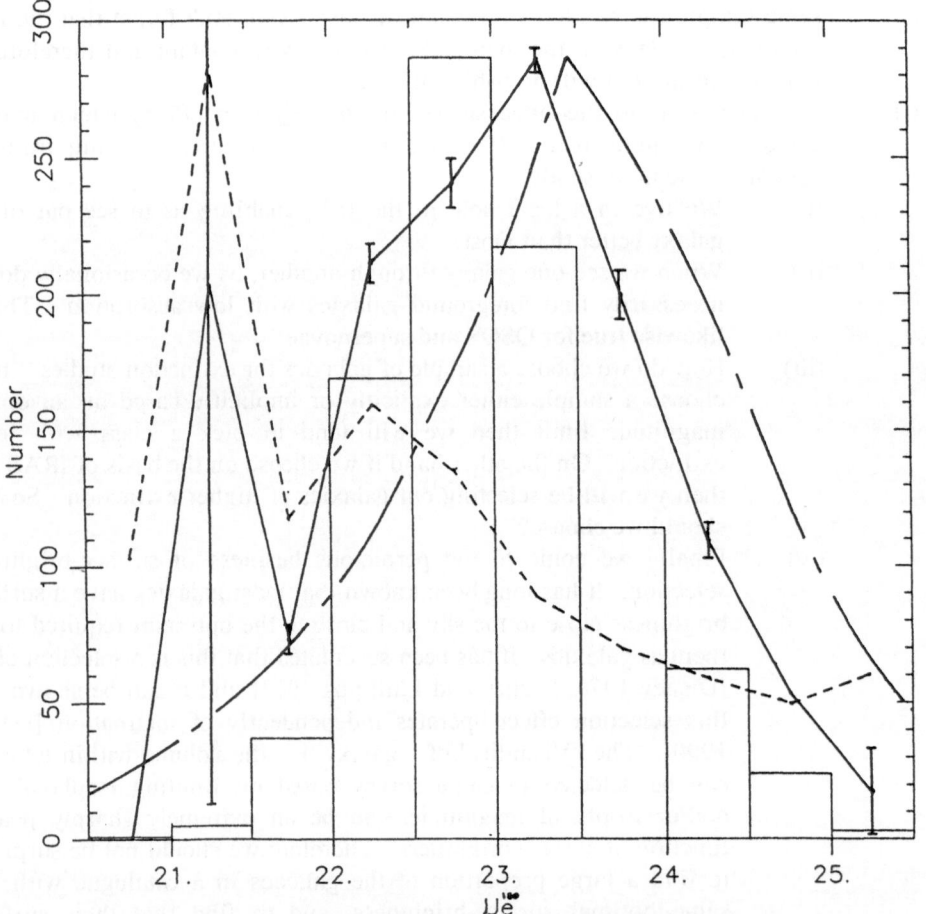

Fig 5:   *Numbers of ESO spirals plotted against surface brightness. Heavy line shows relative median volumes from which each SB category is drawn.*

5.   Future Observational Opportunities

One can say of this topic, as one cannot for most extra-galactic controversies, that it will be largely settled over the next 5 years. K-band imaging, using the new IR arrays, will enable us to pick-up the obscured stars from the ground [Pelletier et al 1994]. The coolest dust, which may yet turn out to be the major absorber, will be harder to find. Nevertheless ISO, which operates out to 250μ, will see down to 14°K. SCUBA, the microwave array to be fitted on the JCMT, will enable us to map galaxies out to 1000μ or 5°K and it may also be possible to do this now using an infra-red interferometer such as IRAM. As a preliminary it will be very interesting to see how much cold dust COBE turns up in our own galaxy.

6.   The Main Questions

There are a number of distinct questions buried within the title of this workshop. Here I gather many of them together, marshalled under 5 separate headings: extinction, opacity, reddening, dust-mass and FIR luminosity.

EXTINCTION Q1   "Given the Morphological Type and Luminosity L_B of a class of galaxies what will be their average (and s.d.) extinction?" where "extinction" is here defined as: "The total fraction of all the original star-light (at all wavelengths) absorbed and re-emitted by smoke."   Q2 "What is the best diagnostic of the total extinction in a particular galaxy?" (e.g. L_{CO}/L_B or L_{IRAS}/L_B or $L_{800\mu}/L_B$.)   Q3 "For a galaxy of a given class how should our estimate of its total extinction depend on its apparent inclination?" Q4 "Given that the inner parts of galaxies are probably more absorbed than the outer parts, how many scale-lengths out will it be before the local extinction falls to less than 50%, less than 10%...?" Q5 "Can social factors (e.g. interactions, clustering) change the extinction in an individual galaxy, and if so by how much for how long?" OPACITY Q6 "What is the average cross-section [pc^2 per solar luminosity in B] of galaxy discs to the absorption of radiation [as a function of wavelength] from the universe beyond?"   Q7 "Given that the inner parts of galaxies are probably more opaque than the outer parts, how far out (in optical scale-lengths) must one go in a particular type of galaxy before the face-on optical depth τ_B has fallen below 1, below 0.5, below 0.1...?"   Q8 "How patchy is the opacity within galaxies, i.e. how many holes/clouds are there in them?" REDDENING: Q9 "How is the spectral energy distribution (SED) of a given type of galaxy affected by internal absorption?" Q10 "How will the reddening depend on the aperture used to define the SED?" Q11 "What is the best (simplest?) diagnostic of the reddening within a particular galaxy?" Q12 "Are the extinction and the reddening simply related?" Q13 Is reddening (and extinction) redshift dependent?" DUST MASS: Q14 "What is the ratio of dust-mass to gas-mass (HI? CO? H_2?...) in local galaxies - and is it reasonably constant?" Q15 "If it is variable then what is the best diagnostic of total dust-mass in a particular galaxy?" Q16 "What is the temperature of most (by mass) of the dust?" FIR LUMINOSITY: Q16 "In galaxies of a given type what proportion of the FIR luminosity arises from extinction the general interstellar radiation field - and not from hidden star-bursts or shrouded AGN's?" Q17 "Do hidden star-bursts really exist - and if so how common are they? Q18 "If star-bursts do really exist, can they remain hidden without supposing a significant increase in the general interstellar absorption - above and beyond that in GMC's?" Q19 "Is high FIR luminosity a transient phenomenon in a particular galaxy - and what triggers it?

7.   In Conclusion

The opacity in spiral discs may not be the most fundamental problem in astronomy, but many of the fundamental problems wait upon its solution.   Because so much of the universe may be cold (<10°K) the wavelength window between 100μ and 1,000μ is the most important of all, and yet the hardest one to peer through.   And even when the observations are in, radiative transfer problems are never the easiest to solve.   At his retirement party Walter Baade was asked the question: "If you had your time over again would you still want to be an astronomer?"   After pausing some time for thought he replied "Only if the ratio of total to selective absorption is everywhere the same".   For whole galaxies at least Baade would have been disappointed.

REFERENCES

1.   Bosma, A., et al, 1994, Pre-print No 131, Univ. Marseille.
2.   Bruzual, G., Magris, C., Calvet, N., 1988 *Ap.J.*, **333**, 673.
3.   Cox, P., and Metzger, P.G., 1987, *Star Formation in Galaxies*, (NASA publ. 2466), p23.
4.   Davies, J.I., 1990, MNRAS, **244**, 8.
5.   Davies, J.I., 1991, Dynamics of Disc Galaxies, ed. Sundelius, Göteborg, p65.
6.   Davies, J.I., et al, 1994, MNRAS, **260**, 491.
7.   Della Valle, M., Panagia, N., 1993, A.J., 104, 696.
8.   Disney, M.J., 1976, Nature, **263**, 573.
9.   Disney, M.J., Sparks, W.B., 1982, Observatory, **102**, 231.
10. Disney, M.J., Davies, J.I. and Phillipps, S., 1989, MNRAS, **239**, 939.
11. Disney, M.J., Phillipps S., 1983, MNRAS, **205**, 1253.
12. Dressler, A., 1980, *Ap.J.*, **236**, 351.
13. Eales, S., et al, 1989, *Ap.J.*, **339**, 859.
14. Fall, S.M., and Pei Y.C., 1993, *Ap.J.*, **402**, 479.
15. de Jong et al, 1984, *Ap.J.*, **278**, L67.
16. Hartwick, F.D.A. and Schade, D., 1990, AR A and A, **28**, 437.
17. Holmberg, E., 1958, Medd. Lund. Astron. Obs., Ser. 2, No 6.
18. Holmberg, E., 1975, *Stars and Stellar Systems IX*, (Univ. Chicago), p123.
19. Huizinga, E., 1994, PhD Thesis, Univ. Groningen.
20. Jura, M., 1980, *Ap.J.*, **238**, 499.
21. Lonsdale Persson, C.J., and Helou, G., 1987, *Ap.J.*, **314**, 513.
22. Miller, D.L. and Branch, D., 1990, A.J., **100**, 530.

23. Moorwood, A. et al,
24. Ostriker, J.P., Heisler, J., 1984, *Ap.J.*, **278**, 1.
25. Pelletier, R. et al, 1994, preprint, Kapteyn Inst.
26. Roberts, M.S., 1975, *Stars and Stellar Systems IX*, (Univ. Chicago), p309.
27. Soifer, et al, 1984, *Ap.J.*, **278**, L71.
28. Saunders, W., et al, 1990, MNRAS, **242**, 318.
29. Valentijn E., 1990, *Nature*, **346**, 153.
30. de Vaucouleurs et al, 1976, Sec. Ref. Cat. of Bright Gals, Univ. Texas.
31. Witt, A., Thronson Jr, H.A. and Capuano Jr, J.M., 1992, *Ap.J.*, **393**, 611.
32. Wright, E.L., 1990, *Ap.J.*, **353**, 411.

INTERSTELLAR GRAIN EVOLUTION AND TEMPERATURES IN SPIRAL GALAXIES

J. MAYO GREENBERG and AIGEN LI
Huygens Laboratory
Leiden University
P.O. Box 9504
2300 RA Leiden
The Netherlands

ABSTRACT. The cyclic evolution of a multimodal interstellar grain population is related to the dynamics of interstellar clouds moving with respect to the density wave pattern of a spiral galaxy. The properties of the dust in the arms and interarms are described in terms of dust in molecular and diffuse clouds. The temperatures characteristic of the "large" tenth micron core-mantle particles are calculated for typical situations to range from 6K to 15K in cool molecular clouds and diffuse clouds respectively. These particles account for some 80% to 90% of the total mass of the dust. The small particles and large molecules in the dust population all emit radiation at higher temperatures than those of the tenth micron particles. The relative proportion of small to large particles is larger in diffuse clouds than in molecular clouds and they are thus expected to be relatively more abundant in the interarm regions than in the arms.

1. Introduction

The properties and evolution of interstellar dust are intimately related to the dynamics and evolution of galaxies. Some 25 years ago I attempted to provide a justification of the appearance of dust concentrations in spiral galaxies based on the density wave picture (Greenberg, 1970). Since then much new has been learned about the dust and, in particular, how it evolves and how its growth and destruction depend on its local environment. The most basic properties of the dust components are their chemical composition and their morphology. These depend on the amounts of the elements which have been created during the course of the cosmic chemical evolution of the galaxy. The distribution of the dust is governed in the broadest sense by the distribution of the clouds in which it is suspended. Partial separation of dust from gas can occur locally and also on a galactic scale as a result of radiation effects (Greenberg and Lind, 1967; Greenberg et al, 1987a; Ferrini et al, 1988). But the observed high degree of correlation of the amounts of dust and gas seems to be the prevailing characteristic result of dust/gas interactions; i.e. the coupling of dust and gas is a strong feature (Spitzer, 1978). On the other hand there exist several different dust populations not all equally correlated and the relative proportions as well as properties of the different dust populations may vary considerably.

There is abundant evidence for the cyclic evolution of the interstellar dust between low density clouds and molecular clouds which suggests how one may develop a theoretical relationship between dust and disk galaxies. There are both quantitative and qualitative differences

J. I. Davies and D. Burstein (eds.), The Opacity of Spiral Disks, 19–31.
© 1995 *Kluwer Academic Publishers.*

to be expected in and out of clouds *and* in and out of spiral arms. The emphasis in this paper is to review grain growth and evolution as observed in the Milky Way and to derive some of the dust properties as related to dust distribution and dust temperatures which may be applied to the observation of dust emission and obscuration in disk galaxies. The galaxy M51 will be used for illustrative purposes even though it may not be entirely typical. A few preliminary new results will be presented. This is a developing field and some of the basic theoretical problems which require further study will be indicated.

2. Observational constraints on dust

The principal observational keys used to define the properties of dust are the following:
a.   wavelength dependence of extinction (and polarization);
b.   correlation of gas and dust;
c.   cosmic abundance of the elements;
d.   infrared spectra.
We shall briefly summarize what these are and how they are interrelated.

2.1. WAVELENGTH DEPENDENCE OF EXTINCTION AND POLARIZATION

The average extinction by interstellar dust in the Milky Way is shown in Figure 1 as a decomposition into 4 major components (Jenniskens and Greenberg, 1993 and references therein; Greenberg and Mendoza-Gómez, 1993). From the most elementary considerations of the theory of scattering by small particles (see van de Hulst, 1957) it may be shown that those particles

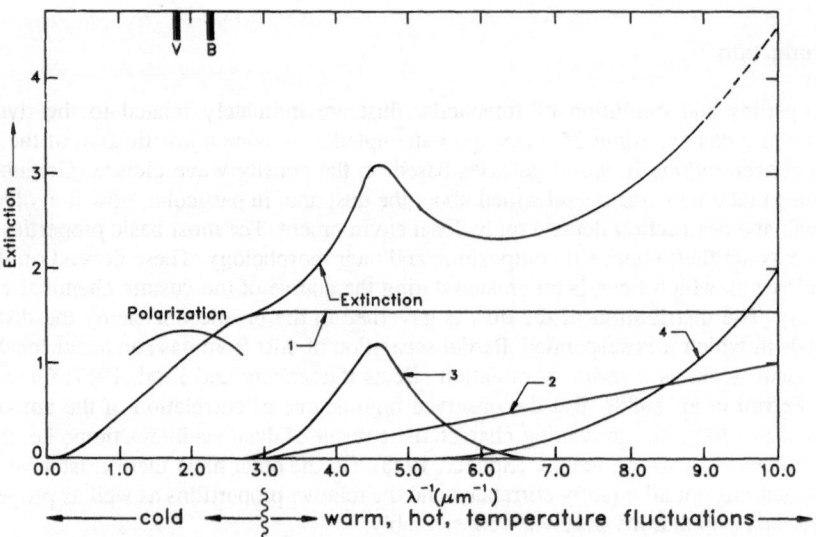

Figure 1. Decomposition of the average interstellar extinction curve into 4 components. Part 1 corresponds to a population of particles with mean size ~0.1μm. Part 2 may be due to small surface perturbation on population 1 particles. Part 3 is due to a population of small (≤0.01μm) carbonaceous particles (*once* thought to be graphite but this is no longer accepted). Part 4 is suggested as due to a population of either very small particles or large molecules - PAH's.

responsible for the rise in the extinction in the visual can *not* provide the continued rise in the extinction in the far ultraviolet because for large values of size divided by wavelength their extinction saturates (Greenberg,1978). Combining the wavelength dependence of the polarization with that of the extinction ties down the mean size of the particles responsible for both. Clearly these particles are not spherical because the polarization results from *differential extinction* by aligned particles along the line of sight and not by scattering. For the purposes of *this* paper it will be simpler and sufficient to consider equivalent spherical particles.

There must be a multimodal size distribution consisting of particles of distinctively different mean sizes in order to provide the full extinction curve as well as varying correlations in different environments (Jenniskens and Greenberg, 1993; Greenberg and Chlewicki, 1983). We first discuss in this paper the tenth micron "large" particles responsible for the "visual" extinction and polarization. These particles contain the major mass fraction of the dust-perhaps 80- 90%.

2.2. CORRELATION OF GAS AND DUST

Wherever it is possible to simultaneously count the hydrogen atoms along the line of sight to a star with its observed extinction one can derive a mean correlation factor (Spitzer, 1978)

$$N_H/A_V = 1.9 \ 10^{21} \ cm^{-2} \tag{1}$$

where N_H = column density of hydrogen, A_V = total extinction in the visual, and where the ratio of total to selective extinction, $R = A_V/(A_B-A_V)$ is taken as 3.1. The coefficient in Eq. 1 is applicable to so-called diffuse clouds but is probably reduced in molecular clouds where $R > 3.1$ indicates that the particles are larger. A *true* correlation is assumed to exist between the relative *number* densities of dust and hydrogen (in all forms) *and* since the extinction depends not only on the number of particles but also their (area), *and* since the extinction *per particle* increases with size, it follows that the coefficient in Equation (1) is lower in dense than in diffuse clouds.

2.3. COSMIC ABUNDANCE OF THE ELEMENTS

The principal elements which can be assumed for the ingredients of solid particles in interstellar space are C, N, O, (S), Mg, Si, Fe. We distinguish sulphur from the others because it plays only a minor role in the discussion of cosmic abundance constraints which follows. The C, N, O group (which we call the organics) constitutes about 1:1000 relative to H and the Mg, Si, Fe group (which we call the rockies) constitutes about 1:10,000. Table 1 lists some of the principal abundances (Cameron, 1982). Other published values for the cosmic (solar system) abundances differ from this and with each other but, generally speaking, not by enough to alter our main conclusions. Two cautions should be kept in mind: (1) current cosmic abundances are different from solar system abundances; (2) cosmic abundances of the condensables in our galaxy and in others are expected to increase toward the galactic centers (Greenberg and Hong, 1974; Sodroski et al.,1994).

Table 1. Relative "cosmic" abundances of the most common elements (Cameron,1982)

H	1	Mg	0.399(-4)
He	0.068	Si	0.376(-4)
C	4.17(-4)	S	0.188(-4)
N	0.87(-4)	Fe	0.388(-4)
O	6.92(-4)		

2.4. CONSTRAINT ON DUST COMPOSITION

The wavelength dependence of extinction leads to a mean particle *size*. The total extinction leads to a required mean particle *area* (per unit area) along the line of sight. Combining the mean size and the mean area leads to a mean required dust *volume* per unit area along the line of sight. The total volume which can be provided by the limited abundance of the rockies (as constituents of silicates) relative to hydrogen is *less* than that *required* to provide the extinction so that even though there is ubiquitous spectroscopic evidence for the SiO stretch at 9.7 μm for silicates (consisting of the rocky elements) the organics are *needed* to supplement the amount of solid material (see e.g. Greenberg,1978). Since oxygen is the most abundant condensable element it is an obvious candidate and, in fact, solid H_2O is indeed often observed as a substantial infrared absorption feature at 3μm relative to the silicate feature making it an important constituent in *molecular* clouds (see Whittet 1992 for a recent survey). *But* in diffuse clouds no ice is observed (see Greenberg, 1973,1982; Sandford et al 1991 for discussions and explanation). Instead one observes a feature at 3.4 μm which is characteristic of *complex* organic molecules whose CH stretches due to CH_2 and CH_3 groups make up the absorptions in this feature. It is certain that in diffuse clouds the abundance of the carbon exceeds that of the oxygen in the solid phase even though it is the less abundant cosmically. The organics constitute a major fraction of the dust material in diffuse clouds, being comparable in mass with the silicates (Greenberg, 1982).

The volume of material in each of the dust components delineated by the decomposition in Figure 1 may be simply deduced from their sizes. The particles responsible for the visual extinction have a mean spherical equivalent radius a ≈ 0.1μm (Xing, 1993; Chlewicki, 1985). For such a size and with physically acceptable optical properties the wavelength dependence of the extinction saturates in the ultraviolet. The particles responsible for the hump and the far ultraviolet extinction may then be shown to have mean sizes ≤ 0.01μm (Greenberg and Chlewicki, 1983; Greenberg, 1984). If the extinction for a *given* size particle is measured at a wavelength comparable to *this* size its extinction efficiency , $Q_{ext} = (C_{ext}/$geometrical area $) \approx 1$ and, in this case, the volume per unit area along the line of sight is Vol ~aΔA(a) where ΔA(a) is the *decomposition* extinction due to the population of particles of size a. Since the effective area of each population is defined by its contribution to the extinction and since the *extra* hump extinction and the *extra* FUV extinction are both comparable with the visual extinction (see Fig. 1) one may conclude that there is about 10 times as much mass in the large particles as in each of the small ones. Thus the *visual* extinction is an indicator of about *80% of the total mass* in the form of dust.

3. A cyclic model for dust.

The picture which can be deduced from a wide range of theoretical and observational arguments is that dust grains evolve by cycling between molecular and diffuse clouds (Greenberg, 1982, 1984, 1986). *Initially* dust starts out as small silicate particles which are produced in the atmospheres and blown out of cool evolved stars by radiation pressure. These particles serve as condensation nuclei for the condensable organics in molecular clouds and, as a result of condensation, surface reactions, and a variety of photo-induced chemical reactions within the low temperature accreted "ices", leads to the formation of complex organic matter. When the molecular cloud dust is injected back to the diffuse cloud medium after star formation, the complex organics are maintained while the ices are eroded away by various destructive processes leaving dust which consists of silicate cores with organic refractory mantles. During the diffuse cloud phase some of the organic refractory mantles are broken off or eroded away providing a source of the small particles / large molecules which are observed via their extinction in the near and far ultraviolet (Greenberg, 1986; Jenniskens et al, 1993). A typical "large" dust grain spends about equal time in diffuse and dense clouds as deduced from the roughly equal amounts of *gas* in each (Burton,1992) and because of the dust/gas correlation. The mean time spent in each is ~ 5×10^7 years so that a typical large grain cycles back and forth between molecular and diffuse clouds every 10^8 years. It is worth noting that *no* stellar source can produce *silicate or carbonaceous solid particles* at a rate sufficient to compensate for their rate of destruction in diffuse clouds (Greenberg, 1982). Silicate particles survive in diffuse clouds because their mantles of organic refractories provide an "ablation shield". The PAH's are exceptional in that they are destroyed in molecular rather than in diffuse clouds(Mendoza-Gómez et al., 1995, Greenberg et al, 1993). This implies that the *major* fraction of all organic matter whether as grain mantles or as small carbonaceous particles/large molecules are created as a result of processes *originating* in molecular clouds, primarily ultraviolet processing of simple molecules in the icy grain mantles. In the diffuse clouds the organic constituents are further processed so that the organic mantles are "carbonized" (O, N, H are photodetached). This may be inferred by comparing the infrared absorption spectrum of processed laboratory organics with first generation organics and with the diffuse cloud-dust spectrum (Greenberg et al., 1995). The mean *total* lifetime of a grain is limited to ~ 5×10^9 years because this is the mean turnover time of the *entire* interstellar medium resulting from the rate of new star formation; i.e., the interstellar medium is used to create new stars.

A spherical representation for the core-mantle particles in diffuse and molecular clouds is shown in Fig. 2. The spherical shape is used for simplicity and is scaled to give approximately

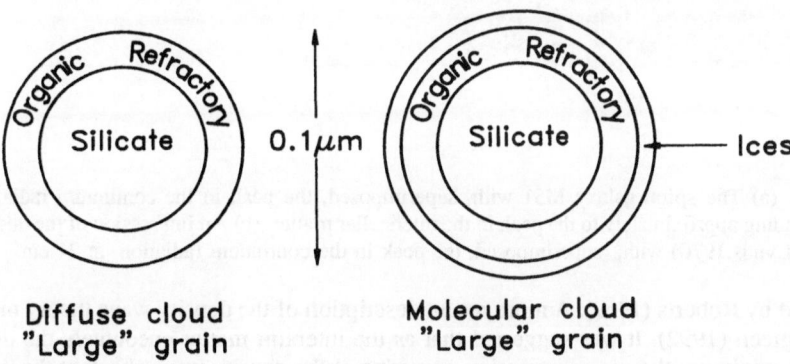

Figure 2. Schematic spherical representation of an average large interstellar grain in diffuse regions and in molecular clouds. Diffuse cloud grain: silicate radius a_{sil} = 0.07μm and an organic refractory mantle a_{OR} = 0.1μm. Molecular cloud grain: a_{sil} = 0.07μm; a_{OR} = 0.1μm; a_{ice} = 0.105 μm.

equal masses for the silicate core and organic refractory mantles which are respectively responsible for the strength of the 9.7 μm and 3.4 μm absorption features in diffuse cloud dust. The extra thickness of ice mantles in molecular clouds is not much more than only Δa = 0.005μm even though thicknesses as high as 0.05 micron are possible if all remaining condensables are accreted.

4. Galactic distribution of dust.

When one observes disk galaxies edge-on or face-on, the overall impression is that the dust which gives the major extinction is concentrated in the plan of the disk about like the gas and that it is concentrated in the inner edges of spiral arms (see Fig. 3). A picture of how and why the latter occurs was proposed some 25 years ago (Greenberg, 1970) based on the density wave model (Lin and Shu, 1964) as

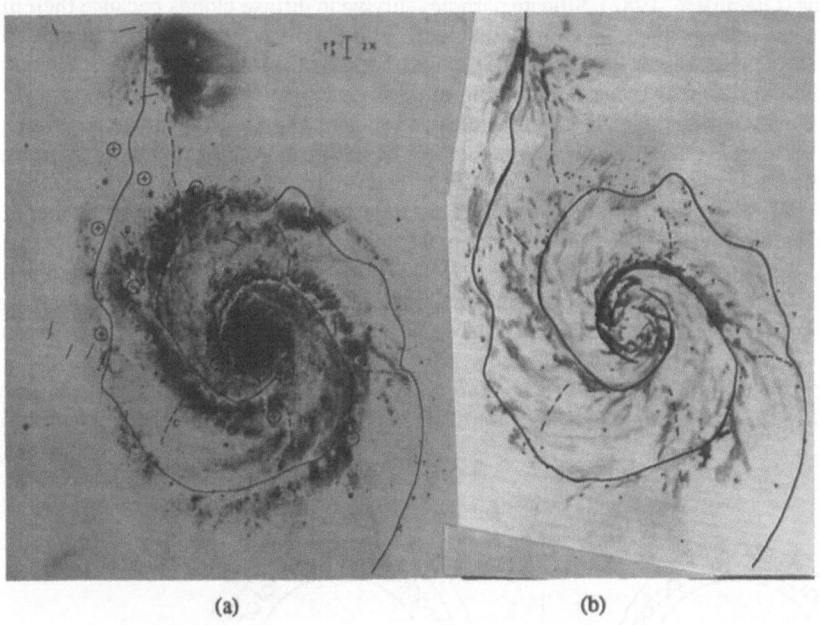

(a)   (b)

Figure 3. (a) The spiral galaxy M51 with, superimposed, the peak in the continuum radiation at 21cm corresponding approximately to the peak in the interstellar matter ,(b) An impression of the dust distribution in M51 (Lynds,1970) with, superimposed, the peak in the continuum radiation at 21 cm.

described by Roberts (1969). A more recent description of the density wave theory may be found in Elmegreen (1992). It was suggested that as the interarm matter encounters the density wave potential minimum the gas compression occurring at the density wave shock at the inner edge of an arm leads not only to a larger number density of grains as correlated with the increased gas density, but also an increase of size by accretion and together these lead to a non-linear amplification of the local extinction at inner edges of spiral arms. However the "effective" opacity may be less than directly deduced from this enhancement because it is not a smoothly occurring phenomenon but is rather concentrated in the molecular clouds leading to star formation. As is well known, a patchy distribution of highly concentrated dust (van der Hulst et al,1988) with optical depths τ_i in areas A_i will lead to less average extinction than deduced from the *average* dust density in the clouds because

$$\Sigma\ A_i exp(-\tau_i)\ /\Sigma A_i\ \geq exp(-\Sigma\tau_i A_i\ /\ \Sigma A_i\ ) \tag{2}$$

and equality applies only if $\tau_i \ll 1$ for all the clouds. This means that the *apparent* contrast in the dust opacity between the arm region and the interarm region is *less* than would be deduced from the overall mean dust density and size. Probes of dust concentration as deduced from far infrared emission could, if properly interpreted, provide a better measure of the contrasting volumes of dust in the arm and interarm regions.

We should note that not only does the mean size of the core-mantle particles change with accretion in molecular clouds but the chemical/morphological properties as well. In the diffuse cloud dust the tenth micron particles consist only of a silicate core and organic refractory mantle while in the molecular cloud there is an additional frozen frost of ices with relatively little infrared absorptivity. If the interarm dust enters the inner edge of the spiral arm as diffuse cloud dust it will grow at a rate (Greenberg, 1970)

$$\mathring{a} = 3.43 \ (1\text{-}\delta)10^{-22} \ n_H \ cm \ s^{-1} \qquad (3)$$

where δ is the fractional abundance of the cosmically abundant O, C, N species in the diffuse cloud dust: $\delta \approx 0.3$ (averaged over O, C, and N). In molecular clouds, formed with $n_H > 10^3 \ cm^{-3}$ in the spiral arm, the increase in grain size is $\Delta a \approx 0.01 \ \mu m$ in 10^5 yrs. However, because of desorption of these volatile mantle components by grain explosions which also occur every $\sim 10^5$ years we generally anticipate *outer* mantle thicknesses of $< 0.01 \ \mu m$. Thus the optical depth per dust grain in regions of star formation may be increased by only about 25% except immediately surrounding protostars or very young stars. As noted above, the total optical depth will not be increased proportional to the mean density increase because of saturation extinction effects in dense clouds with optical depths > 1. In Fig. 4 there is shown a simplistic and schematic representation of the distribution of diffuse and (giant) molecular clouds. When the diffuse clouds enter the potential minimum at the inner edge of the spiral arm the compression leads to coagulation and growth into giant molecular clouds so that GMC's are strongly concentrated in

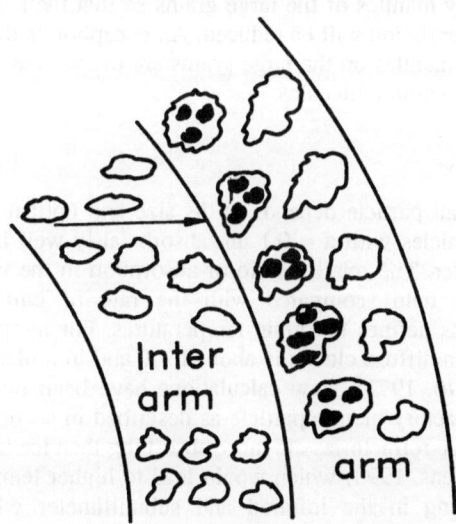

Figure 4. Cloud distributions and types in arm and inter arm. Filled regions represent molecular clouds, open regions represent diffuse clouds. Arrow indicates material motion relative to density wave pattern.

spiral arms (see Roberts et al, 1990). As the result of star formation some dissipation occurs during passage in the arm. When the interstellar clouds emerge at the trailing edge of the arm, they are further dissipated and dispersed so that, while some clouds persist, the major fraction of the dust is of the diffuse cloud type. It is for this reason that while CO is more concentrated in the arms it persists in the interarm. The CO contrast between arm and interarm in M51 is about 5 (García-Burillo et al, 1993). The arm averaged dust density does not necessarily reflect the large contrast exhibited by the strong dust lanes. The difference is mostly in the localized high concentration in the arm-interarm region as compared with the more spread out distribution in the interarm region. The arms traced by CO and 21 cm continuum nonthermal emission coincide and have the same width inside M51 at least within the co-rotation radius (García-Burillo et al, 1993; see this reference also for other aspects of molecular structure in M51), The question of correlation of dust with CO within dense and diffuse clouds is important in order to relate dust morphological properties with growth and destruction processes as well as with local differences in radiation fields which determine grain temperatures and submillimeter emissions. While a substantial fraction of the dust *is* correlated with CO, there must be regions with dust but no CO. How much dust is associated with CO and how much not can best be determined by tracing the submillimeter emission by the dust with sufficiently high resolution. According to Garcia- Burillo et al (1993) the best fit for CO as an interarm gas is obtained for *small* clouds (~ 1.8 pc) with $n_H \approx 500$ cm^{-3} which implies that the interarm dust whether in or out of clouds is almost uniformly exposed to a galactic interstellar radiation field with its full visible and ultraviolet flux. The dust in the small diffuse clouds is not expected to have any ice mantles and the small particle populations should be characteristic of the average diffuse cloud extinction. In the following section a few results will be presented for grain temperatures in diffuse and dense (ultraviolet deficient) clouds to illustrate the range of the different temperatures to expect. A final and very critical comment must be made on dust properties in and out of clouds; namely, that the small particle/large molecule populations in dense clouds are expected to be depleted by at least partial accretion in the icy mantles of the large grains so that their relative contribution to the IR and long wave length emission will be reduced. An exception is the immediate neighborhood of hot young stars where mantles on the large grains are evaporated releasing the trapped small particles along with the volatile molecules.

5. Grain temperatures.

The temperature of a small particle depends on its size and optical properties *throughout* the *entire* spectral range. Particles with a ≈ 0.1 μm absorb fairly well in the ultraviolet but rather poorly in the submillimeter. The relative ratio of absorption in the visual and ultraviolet (from the interstellar radiation field) compared with the rate of emission at far infrared and submillimeter wavelengths defines the grain temperatures. For *ice* particles of a = 0.1 μm the equilibrium temperature in diffuse clouds is about 15 K and in molecular clouds is about 10 K or lower (Greenberg, 1970, 1971). New calculations have been made for a typical (average) silicate core- organic refractory mantle particle as described in section 3 for diffuse cloud dust. It is known that the organic refractories are more absorbing than ice in the visual (Chlewicki and Greenberg, 1990; Jenniskens, 1993) which would lead to higher temperatures. But organics are also much more absorbing in the infrared and submillimeter which leads back to lower temperatures. Actually the silicates dominate in the emissivity at submillimeter wavelengths while the organics dominate in the visual. We approximate the complex index of refraction m= m' - i m" for organic refractories in the region λ < 1μm by the values given in Chlewicki and Greenberg (1990). For 500μm > λ > 1μm we represent the higher absorptivity characteristic of photoprocessed organics (Jenniskens, 1993) by m" = 0.15 λ^{-1}. The real and imaginary parts of the indices of refraction of the organic refractories are calculated from the Kramers-Kronig relations.The optical properties of the silicates are as in Greenberg and Hage (1990) and similar to those of Draine (1985). The core mantle particle is chosen to have an outer organic mantle of radius 0.1 μm with a silicate core radius of a_{sil} = 0.07 μm. The grain temperature is derived from equating the emission at T_d with the absorption from the external radiation field

$$\int C_{abs}(\lambda)B(T_d,\lambda)d\lambda = \int C_{abs}(\lambda)R(\lambda)d\lambda \tag{4}$$

where $R(\lambda)$ is the wavelength distribution of the radiation field and $B(T_d,\lambda)$ is the Planck function for a temperature T_d. In Fig. 5 we see that for such a particle the equality between emission and absorption from a diffuse region radiation field of 10^4 K diluted by 10^{-14} (Eddington, 1926) occurs at T_d = 14.5 K. Using more recent diffuse radiation fields (Greenberg, 1978; Mezger et al,1982) the temperature is T_d = 13.7 K or 15K. For a=0.1 μm graphite the temperature is substantially higher at T_d=25.2 K.

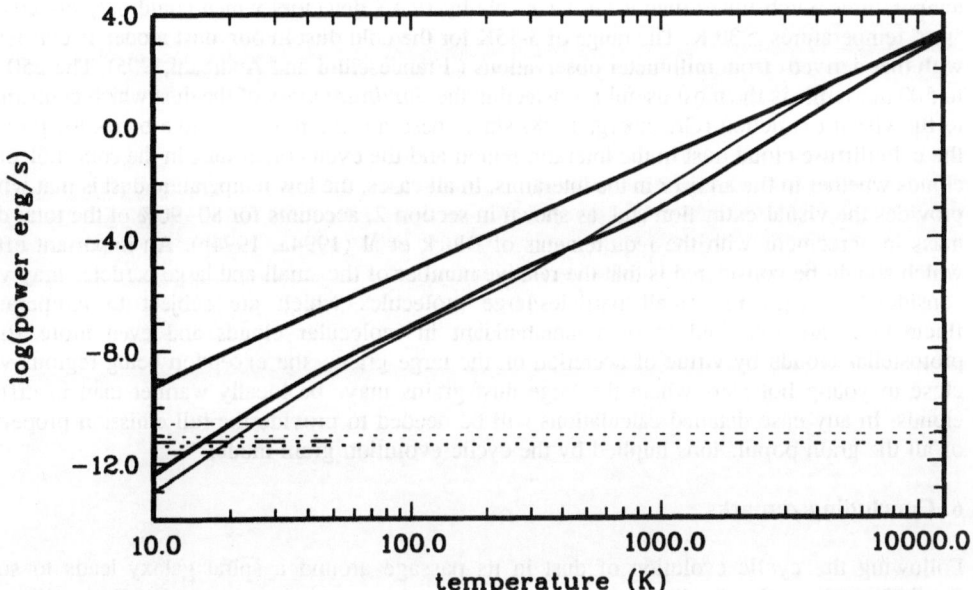

Figure 5. Absorption/emission of radiation by a 0.1 μm size particle as a function of temperature. Upper thick solid line: absorption by a black body. Lower thick solid curve: absorption by a silicate core - organic refractory mantle (a_c= 0.07μm, a_m = 0.10μm) particle. Thin solid curve: absorption by a graphite particle. Upper and lower dotted horizontal lines respectively: absorption by a graphite and core-mantle particle in a T = 10^4K radiation field diluted by 10^{-14}. Dashed line: absorption by a c-m particle in the interstellar field as in Greenberg (1978). Upper and lower dash-dot lines respectively: absorption by a graphite and c-m particle in the diffuse radiation field as in Mezger et al (1982) Intersections of horizontal lines with absorption curves define the grain temperatures.

More detailed calculations are in progress for more grain components and for dust in molecular clouds. For example, in a cloud whose optical depth is 4 the right hand side of Eq. 2 is reduced by at least e^{-4} and the resulting large particle temperature is reduced by (e^{-4})$^{1/4}$. Consequently we expect that the temperatures of the large dust grains in molecular clouds - with additional mantles of ices and in weaker ultraviolet radiation environments - will be as low or *lower than* T = 6 K. The limiting temperature for any particle is no lower than ~ 3 K which would be the case for a black body which radiates and absorbs perfectly and equally efficiently at all wavelengths . A simplistic way of showing the size effect on grain temperature, valid for grains ≥ 0.1 μm, is based on the fact that emission at long wavelengths is ~ a^3 while absorption at short wavelengths in a high (equivalent) temperature, T_R , radiation field is ~ a^2. Equating emission with absorption : $a^3T_d^4 \sim a^2T_R^4$: gives $T_d \sim a^{-\frac{1}{4}}$. For small and very small particles the situation is more complex although generally their temperatures are higher than those for "large" particles.

The smaller particles, say a ≤ 0.01 μm, are expected to be at higher temperatures, perhaps as high as 35 K as is obtained for small graphite particles (Greenberg, 1970) while *very* small particles (or large molecules) will exhibit *effective* temperatures *much* higher as a result of temperature fluctuations (Greenberg, 1968; Greenberg and Hong, 1974; Greenberg, 1971; Sellgren, 1984).

The general consequence of these results is that the **major** mass of the dust is at temperatures much lower than were detectable by IRAS detectors which could only effectively "see" temperatures ≥ 30 K. The range of 5-15K for the cold dust in our dust model is consistent with that derived from millimeter observations (Franceschini and Andreani,1995). The 250 μm to 500 μm range is the most useful for detecting the **dominant** mass of the dust which contributes to the visual extinction (Greenberg, 1978) since these are the most sensitive bands for probing the cold diffuse cloud dust in the interarm region and the even colder dust in the cool molecular clouds whether in the arms or in the interarms. In all cases, the low temperature dust is that which provides the visual extinction and, as shown in section 2, accounts for 80- 90% of the total dust mass in agreement with the requirements of Block et al (1994a, 1994b). An important effect which should be considered is that the relative number of the small and large particles may vary considerably; e.g., the small particles/large molecules which are subject to temperature fluctuations are expected to be underabundant in molecular clouds and even more so in protostellar clouds by virtue of accretion on the large grains, the exception being regions very close to young hot stars where the large dust grains may be locally warmer than in diffuse clouds. In any case detailed calculations will be needed to provide the full emission properties of all the grain populations implied by the cyclic evolution grain model.

6. Concluding remarks

Following the cyclic evolution of dust in its passage around a spiral galaxy leads to some predictions about the distribution of the different sources of infrared emission.The multimodal dust population in spiral arms and especially in the inner edges of arms is expected to be deficient in the high temperature small particle component. A large fraction of the "large" cold particles in the molecular and giant molecular clouds, except in the immediate vicinity of hot young stars, are at *very* low temperatures, possibly as low as 5K. In diffuse clouds, which are presumed to dominate the interarm region, the large dust grains are at temperatures of about 15K while only the small grains provide the high temperature emission; i.e., the IRAS detectors saw *only* the small particles which constitute a *minor* fraction of the mass of the interstellar dust.

7. Acknowledgements

We wish to acknowledge helpful discussions with Butler Burton, Frank Israel and Leo Blitz. We also thank Osama Shalabiea for a careful reading and suggestions for clarification. Most important, one of us (JMG) thanks the organizers of the meeting for inviting me to present my ideas on dust as it may affect observations of disk galaxies in the visible, and far infrared.

8. References

Block, D.L., Bertin, G., Stiction, A. et al (1994a) 2.1μm images of the evolved stellar disk and the morphological classification of spiral galaxies, Astron. Astrophys. **288**, 365-382.
Block, D.L., Witt, A.N., Grosbol, P. et al (1994b) Imaging in the optical and near-infrared regimes, II. Arcsecond spatial resolution of widely distributed cold dust in spiral galaxies, Astron. Astrophys. **288**, 383-395.
Burton, W.B. (1992) Distribution and observational properties of the ISM, in: W.B. Burton, B.G. Elmegreen, R. Genzel (eds), The Galactic Interstellar Medium, Springer, Berlin, pp. 1-155.
Cameron, A.G.W. (1982), in: C. Barnes, R.N. Clayton and D.N. Schramm (eds),Elements and Nuclidic Abundances in the Solar System, Cambridge Univ. Press, p. 23.
Chlewicki, G. (1985) Observational constraints on multimodal interstellar grain populations. Thesis Leiden University.
Chlewicki, G., Greenberg, J.M. (1984) General constraints on the average scattering characteristics of interstellar grains in the ultraviolet, Mon. Not. RAS **210**, 791-801.
Chlewicki, G., Greenberg, J.M. (1990) Interstellar circular polarization and the dielectric nature of dust grains, Astrophys. J. **365**, 230-238.
Draine,B.T. (1985) Tabulated optical properties of graphite and silicate grains, Astrophys. J. Suppl. **57**, 587-594.
Eddington, A.S. (1926) Diffuse matter in interstellar space (Bakerian Lecture) Proc. Roy. Soc. **111A**, 424-456.
Elmegreen, B.G. (1992) Large scale dynamics of the interstellar medium, in: W.B. Burton, B.G. Elmegreen, R. Genzel (eds), The Galactic Interstellar Medium, Springer, Berlin, pp. 157-274.
Ferrini, F., Barsella, B., Greenberg, J.M. (1988) Dust grains in galactic haloes, in: M.E. Bailey and D.A. Williams (eds), Dust in the Universe, Cambridge Univ. Press, pp. 513-519.
Franceschini, A., Andreani, P. (1995) Millimeter observations of a complete sample of IRAS galaxies: dust emission and absorption in spirals Astrophys. J. Let. (in press).
García-Barely, S., Guélin, M., Cernicharo, J. (1993) CO in Messier 51: I. Molecular spiral structure, Astron. Astrophys. **274**, 123-147.
Greenberg, J.M. (1968) Interstellar grains, in: B.M. Middlehurst and L.H. Aller (eds), Nebulae and Interstellar Matter, University of Chicago Press, pp. 221-364.
Greenberg, J.M. (1970) Interstellar grains and spiral structure, in: H.J. Habing (ed), Interstellar Gas Dynamics, Reidel, Dordrecht, Holland, pp. 305-315.
Greenberg, J.M. (1971) Interstellar grain temperatures, effects of grain materials and radiation fields, Astron. Astrophys. **12**, 240-249.
Greenberg, J.M. (1973) Chemical and physical properties of interstellar dust, in: M.A. Gordon and L.E. Snyder (eds), Molecules in the Galactic Environment, Wiley, pp 94-124.
Greenberg, J.M. (1978) Interstellar dust, in: J.A.M. McDonnell (ed), Cosmic Dust, Wiley, pp. 187-294.

Greenberg, J.M. (1982) Dust in dense clouds. One stage in a cycle, in: J.E. Beckman and J.P. Phillips (eds), Submillimetre Wave Astronomy, Cambridge University Press, pp. 261-306.
Greenberg, J.M. (1984) Structure and evolution of interstellar grains, Sci. Amer. **250**, 124-135.
Greenberg, J.M. (1986) Dust in diffuse clouds: one stage in a cycle, in: F. Israel (ed.), Light on Dark Matter (Proc. IRAS Symp., Noordwijk 10-14 June 1985), Reidel, Dordrecht, pp. 177-188.

Greenberg, J.M., Chlewicki, G. (1983) A far ultraviolet extinction law: what does it mean?, Astrophys. J. **272**, 563-578
Greenberg, J.M., Ferrini, F., Barsella, B. and Aiello, S. (1987a) Is there dust in galactic haloes? Nature **327**, 214-216.
Greenberg, J.M., de Groot, M.S. and van der Zwet, G.P. (1987b) Carbon components of interstellar dust, in: A. Léger, d'Hendecourt, L.B., and N. Boccaro (eds), Polycyclic aromatic hydrocarbons and astrophysics, Reidel, pp. 177-181.
Greenberg, J.M., Hage, J.I. (1990) From interstellar dust to comets: a unification of observational constraints, Astrophys. J. **361**, 260-274.
Greenberg, J.M. and Hong, S.S. (1974) The chemical composition and distribution of interstellar grains, in: F.J. Kerr and S.C. Simonson III (eds), Galactic Radio Astronomy, Reidel, Dordrecht, pp. 153-177.
Greenberg, J.M. and Lind, A.C. (1967) Some problems of interstellar grains, in: J.M. Greenberg and T.P. Roark (eds), Interstellar Grains, NASA SP-140, 217-227.
Greenberg, J.M., Mendoza-Gómez, C.X. (1993) Interstellar dust evolution: a reservoir of prebiotic molecules, in: Greenberg, J.M., Mendoza-Gómez, C.X., and Pirronello, V. (eds), The Chemistry of Life's Origins, Kluwer, Dordrecht, pp. 1-32.
Greenberg, J.M., Mendoza-Gómez, C.X., de Groot, M.S., Breukers, R. (1993) Laboratory dust studies and gas-grain chemistry, in: T.J. Millar, D.A. Williams (eds), Dust and Chemistry in Astronomy, IOP publ., pp. 265-288.
Greenberg, J.M., Shalabiea, O.M., Mendoza-Gómez, C.X., Schutte, W., Gerakines, P.A. (1995) Origin of organic matter in the protosolar nebula and in comets. To appear in Adv. Space Res.
Jenniskens, P. (1993) Optical constants of organic refractory residue, Astron. Astrophys. **274**, 653-661.
Jenniskens, P., Baratta, G.A., Kouchi, A., de Groot, M.S., Greenberg, J.M., Strazzulla, G. (1993) Carbon dust formation on interstellar grains. Astron. Astrophys. **273**, 583-600.
Jenniskens, P., Greenberg, J.M. (1993) Environment dependence of interstellar extinction curves, Astron. Astrophys. **274**, 439-450.
Lynds, B.T. (1970) The distribution of dark nebulae in late-type spirals, in: W. Becker, G. Contopoulos (eds), The Spiral Structure of our Galaxy, Reidel, Dordrecht, pp. 26-34.
Mendoza-Gómez, C.X., de Groot, M.S., Greenberg, J.M. (1995) The fate of polycyclic aromatic material in space. Astron. Astrophys. in press.
Mezger, P.G., Mathis, J.S., Panagia, N. (1982) The origin of the diffuse galactic far infrared and sub-millimeter emission, Astron. Astrophys. **105**, 372-388.
Roberts, W.W. (1969), Large-scale shock formation in spiral galaxies and its implications on star formation, Astrophys. J. **158**, 123-143.
Roberts, W.W., Lowe, S.A., Adler, D.S. (1990) Simulations of cloudy, gaseous galactic disks, Ann. New York Acad, Sci, **596**, 13--144.
Sandford, S.A., Allamandola, L.J., Tielens, A.G.G.M., et al (1991) The interstellar CH stretching band near 3.4 microns: constraints on the composition of organic material in the diffuse interstellar medium, Astrophys. J. **371**, 607-620.
Sellgren, K. (1984) The near-infrared continuum emission of visual reflection nebulae, Astrophys. J. **277**, 623-633.

Sodroski ,T.J. et al., (1994), Large-scale characteristics of interstellar dust from *COBE* DIRBE observations, Astrophys. J. 428,638-646.

Spitzer, L. (1978) Physical processes in the interstellar medium, Wiley, New York.

Van de Hulst, H.C. (1957) Light Scattering by Small Particles, Wiley, New York.

Van der Hulst, J.M.,, Kennicutt, R.C., Crane, P.C., Rots, A.H. (1988) Radio properties and extinction of the H II regions in M51, Astron. Astrophys. **195**, 38-52.

Whittet, D.C.B. (1992) Dust in the Galactic Environment, IOP publ., London.

Xing, Zhangfan. (1993) Fundamentals of electromagnetic scattering theory, with applications to cosmic dust. Thesis Leiden University.

RADIATIVE TRANSFER MODELS

GUSTAVO BRUZUAL A.
Landessternwarte Heidelberg-Königstuhl
69117 Heidelberg
Germany

1. Introduction

The intensity of the radiation leaving a dusty interstellar medium, in which stars and dust grains coexist, is given by the solution to the radiative transfer equation (RTE). For the simple case of a plane parallel slab, used as an approximation to a galaxy disk, the RTE is

$$\mu \frac{\partial I_\lambda}{\partial z} = -\kappa_\lambda I_\lambda + \epsilon_\lambda^* + \kappa_\lambda \frac{a_\lambda}{4\pi} \int I_\lambda \Phi_\lambda(cos\Theta) d\Omega, \qquad (1)$$

where $\theta = cos^{-1}\mu$ is the angle between the direction perpendicular to the plane and the direction of the light beam, ϵ_λ^* is the emissivity of the stellar sources, κ_λ is the absorption coefficient, and a_λ is the dust albedo (ratio of the scattering to the absorption coefficient). The term

$$\kappa_\lambda \frac{a_\lambda}{4\pi} \int I_\lambda \Phi_\lambda(cos\Theta) d\Omega \qquad (2)$$

measures the contribution to the emissivity due to scattering by dust particles (assuming coherent scattering $\lambda_{out} = \lambda_{in}$). Θ is the scattering angle and

$$\Phi_\lambda(cos\Theta) = \frac{(1-g_\lambda^2)}{(1+g_\lambda^2 - 2g_\lambda cos\Theta)^{3/2}} \qquad (3)$$

is the Henyey & Greenstein (1941) scattering phase function, characterized by the asymmetry parameter $g_\lambda = < cos\Theta >$.

From the mathematical point of view the radiative transfer problem (RTP) is well formulated. The RTE can be easily written for more complex geometries. In order to solve the RTE we must specify: (a) The intrinsic properties of dust grains, a_λ and g_λ (see Witt et al. 1994 for a recent

33

J. I. Davies and D. Burstein (eds.), The Opacity of Spiral Disks, 33–41.
© *1995 Kluwer Academic Publishers.*

determination of these quantities for galactic dust grains). (*b*) The total amount of dust, measured by the optical depth of the slab, $\tau_\lambda = \kappa_\lambda L$, where L is the physical thickness of the slab. (*c*) The geometrical distribution and degree of mixing of the stellar and dust component in a galaxy disk. (*d*) An algorithm (numerical, analytical, approximation) to solve the RTE for the specified geometry. (*a*) is relatively well known (Witt et al. 1994), (*d*) may be the least of a problem, and (*b*) and (*c*) are related and constitute the major source of uncertainty in modeling the radiative transfer in galaxy disks.

2. Existing models

Three major efforts to understand the systematic properties of dusty galaxies, including the effects of scattering of stellar light by dust grains, are reported in the recent literature. These are, in chronological order: Bruzual, Magris, & Calvet (1988, hereafter BMC88); Disney, Davies & Phillips (1989, hereafter DDP89); and Witt, Thronson & Capuano (1992, WTC92 hereafter). Simultaneously, a number of more restricted applications have appeared. Some of them are: Kylafis & Bahcall (1987); Davies (1990); Davies, Phillips, Boyce, & Disney (1993); Di Bartolomeo, Barbaro, & Perinotto (1993); Calzetti, Kinney, & Storchi-Bergmann (1994). That the RTP in galaxy disks is a current one is revealed by the fact that a large number of these authors is attending this conference.

The papers by BMC88, DDP89, and WTC92 concentrate on different aspects of the RTP (according to the interests of the authors) and use different assumptions about the distribution of dust and stars in galaxies, as well as different methods to solve the RTE.

2.1. BMC88

This paper assumes that galaxy disks can be approximated by a *plane parallel slab* (infinite disk) in which stars and dust are mixed homogeneously. A numerical technique, borrowed from the theory of model stellar atmospheres, is used to solve the RTE.

2.2. DDP89

The authors consider three different geometrical configurations: *slab* (uniformly mixed), *sandwich* (obscuring layer narrower than emitting layer), and *triple exponential* models (as a more realistic representation of spiral galaxies). The RTE is solved by means of an analytical approximation, which is good for geometrically thin disks.

2.3. WTC92

By means of 3-D Montecarlo simulations the authors study the RTP for spherically symmetric galaxies under five sets of assumptions: *dusty galaxies* (stars and dust fill uniformly the sphere); *cloudy galaxies* (uniform spherical stellar distribution within which is embedded a smaller uniform sphere of dust); *starburst galaxies* (stellar distribution declines from the nucleus following r^{-6} embedded within a sphere of uniformly distributed dust); *dusty galactic nuclei* (central dust-free sphere of stars surrounded by a spherical star-free shell of dust); *elliptical galaxies* (spherical stellar distribution following r^{-3}, dust extends up to $\frac{2}{3}$ of galaxy radius and follows r^{-1}).

In all of these three papers the results are compared with those obtained with the *mythological* overlying screen approximation (WTC92). All authors recommend not to use this *model*.

3. Summary of results from existing models

Despite the different geometrical distributions assumed and the different solution techniques used, the basic conclusions derived from the papers mentioned above are in agreement and can be summarized as follows:

• The attenuation suffered by the light emitted by diverse sources in the galaxy is less than the exponential attenuation produced by non-dispersive dust with the same optical depth placed in front of the sources (absorbing screen).

• Light scattering acts as an extra source of radiation, since photons are not only removed but added to the beam.

• If the surface brightness profiles of spiral galaxies do not depend on the inclination μ, then the disks of these galaxies must be optically thick with $\tau_V \sim 1$.

• In their central parts, galaxy disks are optically thick, and in many cases we may be seeing no more than the top surfaces of galaxies, at least to $\mu_B = 24$. An optically thick sandwich will appear to behave, in the optical, like an optically thin disk. The light will be dominated, in the optical, by a thin upper crust of stars comprising even as much as $\frac{1}{6}$ of the total. We only see the cream of the pudding; we do not see if there are raisins in it. No matter how large the optical depth, they behave in the optical as optically thin.

• Optical tests of the optical depth and internal extinction will give wrong values, systematically low, because optical detectors will preferentially pick up radiation from the least obscured patches and parts of galaxies.

• For optically thick sandwich models, the ratio of FIR to optical luminosity is almost independent of inclination. Thus, it is wrong to argue that because this ratio is independent of inclination, then galaxies are optically thin (DDP89).

- Galaxy data (e.g. $\frac{M}{L}$ ratios) which have been derived assuming that their disks are optically thin may need to be revised, or used with caution.

- The range of observed $L(IRAS)/L(B)$ is explicable by pure internal extinction in galaxies that are not undergoing starbursts for suitable chosen sandwich models.

- It is not true that an observed lack of reddening means that little dust exists in a galaxy. Heavily reddened stars are also heavily obscured and do not contribute to the emerging light. For small optical depths, scattered light is bluer than the dust-free source color, for all geometries (WTC92).

- $\tau_{eff} \ll \tau_V$, where $\frac{I}{I_o} = e^{-\tau_{eff}}$, and τ_V is the total optical depth (as seen against a background source). Most observational techniques derive τ_{eff}, whereas the total amount of dust is measured by τ_V.

- For the same amount of dust, models with scattering produce much less reddening. The calculated reddening vectors diverge significantly in direction and magnitude from the *standard reddening law* for all geometries and all wavelength bands explored by WTC92. There is no *standard reddening law*.

- For some plausible geometries blue composite colors are produced by both dust-free and extremely dust-rich environments.

- WTC92 suggest to use colors derived from dust-fre SED's in cosmological applications, rather than using corrections which ignore scattering by dust (Wang 1991).

- Calzetti et al. (1994) conclude that their sample of starburst galaxies can be understood only if the obscuration law in these systems is more gray than the Milky Way extinction law, and if the 2175 Å feature is absent. This conclusion is reached by means of an approximate analytical solution to the RTE for a simple plane parallel geometry. It should be investigated if their conclusion is model dependent.

4. Spectral evolution and radiative transfer

Elsewhere in this volume, Magris & Bruzual (1994, hereafter MB94) have described a radiative transfer model of galaxy disks which takes into account the relative differences in the vertical scale height of early-, intermediate-, and late-type stars and dust grains. The population synthesis models of Bruzual & Charlot (1993, hereafter BC93) are used to describe the evolving spectrum of each stellar component. Despite its geometrical simplicity (each component is assumed to fill an infinite disk; the distribution of dust and stars is homogeneous inside each slab) this model provides a unique opportunity to explore the combined effects of spectral evolution

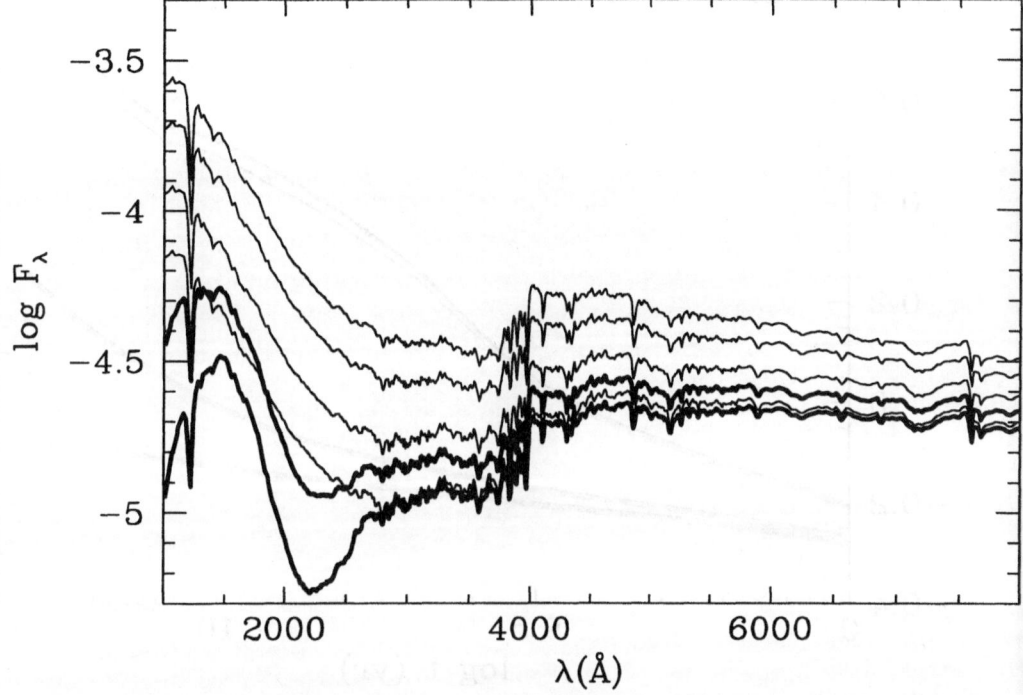

Figure 1. Evolving SED of a BC93 model in which stars form following the Salpeter (1955) IMF according to the SFR $\Psi(t) = 1 M_\odot \tau^{-1} \exp(-t/\tau)$, for $\tau = 6$ Gyr. The SEDs are shown at the age of 8, 10, 13 and 16 Gyr (thin solid lines, top to bottom). The heavy lines represent the emerging radiation from a disk with $\tau_V = 1$ (top heavy line) and $\tau_V = 2$ (bottom heavy line) at the model age of 13 Gyr, according to MB94.

and radiative transfer in a galaxy disk. Here I discuss some of the observational implications of the Magris & Bruzual (1994) model. All the results from this models are shown for a face-on system ($\mu = 1$).

Fig. 1 shows clearly how, depending on the wavelength range, the effects of dust can mimic spectral evolution. The $\tau_V = 1$ 13-Gyr line resembles the 14.5-Gyr dust-free SED longward of 3000 Å. The $\tau_V = 2$ 13-Gyr line is almost identical to the 16-Gyr dust-free SED in the same wavelength range. Below 3000 Å the effects of dust are stronger and the SED becomes fainter relative to optical wavelengths. Thus in the UV the $\tau_V = 1$ 13-Gyr line matches the level of the 16-Gyr dust-free SED.

Fig. 2 shows that the amount of reddening in a given color in the MB94 models depends on the SFR. In the rest frame U-B the reddening by dust is higher for models representing early type galaxies than for the later types. In the reddest model in Fig. 2 the stellar population is dominated by the

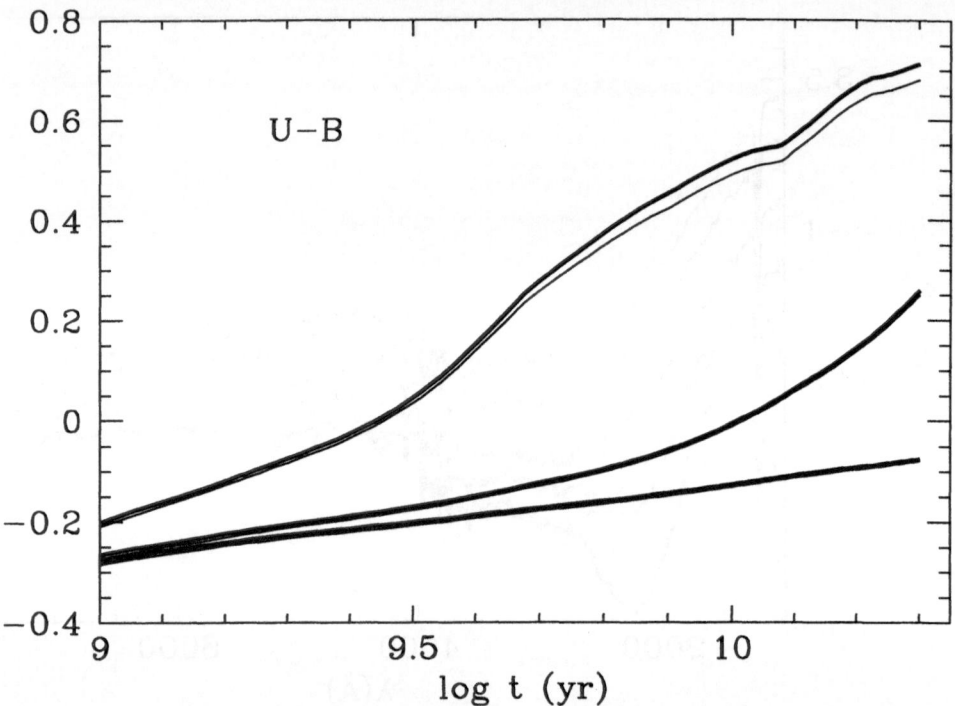

Figure 2. Evolution in the galaxy rest frame of the U-B color for BC93 models with SFR $\Psi(t) = 1M_{\odot}\tau^{-1}\exp(-t/\tau)$, for $\tau = 1$, 6, and 100 Gyr (corresponding to a constant SFR) assuming the Salpeter IMF. The thin lines represent the dust-free models. The heavy lines are from MB94 for $\tau_V = 1$. The effects of dust are not noticeable in the scale of this plot for the two bluer models.

"Rest" stellar group (in MB94 notation), which, as indicated by MB94, is the one more affected by the presence of dust in their particular model discussed here. In the bluer models the contribution of the Rest population is not significant and scattering by dust grains compensates the reddening expected from non-dispersive dust grains.

The dependence of color with redshift is shown in Fig. 3 for several bands. As expected from Fig. 1, at low z the colors of the dusty disks match those of the dust-free systems (little reddening at optical wavelengths is seen in Fig. 1). Once the rest-frame UV below 2500 Å enters the filter at the left of the filter pair, the color of the dusty disk becomes significantly redder than the dust free color. Below 2000 Å the slope of the dusty disk SED is more negative (bluer) than that of the dust free disk (Fig. 1). When both filters in the pair see the region below 2000 Å the dusty disks look to the observer bluer than the dust free disks.

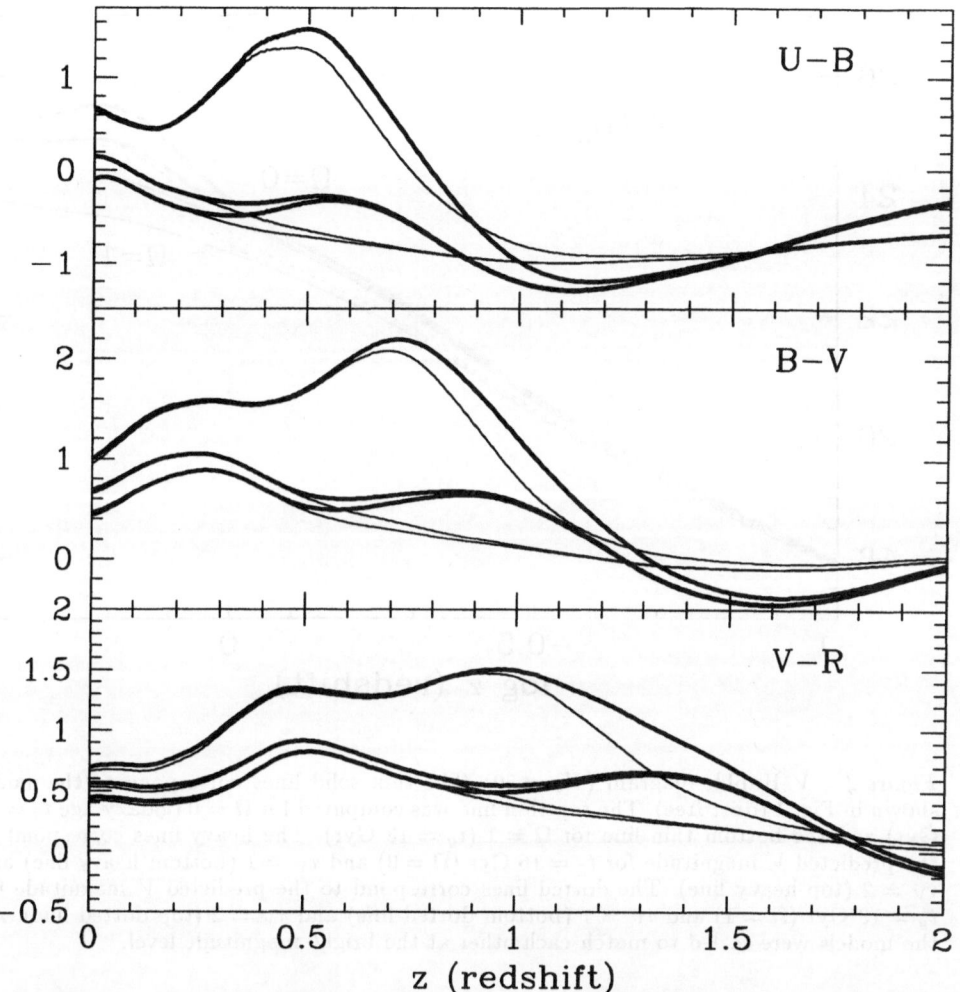

Figure 3. Color vs. redshift lines in the observer frame for the same models of Fig. 2. The thin lines correspond to the dust-free SEDs and the heavy lines to the MB94 results for $\tau_V = 1$. The colors are shown for $H_o = 50$, $\Omega = 0$, and galaxy age $t_g = 16$ Gyr.

The effects of dust in disks is seen quite dramatically in the V-Hubble diagram of Fig. 4. In an $\Omega = 1$ universe dust in disks will bias the determination of Ω towards lower values. The difference between the dust free and dusty disk lines depends on z, and hence cannot be accounted for by a constant shift in the magnitude scale. Dust thus introduces another source of uncertainty in the determination of Ω. Do standard candles have dust?

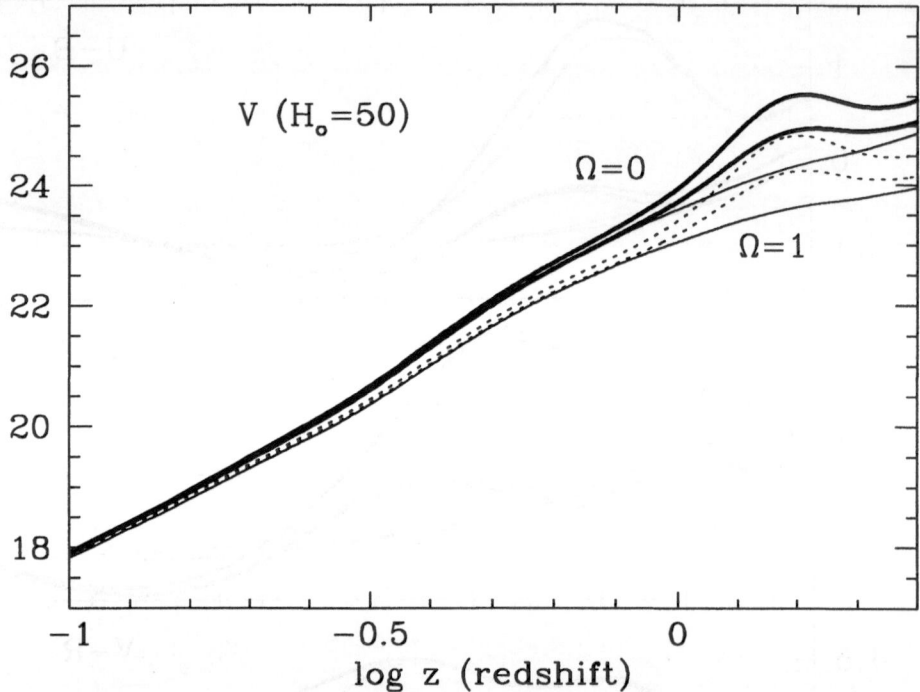

Figure 4. V-Hubble diagram ($H_o = 50$). The thin solid lines correspond to the model shown in Fig. 1 (dust free). The top thin line was computed for $\Omega = 0$ (galaxy age $t_g = 16$ Gyr) and the bottom thin line for $\Omega = 1$ ($t_g = 13$ Gyr). The heavy lines correspond to the predicted V magnitude for $t_g = 16$ Gyr ($\Omega = 0$) and $\tau_V = 1$ (bottom heavy line) and $\tau_V = 2$ (top heavy line). The dotted lines correspond to the predicted V magnitude for $t_g = 13$ Gyr ($\Omega = 1$) and $\tau_V = 1$ (bottom dotted line) and $\tau_V = 2$ (top dotted line). All the models were scaled to match each other at the bright magnitude level.

5. Conclusions

Considerable effort has been put into building *naive* models for the RTP in disk galaxies. Qualitatively all these models reach the same conclusions, despite the different geometries and solution algorithms used:

- Galaxy disks may be optically thick at optical wavelengths and still appear to us optically thin.

- For the same total amount of dust, models which include scattering produce much less reddening than the overlying screen model. All authors recommend *not to use* this model.

- There does not seem to exist a *standard reddening law*. It is model dependent.

- The combined effects of spectral evolution and radiative transfer may complicate the simple interpretation of observational results. For instance, the colors of local Scd-Sdm galaxies can be reached by earlier type systems at some range in z (Fig. 3).

Despite these efforts there has been it little absorption of these theoretical results in the observational work. Possible reasons are:

- Models are too naive.

- Smarter observational techniques may have to be implemented to establish once and for all the role of dust in spiral disks.

- The problem is quite complex and it may be too optimistic to expect that a given class of galaxies can be characterized uniquely according to its radiative transfer properties.

References

1. Bruzual A., G., & Charlot, S. 1993, ApJ, 405, 538 (BC93)
2. Bruzual A., G., Magris C., G., & Calvet, N. 1988, ApJ, 333, 673 (BMC88)
3. Calzetti, D., Kinney, A. L., & Storchi-Bergmann, T. 1994, ApJ, 429, 582
4. Davies, J. I. 1990, MNRAS, 245, 350
5. Davies, J. I., Phillips, S., Boyce, P. J., & Disney, M. J. 1993, MNRAS, 260, 491
6. Di Bartolomeo, A., Barbaro, G., & Perinotto, M. 1993, Mem. Ital. A. S., in press
7. Disney, M. J., Davies, J. I., & Phillips, S. 1989, MNRAS, 239, 939 (DDP89)
8. Henyey, L. G., & Greenstein, J. L. 1941, ApJ, 93, 70
9. Kylafis, N. D., & Bahcall, J. N. 1987, ApJ, 317, 637
10. Magris C., G., & Bruzual A., G. 1994, this volume
11. Salpeter, E.E. 1955, ApJ, 121, 161
12. Wang, B. 1991, ApJ, 383, L37
13. Witt, A. N. et al. 1994, ApJ, in press
14. Witt, A. N., Thronson, H. A., & Capuano, J. M. 1992, ApJ, 393, 611 (WTC92)

Question
Jörsäter

You didn't include any scattering in your models, did you?

Answer
Bruzual

That means that these models still don't tell us how complex reality is.

RADIATIVE TRANSFER IN DUSTY GALAXIES

A. N. Witt
The University of Toledo
Toledo, OH 43606 USA

ABSTRACT. The efficient conversion of UV/optical stellar radiation into mid- and far-infrared thermal emission by dust observed to occur in disk galaxies suggests that interstellar dust must be present in such systems in substantial quantities. The associated optical depths could well be in a range high enough to render large fractions of the disks of late type galaxies optically thick. Yet, the UV/optical spectral energy distributions of disk galaxies show little evidence of reddening effects. Four effects will be discussed which counteract reddening and have the ultimate result of substantially reducing the effective optical depth of widely distributed dust in galaxies. These effects are increased star formation made possible by large amounts of molecular gas associated with optically thick dust, scattering by dust, the distribution of stars in optical depth space, and finally the clumpy structure of the interstellar medium. As a consequence of these four phenomena, substantial amounts of dust can be present in galaxies without producing strong effects in the UV/optical spectral energy distribution.

1. Introduction

Interstellar dust provides the principal source of continuous opacity in disk galaxies at wavelengths longward of the Lyman limit. It is an interesting fact of nature that the wavelength regions of maximum opacity of typical interstellar grains and of maximum emissivity of the photospheres of the dominant stellar sources in galaxies essentially coincide, providing the basis for the effective interaction of dust and the interstellar radiation field. Toward longer wavelengths, the dust opacity declines as $\lambda^{-\beta}$, with $1 \leq \beta \leq 2$, while the stellar emissivity drops as λ^{-4}. Hence, if enough dust is present to intercept a significant fraction of the optical and UV interstellar radiation energy, the differences in the λ-dependence of stellar emissivity and dust opacity assure that the mid- and far-infrared emission from heated dust totally dominates the galaxies' spectra in these spectral regions ($\lambda \geq 10\mu m$). Thus, it was the detection of approximately 20,000 galaxies at far-IR wavelengths in the IRAS survey that provided

43

J. I. Davies and D. Burstein (eds.), The Opacity of Spiral Disks, 43–53.
© 1995 *Kluwer Academic Publishers.*

a clue to the importance of dust in galaxies on a global scale. The qualitative similarity of the mid- and far-infrared spectra of a large number of galaxies, including our own (e.g. Rowan-Robinson 1992), suggests that essentially the same dust opacity law applies in all, and the physics of the emission process is the same.

Unfortunately, the detection of radiating dust at the IRAS bandpasses alone does not form a reliable basis for determining the *amount* of dust present, without a simultaneous knowledge of the temperature distribution of such dust. Hence, major dust-related issues remain open and are important issues for discussion, e.g. the opacity of spiral disks (Disney, Davies, & Phillipps 1989; Valentijn 1990; Burstein, Haynes, & Faber 1991; and Bosma et al. 1992); the source of dust heating in disk galaxies (star forming regions; interstellar radiation field) (e.g. Engargiola 1991); the attenuation of optical luminosity of galaxies by dust (e.g. Davies 1990); the effect of internal obscuration on galaxy number counts and redshift distributions (Wang 1991); the possible obscuration of the distant Universe by the collective opacity of foreground galaxies (Wright 1990); and the effect of internal dust on the optical spectral energy distributions (SED) of galaxies (Evans 1992; Witt, Thronson, & Capuano 1992). In what follows, I shall primarily discuss various radiative transfer aspects which address the SEDs of dusty disk galaxies.

2. Why do Dusty Disk Galaxies Appear Unreddened?

In Fig. 1, I show the (B-V) colors, corrected for foreground extinction by Galactic dust, of a large number of disk galaxies with a wide range of ratios of infrared to optical luminosity ($0.01 < L(IR)/L(Opt) < 20$; Young et al. 1989; Soifer et al. 1987). If increasing $L(IR)/L(Opt)$ indicates increasing dustiness, no obvious trend of color with amount of dust is discernable. Indeed, up to $L(IR)/L(Opt) \simeq 0.5$, the average (B-V) color appears to become *bluer* by 0.2 to 0.3 mag with increasing $L(IR)/L(Opt)$, followed by a gradual reddening trend of similar magnitude for the more extreme dust-rich galaxies.

This remarkable behavior can be attributed to any of four mechanisms, although they likely work in conjunction and re-enforce each other:

- Increased star formation in ISM rich galaxies
- Scattering by dust
- Distribution of stars in optical depth space
- Clumpy structure of the ISM in galaxies

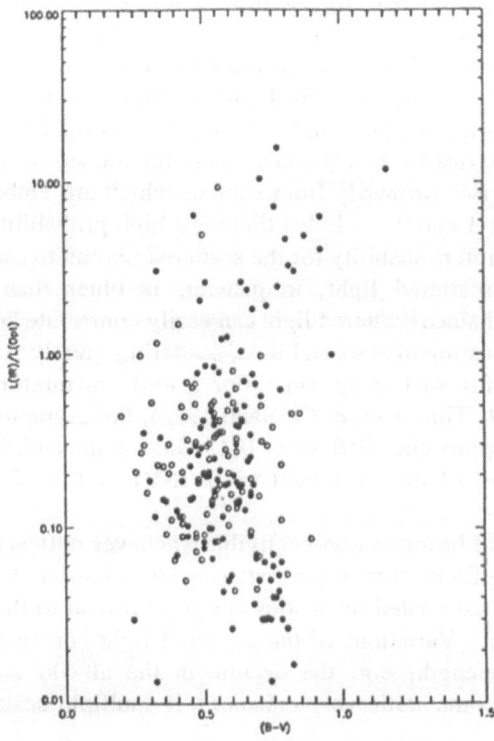

Figure 1. Distribution of the fractional infrared emission
L(IR)/L(Opt) vs. (B-V) color for infrared-bright spiral galaxies.
Open circles were taken from Young et al. (1989), and the
filled circles were taken from Soifer et al. (1987).

2.1. STAR FORMATION

The initial trend toward bluer (B-V) colors for small L(IR)/L(Opt) almost certainly reflects the growing disk-to-bulge light ratio; the effect of more O-B stars adding their light to the system more than balances any potential reddening effects due to the dust. Eventually, however, dust effects must become important, although they are greatly moderated when scattering, distribution of stars in optical depth space, and the clumpy structure of the ISM are included in the consideration of the radiative transfer.

2.2. SCATTERING

The importance of scattering at any wavelength is primarily determined by the dust albedo; a compilation of empirically determined albedo values has been published by Gordon et al. (1994), covering the 550-100 nm range. To these, one could add albedo values at R and I by Witt, Oliveri, & Schild (1990), and at K by Witt et al. (1994). These sources indicate a fairly high albedo in the range 0.6-0.7, for most of the near-IR, optical and UV, with exception of the region around the 217.5 nm UV extinction bump and the far-UV rise ($\lambda < 130$ nm), where the albedo declines to about 0.45.

Scattered light arises primarily from sources which are embedded in dust by an optical depth between 1 and 2, such that there is a high probability for a photon to be scattered and still a high probability for the scattered photon to escape without further interaction. Since scattered light, in general, is bluer than the light of stars responsible for it, and since scattered light can easily contribute 20 to 30% of the total light of a system consisting of stars and dust, scattering greatly reduces the reddening otherwise predicted for such a system. For a uniform mixture of stars and dust ("dusty galaxy"; Witt, Thronson, & Capuano 1992), the asymptotic reddening of the integrated star light approaches E(B-V) = 0.4, whereas the addition of scattered light reduces the reddening of the total light to E(B-V) = 0.2, for an optically thick system.

Multiple scattering becomes non-negligible whenever optical depths exceed about 0.3. The relative magnitude of multiple scattering contributions depends strongly upon the albedo, since the associated intensities are proportional to the second and higher powers of the albedo. Variations of the scattered light contributions due to albedo variations with wavelength, e.g. the decline in the albedo across the 217.5 nm extinction hump, become noticeably enhanced if multiple scattering is taken into account.

Scattering by dust at optical and UV wavelengths is strongly forward directed. The average value of the cosine of the scattering angle, weighted by the phase function, is in the range 0.6 to 0.8, increasing toward shorter wavelengths (Witt et al. 1992). In a system in which stars and dust are thoroughly mixed, this asymmetry of the scattered light pattern is largely self-canceling; thus, model results do not depend sensitively upon the assumed value of the asymmetry parameter (e.g. Gordon et al. 1994).

Very influential for determining the scattered light contribution to the total light of a galaxy is the structure of the interstellar medium. On the one hand, a homogeneous dusty medium with embedded stars is highly effective at scattering, with single scattering being the dominant contribution; on the other hand, a highly clumped medium consisting of externally illuminated optically thick clouds with forward-scattering grains directs most of the scattered photons toward the cloud interiors, where they suffer absorption as their most likely fate. The resultant scattered light contribution to an outside observer may therefore be reduced. The real environment within a disk galaxy is probably intermediate between these extremes. Since most current radiative transfer models for galaxies, which include scattering, are based on the assumption of continuous homogeneous scattering media, their predicted scattered light contribution is generally (but not always) an overestimate.

2.3.   SCATTERING IN FACE-ON DISK GALAXIES

The geometry of a dusty disk galaxy viewed at small inclination angles presents a special condition where, depending upon the dust albedo and the optical thickness of the disk, the amount of light scattered toward the observer may be larger than the amount of attenuation of the line of sight sources, resulting in a net disk brightening compared to the dust-free case.   This is a direct result of the non-isotropy of the illuminating radiation field seen by scatterers in the disk.   In Fig. 2, I show the computed effective attenuation due to internal dust of a doubly exponential disk galaxy as a function of inclination and albedo for a central optical thickness of 5.

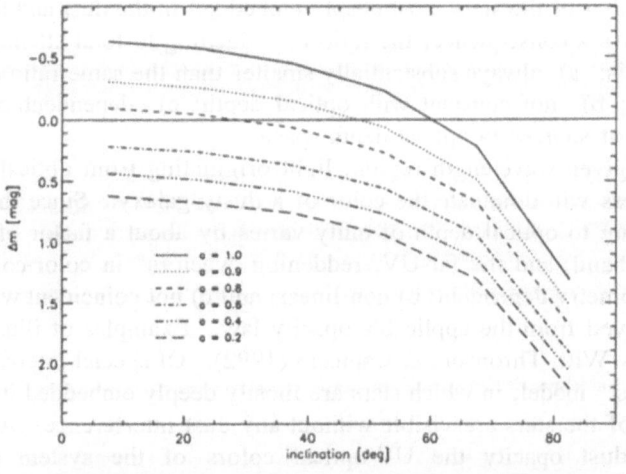

Figure 2.   Model computations of the effective attenuation of a doubly exponential disk galaxy with a central optical thickness of 5., with dust albedos in the range $0.2 \leq a \leq 1.0$. Negative values of Δm designate disk brightening.

The dust-free case is represented by the horizontal solid line at $\Delta m = 0$; negative Δm values indicate disk brightening.   The amplitude of the face-on attenuation correction increases with increasing albedo and is lowest when scattering is ignored.   This may be of importance when face-on magnitude corrections are applied to highly inclined galaxies as part of the Tully-Fisher procedure.

The disk brightening phenomenon is relatively insensitive to the asymmetry of the scattering phase function.   As an example, for a central optical depth of unity and an albedo of 0.6, the net attenuation varies from $\Delta m = -0.06$ for an isotropic phase function to $\Delta m = +0.07$ for a strongly forward throwing phase function, with essentially zero attenuation for a intermediate cases. These conditions may come close to describing near-face-on disk galaxies in the K-band, where the effects of internal dust seem to disappear. (Witt et al. 1994).

2.4. DISTRIBUTION OF STARS IN OPTICAL DEPTH SPACE

The most profound cause for the relative absence of reddening effects in dusty galaxies is the dust-star geometry. In contrast to a single star, which is viewed through a foreground dust screen and whose apparent colors are altered in accordance with the opacity law of the dust, the color of a dusty star system, in which stars are distributed throughout the dust as well as in front of the dust, is determined by the apparent brightest and usually least reddened stars in the ensemble. The attenuation of the total light of the system, on the other hand, is determined by the luminosity and relative number of the stars most deeply embedded in the dust and thus most heavily obscured. As a consequence, the ratio of reddening to total attenuation for a dusty star system is: a) always substantially smaller than the same ratio derived from the opacity law; b) not constant with optical depth; c) dependent upon the detailed distribution of sources in optical depth space.

In any given wavelength region, light originating from optical depths of order unity and less will dominate the color of a dusty galaxy. Since the physical depth corresponding to optical depth of unity varies by about a factor of 30 between the near-IR (K-band) and the far-UV, reddening "vectors" in color-color diagrams are also, a) geometry-dependent; b) non-linear; and c) not coincident with the reddening vectors derived from the applicable opacity law. Examples of illustrative cases are published by Witt, Thronson, & Capuano (1992). Of special interest among these is the "starburst" model, in which stars are mostly deeply embedded in dust, but where some 13% of the stars are visible without any dust interference. As expected, with increasing dust opacity the UV-optical colors of the system are increasingly determined by the light of the unobscured stars, yielding a SED essentially unaffected by reddening in the limit of extremely high dust opacity.

It seems not unreasonable to expect the spatial distribution of different spectral types and different luminosity classes of stars to be different relative to the dust distribution in a star system. Therefore, attempts to derive "effective attenuation laws" by dividing the SED of a dusty galaxy by the SED of a less dusty but otherwise similar galaxy may have more to do with dividing the SED of a B-star by that of an O-star than with deriving anything meaningful about the wavelength dependence of the dust opacity.

The distribution of stars in optical depth space also profoundly influences the ratio of far-IR and optical luminosities. In Fig. 3 this ratio is shown for three different configurations of the same amount of dust and the same number of stars.

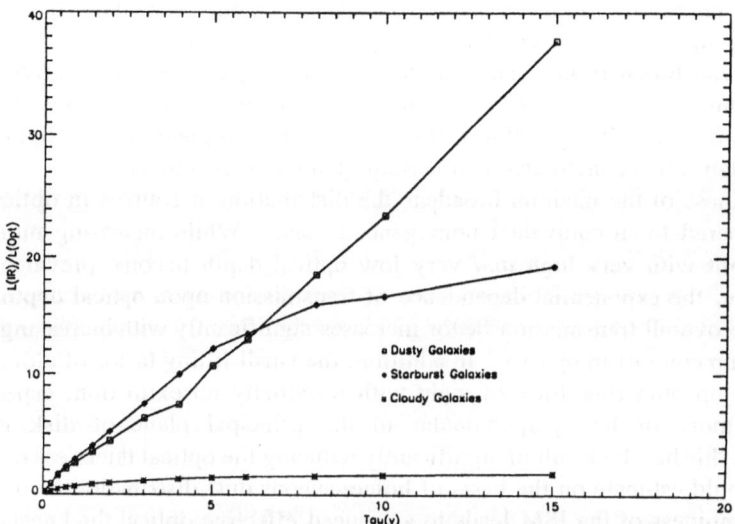

Fig. 3 — The ratio of far-IR to optical luminosity for three
different geometries involving the same number of stars and the
same dust mass.

A "dusty galaxy" modeled by a spherical uniform mixture of stars and dust yield a
ratio L(IR)/L(Opt) which is linearly increasing with optical depth, e.g. L(IR)/L(Opt)
$\approx 0.5 \, \tau_v$. The "starburst galaxy" and the "cloudy galaxy" model have, respectively,
13 % and 67 % of their stars located in dust-free regions. Since most of their light will
escape as optical radiation without being converted into IR, the ratios L(IR)/L(Opt)
will asymptotically approach limiting values representative of these geometries,
independent of the optical depth found in the dust-filled region.

For a dusty disk galaxy, the distribution of sources in optical depth space depends
on the inclination of its axis. If the galaxy is optically thin to its own far-IR emission
and if its optical magnitude varies by two magnitudes (Byun, Freeman & Kylafis
1994) between edge-on and face-on view ($\tau_0 = 5$, a = 0.6 see Fig. 2), its ratio
L(IR)/L(Opt) varies by more than a factor six due to inclination alone. Thus,
L(R)/L(Opt) is not a useful indicator of dust mass, unless the geometrical details, i.e.
the distribution of sources in optical depth space, are known.

2.5. STRUCTURE OF THE INTERSTELLAR MEDIUM

A combination of several astrophysical conditions and processes (e.g. Hartquist 1994
for a recent review), among these the nature of the radiative cooling function of the
ISM, stellar winds, supernova ejecta and shock waves, star formation, stellar mass
loss, and self-gravitation of dense molecular clouds, provide the basis for the presence

of a multi-phase interstellar medium, in which clumps with significant optical thickness occur with a relatively small filling factor (Murthy, Walker, & Henry 1992; Gaustad & van Buren 1993). This fact has been largely ignored in radiative transfer studies of dusty galaxies carried out so far. Still, the work by Boisse (1990) and by Hobson and Scheuer (1993) dealing with the penetration of plane-parallel light through a clumpy slab, allows us to arrive at certain qualitative conclusions.

Clumpiness of the medium broadens the distribution of sources in optical depth space compared to an equivalent homogeneous case. While increasing numbers of lines of sight with very high *and* very low optical depth become prevalent in the clumpy case, the exponential dependence of transmission upon optical depth assures one that the overall transmission factor increases significantly with increasing clump-to-interclump contrast in opacity. In addition, the small filling factor of clouds in the ISM opens up numerous lines of sight with essentially no extinction, especially in directions more or less perpendicular to the principal plane of disk galaxies. Altogether, this has the result of significantly reducing the optical thickness of the disk that one would estimate on the basis of homogeneous dust distribution models.

As clumpiness of the ISM leads to a reduced effective optical thickness, greater fractions of the direct starlight can escape from the system. The effect of clumpiness on the escape of scattered light depends on the optical thickness of the environment to begin with. In the optically thick limit, the probability of scattering is high but the probability of escape is low. A reduction in the effective optical depth will affect the probability of escape before it changes the probability of scattering; thus the amount of scattered light escaping from a clumpy, optically thick medium will be greater than the amount expected from an equivalent homogeneous distribution. In the optically thin limit, the escape probability is near unity but the probability of scattering varies as the optical depth; hence, the amount of scattered light emanating from an optically thin but clumpy medium will be lower than that expected from a homogeneous distribution.

The circumstances described above are modified by the fact that the effective albedo of an externally illuminated cloud is always lower than the albedo of the dust grains making up the cloud, because multiple scattering within the cloud greatly enhances the probability of absorption. This process is further amplified by the strongly forward directed scattering phase function of grains.

The light absorbed in a clumpy system remains fairly constant, as long as the product of the covering factor by clouds and the cloud opacity is reasonably constant. A noticeable change, however, must be expected in the wavelength dependence of attenuation. As the same opacity is provided by fewer and fewer clumps which are increasingly opaque, the attenuation will result in a wavelength-independent blocking of light for those wavelengths and shorter, for which the clumps are optically thick (Natta and Panagia 1984). The resulting effective attenuation law becomes increasingly grey in nature and features such as the 2175Å extinction hump and the far-UV rise are likely to disappear. The color of the residual star light will appear largely unreddened, despite the fact that the galaxy exhibits a large ratio of L(IR)/L(Opt), as shown in Fig. 1.

3. Conclusions

Widely distributed interstellar dust in disk galaxies is highly effective in preventing the escape of stellar radiation from such systems and converting it into thermal radiation emitted by the dust, as shown by the observations of numerous galaxies with ratios L(IR)/L(Opt) > 0.1. Yet, the familiar indicator of high extinction, i.e. reddening, is largely absent from the near-IR/optical/UV continua of the spectra of such galaxies. I have shown how four factors, generally working in combination with one another, are responsible for this phenomenon. These are the increased importance of star formation in systems rich in interstellar matter and three particular aspects of radiative transfer, all leading to reduced effective reddening in the face of large attenuation, namely the role of scattering, the distribution of sources in optical depth space, and the clumpy structure of the interstellar medium. Consequently, substantial quantities of dust, sufficient to render most parts of the disks of spiral galaxies optically thick at visible wavelengths, may be hidden in some galaxies without producing strong optical effects.

The advent of near-IR imaging with arc second spatial resolution offer one opportunity to trace the distribution of such dust, by comparing the surface brightness distribution of disk galaxies in the K-band, where the dust optical depth is low, one-tenth of that in V, and the V-band, where the emission is still dominated by the same stellar population but the dust optical depth is high (Block et al. 1994). A second opportunity for determining the amount of dust present arises through the advent of ISO and SCUBA, allowing one to trace the dust emission throughout the IR and sub-mm spectral regions with enough spatial and wavelength resolution so that the temperature distribution of the radiating dust can be determined reliably. Only then can dust masses be derived with some confidence (Hildebrand 1983).

Radiative transfer models applicable to galaxies in the future should incorporate the effects of multiple scattering, extended dust-star geometries with independent scale lengths and heights, and a cloudy-structured interstellar medium.

4. References

1. Block, D.L., Witt, A.N., Grosbøl, P., Stockton, A., and Moneti, A. (1994) Imaging in the optical and near-infrared regimes. II. Arcsecond spatial resolution of widely distributed cold dust in spiral galaxies, *Astron. & Astroph.* **288**, 383-395.
2. Boissé, P. (1990) Radiative transfer inside clumpy media: the penetration of UV photons inside molecular clouds, *Astron. & Astroph.* **228**, 483-502.
3. Bosma, A., Byun, Y., Freeman, K.C., and Athanassoula, E. (1992) The Opacity of Spiral Disks, *Astroph. J.* **400**, L21-L24.

4. Burstein, D., Haynes, M., and Faber, S. (1991) Dependence of Galaxy Properties on Viewing Angle, *Astroph. J.* **353**, 515-521.
5. Byun, Y.I., Freeman, K.C., and Kylatis, N.D. (1994) Diagnostics of Dust Content in Spiral Galaxies: Numerical Simulations of Radiative Transfer, *Astroph. J.* **432**, 114-127.
6. Davies, J.I. (1990) Missing mass or missing light?, *Mon. Not. R. Astr. Soc.* **245**, 350-357.
7. Disney, M., Davies, J., and Phillipps, S. (1989) Are galaxy disks optically thick?, *Mon. Not. R. Astr. Soc.* **239**, 939-976.
8. Engargiola, G. (1991) Origins of the 12-200 Micron Flux in NGC 6946: Starlight and Continuum Dust Emission from an Sc Galaxy, *Astroph. J. Suppl.* **76**, 875-910.
9. Evans, Rh. (1992) Opacity in Spiral Galaxies, *Ph.D. Thesis*, University of Wales.
10. Gaustad, J.E., and van Buren, D. (1993) The Distribution of Interstellar Dust in the Solar Neighborhood, *Publ. Astr. Soc. Pac.* **105**, 1127-1140.
11. Gordon, K.D., Witt, A.N., Carruthers, G.R., Christensen, S.A., and Dohne, B.C. (1994) The Far-Ultraviolet Dust Albedo in the Upper Scorpius Subgroup of the Sco OB2 Association, *Astroph. J.* **432**, 641-647.
12. Hartquist, T.W. (1994) The Global Structure of the Interstellar Medium, *Astroph. & Space Sci.* **216**, 185-200.
13. Hildebrand, R.H. (1983) The Determination of Cloud Masses and Dust Characteristics from Submillimeter Thermal Emission, *Q. Jl. R. Astr. Sco.* **24**, 267-282.
14. Hobson, M.P., and Scheuer, P.A.G. (1993) Radiative transfer in a clumpy medium — I. Analytical Markov-process solution for an N-phase slab, *M.N.R.A.S.* **264**, 145-160.
15. Murthy, J., Walker, H.J., and Henry, R.c. (1992) The Low Filling Factor of Dust in the Galaxy, *Astroph. J.* **401**, 574-583.
16. Natta, A., and Panagia, N. (1984) Extinction in Inhomogeneous Clouds, *Astroph. J.* **287**, 228-237.
17. Rowan-Robinson, M. (1992) Interstellar Dust in Galaxies, *Mon. Not. R. Astr. Soc.* **258**, 787-799.
18. Soifer, B.T., Sanders, D.B., Madore, B.F., Neugebauer, G., Danielson, G.E., Elias, J.H., Lonsdale, C.J., and Rice, W.L. (1987) The IRAS Bright Galaxy Sample. II. The Sample and Luminosity Function, *Astroph. J.* **320**, 238-257.
19. Valentijn, E. (1990) Opaque Spiral Galaxies, *Nature* **346**, 153-155.
20. Wang, B. (1991) Effects of Internal Absorption on Galaxy Number Count and Redshift Distribution, *Astroph. J.* **383**, L37-L40.
21. Witt, A.N., Lindell, R.S., Block, D.L., and Evans, Rh. (1994) K'-Band Observations of the Evil Eye Galaxy: Are the Optical and Near-Infrared Dust Albedos Identical? *Astroph. J.* **427**, 227-231.
22. Witt, A.N., Oliveri, M.V., and Schild, R.E. (1990) The Scattering Properties and Density Distribution of Dust in a Small Interstellar Cloud, *Astron. J.* **99**, 887-897.
23. Witt, A.N., Petersohn, J.K., Bohlin, R.C., O'Connell, R.W., Roberts, M.S., Smith, A.M., and Stecher, T.P. (1992) Ultraviolet Imaging Telescope Images of the Reflection Nebula NGC 7023: Derivation of Ultraviolet Scattering Properties of Dust Grains, *Astroph. J.* **395**, L5-L8.
24. Witt, A.N., Thronson, H.A., and Capuano, J.M. (1992) Dust and the Transfer of Stellar Radiation within Galaxies, *Astroph. J.* **393**, 611-630.
25. Wright, E.L. (1990) Are High-Redshift Quasars Hidden by Dusty Galaxies? *Astroph. J.* **353**, 411-415.
26. Young, J.S., Xie, S., Kenney, J.D., and Rice, W.L. (1989) Global Properties of Infrared Bright Galaxies, *Astroph. J. Suppl.* **70**, 699-722.

Question
Roberts

Would you like to make some clarifying remarks about extinction, absorption and attenuation.

Answer
Adolph Witt

I suggest that we use consistent terminology during this workshop. When referring to extinction by dust, we mean the reduction of flux by a combination of absorption and scattering caused by dust situated between the observer and a luminous object.

The reduction of surface brightness or change of integrated apparent magnitude of an extended source, such as a galaxy, due to embedded dust might best be referred to as attenuation. This attenuation is caused by internal absorption as well as partial re-direction of flux by scattering, while part of the scattered light may in fact be added to the integrated observed light. The faction of the returned scattered light depends on dust scattering properties and the dust-star geometry.

OPACITY DIAGNOSTICS IN SPIRAL GALAXIES

NIKOLAOS D. KYLAFIS
kylafis@iesl.forth.gr
University of Crete
Physics Department
714 09 Heraklion, Crete, Greece

ABSTRACT. I would like to review the best available observational diagnostics of dust content in spiral galaxies. These come from realistic models of spiral galaxies with immersed dust layers. The radiative transfer, including both scattering and absorption, has been computed for a range of model galaxies in various orientations. Standard galaxy surface photometry techniques were then applied to the numerical data to illustrate how different observables behave under given conditions of dust distribution. These observables are the total magnitude, the color and luminosity distribution, the apparent disk structural parameters and the amplitude of asymmetry between the near and far sides of the galaxy as divided by the apparent major axis. The dependence of the above observables on the orientation of the galaxy with respect to the observer is crucial and is presented here.

1. Introduction

The light emitted by the stars in a spiral galaxy is attenuated not only by the dust in our own galaxy, but also by the dust within the galaxy itself. In spite of numerous studies, the internal extinction within individual galaxies has remained poorly understood. On the other hand, the importance of internal extinction is tremendous. If galaxy disks contain a lot of dust and are therefore opaque, they can hide (in the optical part of the spectrum) all objects that lie behind them. Such a situation would have large implications for Cosmology.

Until recently, internal absorption in spiral galaxies was thought to be small. The conventional view that spiral galaxies are optically rather thin to visible radiation came from a study by Holmberg (1958). He observed 119 spiral galaxies at different inclinations and studied the variation of the mean surface brightness with inclination, interpreting the relation as implying a small amount of absorbing matter. Two major catalogues, the *Second Reference Catalogue of Bright Galaxies* (de Vaucouleurs, de Vaucouleurs, & Corwin 1976) and the *Revised Shapley-Ames Catalog of Bright Galaxies* (Sandage & Tammann 1981), also adopt low values of face-on extinction in the blue.

For the edge-on galaxy NGC 891 there is an accurate determination of the dust distribution (Kylafis & Bahcall 1987). *Despite its prominent dust lane, NGC 891 would be optically thin if it were seen face-on, except for a small part near its*

J. I. Davies and D. Burstein (eds.), The Opacity of Spiral Disks, 55–66.
© 1995 *Kluwer Academic Publishers.*

center where the optical depth would be of order unity.

Recently, there have been suggestions that galaxy disks are optically thick. Disney, Davies, & Phillipps (1989) suggested that the variations of the total or mean surface brightness and color with inclination are not reliable indicators for internal extinction, since these relations are highly model dependent and the patchiness of dust distribution biases the determination of optical depth towards lower values. They also suggested that the optical thickness in the V-band of metal-rich, luminous face-on spiral galaxies may be much larger than 10. If spiral galaxies were so optically thick, we would see light only from one side of them.

Valentijn (1990) looked at a large number of galaxies and reported that the central surface brightness and the brightness at half-light radius do not show any significant change with inclination. He concluded from this that spiral galaxies are opaque across their entire disks including their outermost parts.

Burstein, Haynes, & Faber (1991) and Chołoniewski (1991) took approaches different from Valentijn's and reported that the isophotal diameters do not show the inclination dependence expected for optically thin disks and concluded that spiral disks are optically thick even in the outer regions. The authors of RC3, the *Third Reference Catalogue of Bright Galaxies* (de Vaucouleurs et al. 1991), also adopt the view of optically thick disks and choose to apply no inclination correction to the diameters of spiral galaxies. Other reports in favor of significant optical depth are in Davies (1990), Cunow (1992) and James & Puxley (1993). Peletier & Willner (1992) and Davies et al. (1993) also suggest that the observed constancy of disk central surface brightness (Freeman 1970; Bosma & Freeman 1993) might be the result of significant optical depth in the central region of galactic disks.

There is, however, another group of recent studies with different conclusions. Huizinga & van Albada (1992) used the same database (Lauberts & Valentijn 1989) that Valentijn (1990) used but arrived at the opposite conclusion, namely that the isophotal diameters *increase* with inclination and the optical depth is significant only in the central region. Based on photometric asymmetries and the shape of rotation curves, Byun (1993) also found that the outer parts of disks are nearly transparent (see also Bosma et al. 1992).

The difficulty in the determination of dust content from optical observations arises mainly from the lack of well defined observable indicators. Observations are often compared with simple but very naive models, which are not only inadequate but also misleading.

In order to find the best observable diagnostics for the amount of internal extinction in spiral galaxies, Byun, Freeman, & Kylafis (1994, hereafter referred to as BFK) have constructed more realistic models for disk galaxies with immersed dust layers. The radiative transfer, including both scattering and absorption, has been computed for a range of model galaxies. For other radiative transfer calculations see Bruzual, Magris, & Calvet (1988) and Witt, Thronson, & Capuano (1992).

2. Simulations of Radiative Transfer

BFK have simulated spiral galaxies with realistic distributions for the stars and the dust (see below). The radiative transfer was performed with the code developed by Kylafis & Bahcall (1987), which was modified and vectorized for their needs. A detailed discussion of the approximations made in the calculation of the radiation transport is given by Kylafis & Bahcall (1987).

2.1. MODEL GALAXY

The distribution of stars and dust in real disk galaxies is fairly clumpy on small scales. BFK do not attempt to model such clumpiness, instead they are interested in the large scale effects of dust on the surface photometry. Thus, they adopt simple, smooth distributions for both the stars and the dust.

They take the emissivity of the stars in the disk to be

$$L(R, z) = L(0, 0) \exp(-\frac{R}{h_s} - \frac{|z|}{h_z}) , \tag{1}$$

and the absorption coefficient due to dust as

$$A_\lambda(R, z) = A_\lambda(0, 0) \exp(-\frac{R}{h_d} - \frac{|z|}{h_{dz}}) , \tag{2}$$

where $L(0,0)$ is the stellar emissivity at the center of the disk, $A_\lambda(0,0)$ is the absorption coefficient at wavelength λ at the center of the disk, h_s, h_d are the scalelengths and h_z, h_{dz} are the scaleheights of the stars and dust respectively, R is the (cylindrical) radial distance from the galactic center and z is the vertical distance from the galactic plane.

The dust distribution in each model galaxy is parameterized by the total optical depth through the center of the face-on disk in the visual passband, i.e.,

$$\tau_V(0) = \int_{-\infty}^{\infty} A_V(0, z)dz = 2A_V(0, 0)h_{dz} , \tag{3}$$

where $A_V(R, z)$ is the absorption coefficient in the visual passband at any point (R, z). For calculations in other passbands BFK used the interstellar extinction law by Rieke & Lebofsky (1985) to account for the wavelength dependence of $A_\lambda(0, 0)$.

In order to simulate the Hubble sequence, BFK added stellar bulge components to the disk model. The relative significance of the bulge component is known to correlate with Hubble type (e.g., Simien & de Vaucouleurs 1986). Therefore, they used bulge-to-total light ratios of 0.5, 0.3, 0.1 and 0.0 to represent typical Sa, Sb, Sc and Sd spirals, as suggested by the work of Simien & de Vaucouleurs (1986). These models are designated as BT0.5, BT0.3, BT0.1, and BT0.0 throughout this paper.

For the light distribution in the bulge, BFK used the table given by Young (1976) so that the projected profile follows an $R^{1/4}$ law and they assumed the bulge to be spherically symmetric. The total bulge luminosity depends on its effective radius and surface brightness. The bulge surface brightness relative to the disk decreases along the Hubble sequence but the trend of the relative effective radius with Hubble type is not clear (Kent 1985; Simien & de Vaucouleurs 1986). BFK adopt a constant effective radius for the bulge equal to 40% of the disk scalelength, which is consistent with Kent's (1985) results. The required bulge luminosity for each BT class is then set by adjusting the effective surface brightness μ_{eff} of the bulge.

The model of BFK is a closer approximation to real three-dimensional galaxies. The line-of-sight optical depth and the amount of extinction depend on both radius and inclination. The integrated quantities, such as total internal extinction, can be inferred *after* a model surface photometry for the galaxy has been computed.

The quantity $\tau_V(0)$ that BFK are using is a *local* parameter. The face-on optical depth in the V band at radius R is defined as $\tau_V(R) = \int_{-\infty}^{\infty} A_V(R, z)dz = \tau_V(0)\exp(-R/h_d)$.

2.2. CALCULATIONS

BFK have calculated the apparent two-dimensional luminosity distribution in the B and I passbands for model galaxies with inclinations $i = 0, 30, 50, 70, 80, 85$ and $90°$. Influenced by our own galaxy, they adopted a stellar scalelength $h_s = 4$ kpc, a stellar scaleheight $h_z = 350$ pc and a dust scaleheight $h_{dz} = 0.4h_z$. They assumed that the stars and the dust have the same scalelength, $h_s = h_d$ (cf. Kylafis & Bahcall 1987). The conclusions presented here *do not* depend sensitively on the specific values used for the parameters.

For the model galaxies with bulges the adopted effective radius R_{eff} of the bulge is 1.6 kpc and the effective surface brightness μ_{eff} is given by $\mu_{\text{eff}} - \mu(0)_{\text{disk}} = -0.596, 0.324, 1.790$ in the B band for BT0.5, BT0.3, BT0.1, respectively, to give the required bulge-to-total luminosity ratios of 0.5, 0.3, and 0.1 respectively. BT0.0 is a pure disk galaxy without any bulge.

BFK took the bulge luminosity in the I band to be one magnitude brighter than in the B band by increasing μ_{eff}. Thus, the bulge color is redder than the disk color by $\Delta(B - I) = 1$ and the bulge-to-total luminosity ratios in the I band are changed accordingly. A range of optical depths were used for the simulations, $\tau_V(0) = 0, 0.5, 1, 2, 5$ and 10, covering the range from transparent to fairly opaque.

3. Diagnostics

In what follows we study the effects of dust on 1) the major axis profiles, 2) the elliptically averaged profiles, 3) the disk central surface brightness, 4) the disk scalelength, 5) the minor axis profiles, 6) the total extinction, 7) the face-on correction and 8) the isophotal diameter.

3.1. MAJOR AXIS PROFILES

Several interesting points can be made when there is dust in the galaxy:

a) In galaxies with bulges and $\tau_V(0) \lesssim 2$ *there is hardly any extinction at low inclinations ($i \lesssim 30°$) even in the central part of the profile.* This is because the scattered light compensates for most of the absorbed light. The degree of extinction rapidly increases with inclination.

b) The major axis profiles become significantly flatter by dust extinction. Generally, the larger the amount of dust, the flatter the profile. At low inclinations ($i \lesssim 30°$), this distortion is negligible at most radii unless the optical depth is fairly large $[\tau_V(0) \gtrsim 5]$. At high inclinations ($i > 70°$), however, the major axis profiles are significantly distorted *even in the case where the disk optical depth is small.*

c) The flattening of the major axis profiles causes significant changes in the disk apparent central surface brightness. The disk central surface brightness does not show much of an inclination effect when $\tau_V \gtrsim 1$ (see §3.3). The effect is stronger in galaxies with bulges. This is because the extinction more than compensates for the enhancement by inclination. Therefore, *the disk central surface brightness is not a good diagnostic of dust content in spiral galaxies.*

   d) The flattening of the major axis profiles by dust extinction has a direct effect on the apparent disk scalelength. Its value increases with inclination and dust content. The change in disk scalelength with inclination and optical depth has not attracted any attention in previous observational studies but it is an important diagnostic of dust (see §3.4).

   The major axis B-I color distributions also prove to be useful diagnostics for dust content, if there is no intrinsic disk color variations with radius or if the intrinsic color variation is known. In the simulations of BFK they have assumed no intrinsic color variation across the disk. Nevertheless, in models BT0.5, BT0.3 and BT0.1 the redder bulge causes strong color gradients along the major axis.

   Several interesting remarks can be made regarding the B-I color distributions.

   a) In general, the dust reddens the central parts of the galaxies more efficiently than the outer parts and the distorted color profiles are well correlated with both dust optical depth and inclination.

   b) The disk dominated galaxies (BT0.0 models) show a strong color gradient generated by the dust. For such face-on galaxies, the color gradient is contained in annuli in which $1 \lesssim \tau_V(R) \lesssim 3$. Thus, *if a color gradient is detected in an annulus of a face-on disc galaxy, the optical depth is inferred to be of order unity there.*

   c) For inclined pure disk galaxies (BT0.0 models) the color gradient is contained in radial distances R along the major axis for which *the line of sight optical depth in the V band is of order unity.*

   d) The presence of a strong bulge (e.g., models BT0.5) weakens the color gradient diagnostic.

3.2. ELLIPTICALLY AVERAGED PROFILES

With the exception of early S0 galaxies, most disk galaxies have clumpy disks and the major axis profiles are affected by local irregularities. The low signal-to-noise ratio in the outer part of the disk also prevents one from getting reliable major axis profiles. For this reason BFK have applied the ellipse analysis package GASP (Cawson 1983) to the simulated galaxies in order to get elliptically averaged radial luminosity distributions. This averaged radial luminosity profile is directly comparable to real data because the observational analysis follows the same procedure.

   As expected, the elliptically averaged profiles for highly inclined systems are brighter than the corresponding major axis profiles, while low inclination systems have almost identical profiles to the major axis profiles. This suggests that in comparing data with models one must consistently use one-dimensional or two-dimensional profiles but never a mixture.

3.3. DISK CENTRAL SURFACE BRIGHTNESS

The disk central surface brightness is observationally defined by an extrapolation of the exponential disk toward the center (e.g., Freeman 1970). BFK derived such central surface brightness for BT0.0 galaxies using two-dimensional fitting and concluded that *even a small optical depth can significantly reduce the expected projec-*

tion effect especially in the B band.

The conventional measure of the amount of dust in a disk is the constant C in the empirical equation

$$\mu_{\text{obs}} = \mu_{\text{face-on}} - 2.5 \, C \, \log \, (a/b) \, , \tag{4}$$

where μ_{obs} and $\mu_{\text{face-on}}$ are respectively the observed and face-on surface brightnesses, a/b is the axial ratio, $C=1$ is the no internal extinction limit and $C=0$ is the totally opaque disk. Care must be exercised because this equation is correct *only* in the limits $C = 0$ and $C = 1$ and even then under the assumption of uniform mixture of stars and dust. In all intermediate cases its accuracy is not guaranteed. Valentijn (1990) found that the B-band disk central surface brightness of spiral galaxies in the ESO-LV catalogue (Lauberts & Valentijn 1989) indicates very low C values and thus concluded that disks in spiral galaxies are generally opaque.

The simulations of BFK show that this is not necessarily the case. Valentijn's C values of 0.3 - 0.5 for Sd galaxies correspond to $\tau_V(0) = 1 - 2$ in BFK. As it is shown later (see §3.6), $\tau_V(0) \lesssim 2$ is not much of an optical depth because the *effective* face-on total extinction is negligible. For earlier type galaxies, Valentijn finds smaller C values; about 0 for Sb and 0.1 for Sc galaxies. Note however, that the presence of a bulge makes it difficult to pick up the enhancement of central disk surface brightness. Both major-axis and elliptically averaged profiles given by BFK indicate that the extrapolated disk central surface brightness of large-bulged galaxies *does not change significantly with inclination even at low optical depth.*

3.4. RADIAL SCALELENGTH

The flattening of the radial profiles, when there is dust in the galaxy, results in an *observed* scalelength R_h for the stars which is larger than the *intrinsic* scalelength h_s. BFK have calculated how the observed scalelength relative to its face-on value changes with inclination for BT0.0 galaxies in the B and I bands. By comparing the I-band scalelength distributions of galaxies with low and high inclinations, one could in principle estimate the disk optical depth. This cannot be done with the B-band distributions because the curves are not well separated.

It is important to point out that the scalelength analysis presented above can be quite difficult in real observations. Galaxies cover a large range in size and the intrinsic scatter in scalelength is probably too large for this kind of approach to be directly useful. One possible solution is to normalize each scalelength by (the galaxy luminosity)$^{1/2}$ derived from its inclination-corrected rotation velocity width. This is likely to reduce the scatter significantly since, following from the approximate constancy of the disk central surface brightness (Freeman 1970), the intrinsic scalelength should be closely coupled to the luminosity of the galaxy.

A more direct procedure would be to compare the apparent scalelengths in different passbands for a given galaxy. BFK have computed the inclination dependence of the ratio of the B-band observed scalelength to the I-band one $R_h(B)/R_h(I)$. Note that the model galaxy (BT0.0) has no intrinsic color gradient, so its face-on $R_h(B)/R_h(I)$ is unity in the absence of dust. Likewise the ratio stays unity at all inclinations if galaxies are dust free. With dust, both the face-on ratio and the ratio at any inclination depend on the amount of dust.

It is important to remark that the measurement of $R_h(B)/R_h(I)$ for *face-on galaxies alone* may produce wrong results for the optical depth because any departure from unity can equally well be explained as an intrinsic color gradient.

From analysis of six face-on spirals Schweizer (1976) reported that, although old disks show no significant color gradient, the relative strength of the arm population is generally increasing outward making the galaxy look bluer in the outer part. If this is the case, the galaxy would have $R_h(B)/R_h(I)$ greater than unity and would be regarded as a high extinction case.

Therefore, one should look for the variation of the ratio $R_h(B)/R_h(I)$ with inclination rather than its values at a given inclination. Opaque galaxies [$\tau_V(0) = 5$ and 10] will show a constant or even decreasing ratio with inclination. Galaxies with moderate or small optical depth will exhibit a monotonically increasing ratio at high inclinations.

3.5. MINOR AXIS PROFILES

When a disk galaxy of the form studied here is seen edge-on, the surface brightness is symmetric with respect to the galactic plane. At lower inclinations, however, the symmetry with respect to the major axis is broken. This is because on one side of the major axis most of the light along the line of sight is above the dust while on the other side the opposite is true. The effect is stronger in galaxies with large bulges (BT0.5 and BT0.3). As a consequence, minor axis profiles exhibit a characteristic asymmetry which is entirely due to dust and to our knowledge it was pointed out by BFK for the first time. At a given inclination, the asymmetry depends on the type of galaxy and the optical depth.

Inspection of the Hubble Atlas (Sandage 1961) reveals that most inclined galaxies show such asymmetry between the two sides of the major axis. A direct comparison of the theoretical with the observed minor axis profiles can provide a good estimate of dust content.

We also note that the minor axis asymmetry is accompanied by a displacement of the apparent galactic center (i.e., the peak brightness) toward the far side along the minor axis. This displacement occurs in all BT model galaxies and at inclinations $i \gtrsim 50°$. The amount of displacement depends strongly on the optical depth. We are not aware of any observational reports on the dust-generated displacement of the apparent centers of galaxies. This seems to suggest that *galaxy disks are generally not optically thick near their centers.* We note, however, that in a recent observational study (Byun 1992), there is evidence for this kind of displacement in a few spiral galaxies.

Due to fluctuations in real galaxies, the usefulness of minor axis profiles is limited in a way similar to major axis profiles. To smooth out the effects of fluctuations, BFK defined an asymmetry amplitude as the difference between the integrated brightness for the far and the near halves of the galaxy. This quantity can be easily measured from the observed galaxy images and it is independent of distance and foreground extinction. *It is remarkable that for $\tau_V(0) = 10$, the asymmetry at $i = 85°$ is as much as 0.7 to 0.9 magnitude for all BT model galaxies. If disks are optically thick, this should be readily observable.*

3.6. TOTAL EXTINCTION

One of the most important questions in the study of internal absorption in galaxies is how much light a galaxy loses by the dust and how this varies with inclination.

One of the conclusions of BFK was that by ignoring scattering one makes a big error, especially at low inclinations. For instance, an optical depth of $\tau_V(0)=10$

would make a face-on pure disk (BT0.0) galaxy look 0.83 mag (in B band) fainter if scattering were ignored, but with scattering taken into account it is only 0.4 mag. A moderate optical depth of $\tau_V(0) = 2$ gives a negligible face-on extinction of 0.02 mag, but with scattering ignored it gives a non-negligible 0.3 mag. At small optical depths $[\tau_V(0) \lesssim 1]$, scattering makes a face-on galaxy look slightly *brighter* than when it is dust free. This is because light that would propagate close to the galactic plane is scattered into the face-on direction. This effect quickly disappears with inclination.

3.7. FACE-ON CORRECTION

An important quantity in galaxies is the magnitude correction required to convert the measured total magnitude of an inclined galaxy into its face-on value. The calculations of BFK have revealed the following points:

a) Comparison between pure absorption and scattering-included models shows that the face-on correction is generally *larger* if scattering is considered. This conclusion is *opposite* to that of RC3, where scattering is suggested to be responsible for reducing the amplitude of face-on correction.

b) The simulations show that in pure disk galaxies and in galaxies with small bulges it *is* possible to estimate the dust content from an analysis of the magnitude-inclination relation. For galaxies with large bulges this is not possible because the curves are intermingled.

c) At large optical depths, the amount of face-on correction increases with decreasing bulge size. The correction for an edge-on BT0.5 galaxy is less than half of that for a corresponding BT0.0 one with $\tau_V(0)$=5 or 10. For optically thin galaxies on the other hand $[\tau_V(0)$=0.5 and 1], the situation is reversed at low inclinations ($i \lesssim 80°$).

d) The wavelength dependence of internal absorption is often miscalculated. If the obscuring matter were located between the galaxy and the observer (i.e., as in the screen model of Holmberg 1958, 1975), the amount of internal absorption in one passband could be readily converted into another using some standard extinction law such as that of Rieke & Lebofsky (1985). However, the dust layers in spiral galaxies lie in the mid-plane and are mixed with stars. The extinction term is then no longer a simple function of wavelength. An internal absorption correction derived from a certain passband should not be applied to other passband data even after the application of the standard extinction law (see also Han 1992). In fact the study of BFK reveals that, although the amount of face-on correction in the I band is less than in the B band, the difference in the two corrections is smaller than one might expect from the standard extinction law. For example, for the BT0.0 pure absorption model with $\tau_V(0) = 5$ at $i = 80°$, we see that $\Delta M_B = 0.51$ and $\Delta M_I = 0.4$, so $\Delta M_B - \Delta M_I = 0.11$. Naive use of the standard extinction law would imply $\Delta M_B - \Delta M_I = 0.36$.

3.8. ISOPHOTAL DIAMETER

It is well known that if galaxies were completely free of dust, their surface brightness would increase with inclination due to the fact that as the inclination increases, so does the column density of stars. The apparent isophotal diameter of a galaxy would then increase with inclination, assuming of course that the isophotal level in question is within the stellar disk. As expected, BFK have found that the presence of dust reduces the apparent size of galaxies, *but not as much as one might intuitively expect. The isophotal diameter of a galaxy is still an increasing function of inclination.*

4. Conclusions

The most important diagnostics of dust content in spiral galaxies are:

a) For edge-on spiral galaxies there is a clear correlation between the shape of the isophotes and the dust content. In very optically thick galaxies the isophotes are closed curves in each of the two halves of the galaxy as divided by the dust lane. The amount of dust can be determined from the brightest isophote that surrounds the entire galaxy.

b) The minor-axis profiles in spiral galaxies with inclinations $0 < i < 90°$ show a characteristic asymmetry due to dust. Its shape and the amplitude of asymmetry between the two halves of the galaxy is a direct measure of the dust content.

c) The apparent galactic center of inclined galaxies is displaced from its true position when there is dust present. The amount of displacement depends strongly on the optical depth.

d) The total extinction in spiral galaxies correlates well with the dust content. It is therefore possible to estimate the amount of dust from an examination of the magnitude-inclination relation.

e) A color gradient is predicted in dusty spiral galaxies. This dust generated color gradient should be contained in radial distances along the major axis for which the line-of-sight optical depth in the V band is of order unity.

f) The inferred scalelength of a dusty spiral galaxy is predicted to be different in different bands. The ratio of the inferred B- and I-band scalelengths correlates with the amount of dust in the galaxy and can be used to estimate the dust content.

The calculations of BFK have revealed the following points which are contrary to common belief:

a) The disk central surface brightness is not a good diagnostic of dust content in the galaxy because even a small amount of dust can compensate for the expected increase in central disk surface brightness with inclination.

b) Using the empirical equation (4) to infer the transparency or not of a spiral galaxy can lead to large errors.

c) The face-on correction for an inclined spiral galaxy is generally larger than what simple estimates, which ignore the scattering by the dust, imply.

d) The internal extinction of a galaxy in one band cannot be converted to that in another band by simply using an extinction law. This works only when the dust is between the galaxy and the observer.

e) The apparent diameter of a spiral galaxy is expected to increase with inclination even if there is a large amount of dust in the galaxy.

f) An optical depth of order unity through the center of a face-on spiral galaxy implies that the galaxy is effectively transparent. However, if the same galaxy is seen edge-on it will still exhibit a prominent dust lane.

64

5. References

Bosma, A., Byun, Y. I., Freeman, K. C. & Athanasoula, E. 1992, ApJ, 400, L21
Bosma, A., & Freeman, K. C. 1993, AJ, 106, 1394
Bruzual, G. A., Magris C., G., & Calvet, N. 1988, ApJ, 333, 673
Burstein, D., Haynes, M. P., & Faber, S. M. 1991, Nature, 353, 515
Byun, Y. I. 1992, PhD Thesis, Australian National University
Byun, Y. I. 1993, PASP, 105, 993
Byun, Y. I., Freeman, K. C., & Kylafis, N. D. 1994, ApJ, 432, 114 (BFK)
Cawson, M. 1983, Ph.D. thesis, University of Cambridge
Chołoniewski, J. 1991, MNRAS, 250, 486
Cunow, B. 1992, MNRAS, 258, 251
Davies, J. I. 1990, MNRAS, 245, 350
Davies, J. I., Phillipps, S., Boyce, P. J., & Disney, M. J. 1993, MNRAS, 260, 491
de Vaucouleurs, G., de Vaucouleurs, A., & Corwin, H. G. 1976, *Second Reference Catalogue of Bright Galaxies*, (Austin:University of Texas Press)
de Vaucouleurs, G., de Vaucouleurs, A., Corwin, H. G., Buta, R. J., Paturel, G. & Fouqué, P. 1991, *Third Reference Catalogue of Bright Galaxies*, (New York: Springer-Verlag) (RC3)
Disney, M., Davies, J., & Phillipps, S. 1989, MNRAS, 239, 939
Freeman, K. C. 1970, ApJ, 160, 811
Han, M. 1992, ApJ, 391, 617
Holmberg, E. 1958, Medd. Lunds Astr. Obs., Ser II, Nr. 136
――― 1975, in *Stars and Stellar Systems*, Vol. IX, eds A. Sandage, & J. Kristian (Chicago: University of Chicago Press)
Huizinga, J. E., & van Albada, T. S. 1992, MNRAS, 254, 677
James, P.A., & Puxley, P. J. 1993, Nature, 363, 240
Kent, S. M. 1985, ApJS, 59, 115
Kylafis, N. D., & Bahcall, J. N. 1987, ApJ, 317, 637
Lauberts, A., & Valentijn, E. A. 1989, *The Surface Photometry Catalogue of the ESO-Uppsala Galaxies*, ESO-München
Peletier, R. F., & Willner, S. P. 1992, AJ, 103, 1761
Rieke, G. J., & Lebofsky, M. J. 1985, ApJ, 288, 618
Sandage, A. 1961, *The Hubble Atlas of Galaxies* (Washington: Carnegie Institution of Washington)
Sandage, A. & Tammann, G. A. 1981, *A Revised Shapley-Ames Catalog of Bright Galaxies*, (Carnegie Institution of Washington, Washington)
Schweizer, F. 1976, ApJS, 31, 313
Simien, F., & de Vaucouleurs, G. 1986, ApJ, 302, 564
Valentijn, E. A. 1990, Nature, 346, 153
Witt, A. N., Thronson, Jr., H. A., & Capuano, Jr., J. M. 1992, ApJ, 393, 611
Young, P. J., 1976, AJ, 81, 807.

Question
Neininger

In which way does the clumpiness of the dust influence your results about the opacity? NGC891 is known to have a rather clumpy and thick dust lane which is not directly expected from the long lines of sight involved.

Answer
Kylafis

The fact that NGC891 is seen edge-on means that a lot of the clumpiness along the line of sight is smeared out. Despite the appearance of NGC891, we found that the double exponential is a very good fit to the data. For face-on galaxies, the double exponential misses all the clumpiness. In such cases more detailed models must be constructed.

Question
Madore

You suggest that the central surface brightness is not a sensitive diagnostic of dust opacity; however, your diagrams still do indicate that the extrapolated central surface brightness (based on the exponential fall-off at large radii) does in fact respond sensitively to the dust content.

Answer
Kylafis

Not that sensitively. Note that the surface brightness curves more or less give the same value at $R=0$ for optical depths 0.5 to 10.

Comment
Evans

A comment on what will change the ratios of α_B to α_I (scale lengths) as Kylafis' models didn't go above 1.2 but Pelletier has data where $\alpha_B/\alpha_I > 1.4$. My comment was that increasing ξ (relative scale height of dust to stars) from 0.4 to 0.5 or higher will change this ratio dramatically even for the same central τ.

Answer
Kylafis

We will create such models to check it.

Question
de Vaucouleurs

Please note that in early types (Sa, Sb) there is no, or very little, gas and dust in the centre, the dust is concentrated in a peripheral annulus. Only in the late types (Sc,Sd) is the gas-dust distribution exponential (like the stars).

Also the z scale height of the stars depends on spectral type.

Answer
Kylafis

In future models we will take your remarks into account.

MODELING DUSTY GALAXIES

GLADIS MAGRIS C.
Centro de Investigaciones de Astronomía (CIDA)
Apartado Postal 264, Mérida, Venezuela

AND

GUSTAVO BRUZUAL A.
Landessternwarte Heidelberg-Königstuhl
69117 Heidelberg, Germany

1. Models

We study the effects of dust in the evolving spectral energy distributions (SEDs) of disk galaxies by means of models that take into account the absorption and scattering of stellar radiation by dust grains. We solve the radiative transfer equation in a plane parallel geometry, using the algorithm of Bruzual, Magris & Calvet (1988, hereafter BMC88). The boundary conditions have been modified to include the relative differences in the vertical scale height of early-, intermediate-, and late-type stars and dust grains. The population synthesis models of Bruzual & Charlot (1993, hereafter BC93) are used to describe the evolving spectrum of each stellar component. The wavelength dependence of the dust albedo, a, and the phase function asymmetry parameter, g, was taken from the recent determination by Witt et al. (1994). The parameter $\alpha = h_{dust}/h_{stars}$ measures the dust-to-star scale height ratio for the geometry shown in Fig. 1. When $\alpha < 1$ the dust layer is inside the stellar disk. The distribution of dust and stars is homogeneous inside each slab where they coexist.

2. Results

In the top panel of Fig. 2 we show separately the contribution of main sequence OB stars, main sequence A stars, and the Rest of the stellar population, to the total dust free SED in a BC93 model for an Sb spiral galaxy at the age of 13.5 Gyr. In this model stars form following the Salpeter

J. I. Davies and D. Burstein (eds.), The Opacity of Spiral Disks, 67–70.
© 1995 *Kluwer Academic Publishers.*

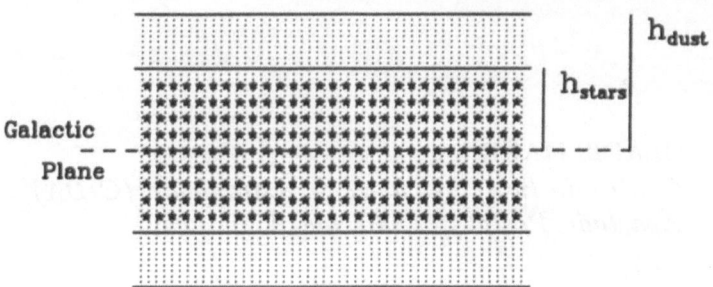

Figure 1. Parameters defining the relative scale height of the stellar and dust layers.

(1955) IMF according to the SFR $\Psi(t) = 1M_\odot\tau^{-1}\exp(-t/\tau)$, for $\tau = 6$ Gyr. The middle panel shows for the same stellar components the SEDs resulting from our solution to the radiative transfer problem for a face-on galaxy of total optical depth $\tau_V = 1$. These stellar groups have different scale heights with respect to the dust component. The typical values of α for spiral galaxies listed in Table 1 (Allen 1973) have been used. In the bottom panel the same SEDs are repeated for the $\tau_V = 2$ case.

TABLE 1. Scale heights (pc)

Stars	h_{stars}	h_{dust}	α
OB	50	120	2.4
A	120	120	1.0
Rest	500	120	0.24

Fig. 3 shows in a magnitude scale the difference between the reddened and the dust free total SED of Fig. 2 for the $\tau_V = 1$ case. The heavy solid line corresponds to the models presented in this paper. For comparison the dotted line shows the results for a homogeneous mixture of stars and dust computed as in BMC88, but for the values of a and g of Witt et al. (1994). The line marked "dust screen" corresponds to a layer of dust located between the source and the observer. The differences between our stratified model and the homogeneous layer model of BMC88 are larger at optical and IR wavelengths. Our models are fainter redward of 4000 Å. In this range and at the age shown in Figs. 2 and 3 the SED is dominated by the Rest stellar group. For the parameters of Table 1, 38% of the stars of this group are located behind a dust screen of $\tau_V = 1$. The scattering of light by the grains in this internal dust layer, plus the contribution from the other 38% of the stars located in front of the layer is not enough to

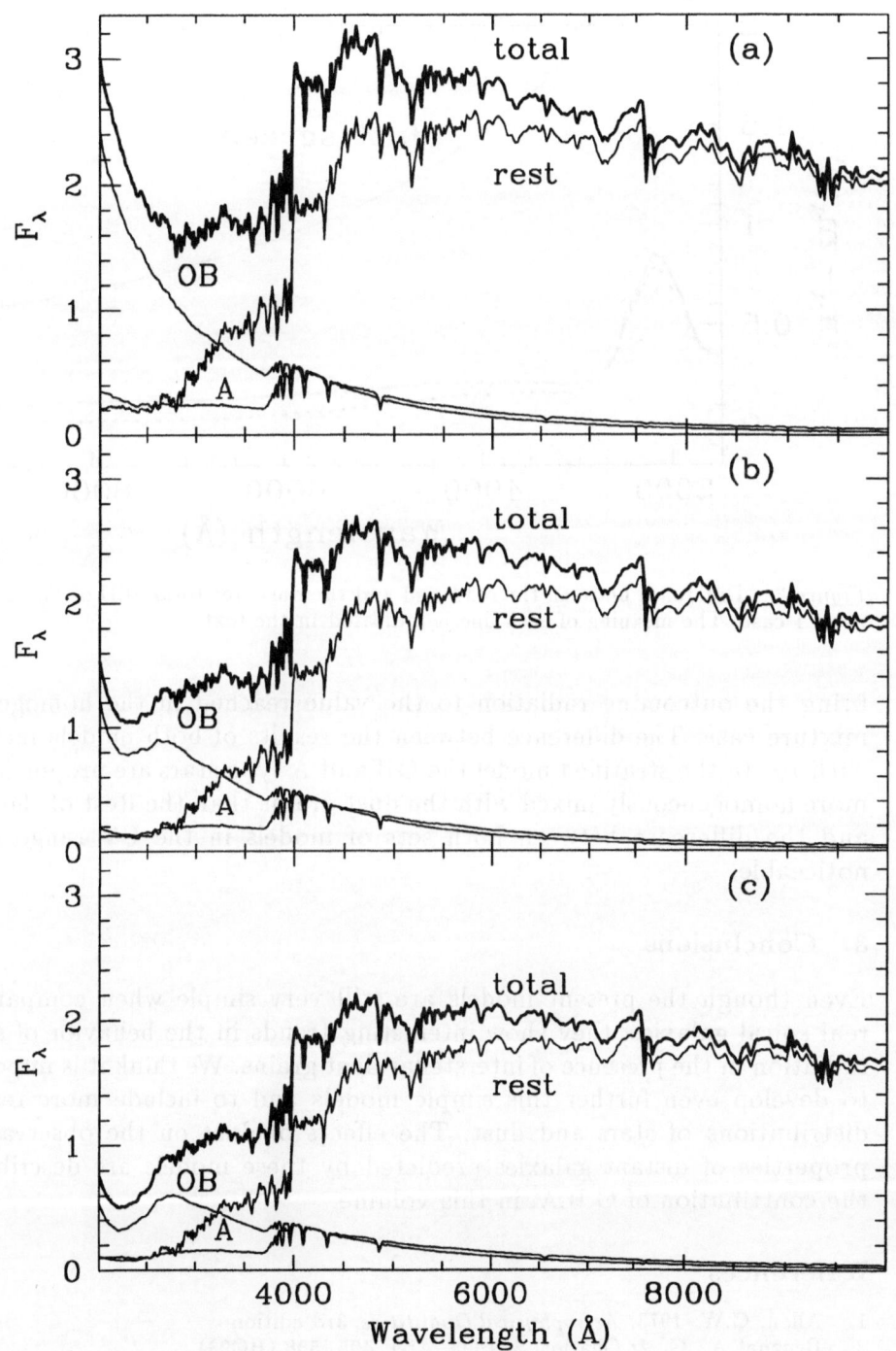

Figure 2. (a) Contribution of MS OB stars, MS A stars, and the Rest of the stellar population, to the total dust free SED in a BC93 model for an Sb spiral galaxy at the age of 13.5 Gyr. (b) Same as (a) but for the case of a dusty disk of optical thickness $\tau_V = 1$

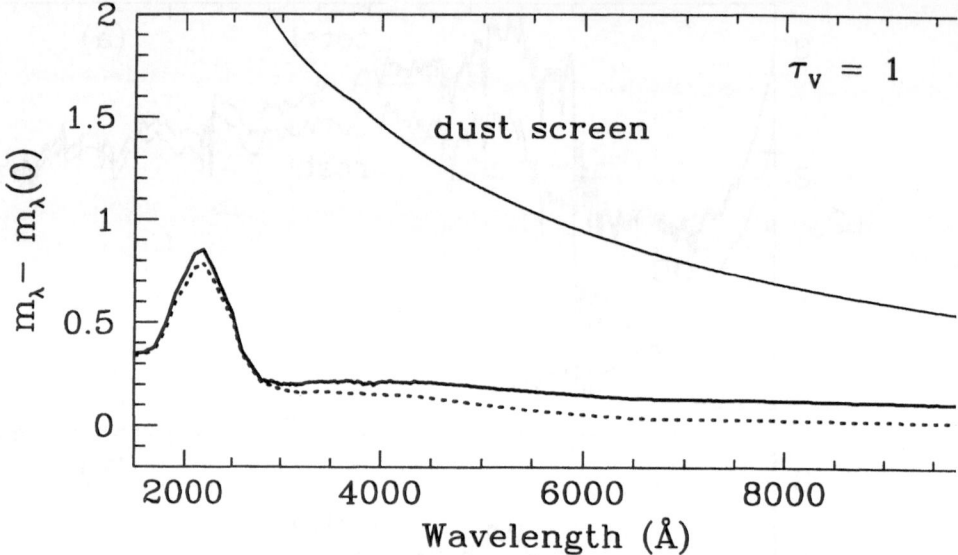

Figure 3. Difference between the reddened and the dust free total SED of Fig. 2 for the $\tau_V = 1$ case. The meaning of each line is indicated in the text.

bring the outcoming radiation to the value reached in the homogeneous mixture case. The difference between the results of both models increases with τ_V. In the stratified model the OB and A type stars are proportionally more homogeneously mixed with the dust grains than the Rest of the stars, and the difference between both sets of models in the UV range is less noticeable.

3. Conclusions

Even though the present models are still very simple when compared to real spiral galaxies, they show interesting trends in the behavior of stellar radiation in the presence of interstellar dust grains. We think it is important to develop even further this simple models and to include more realistic distributions of stars and dust. The effects of dust on the observational properties of distant galaxies predicted by these models are described in the contribution of G.B.A. in this volume.

References

1. Allen, C.W. 1973, *Astrophysical Quantities*, 3rd edition
2. Bruzual A., G., & Charlot, S. 1993, ApJ, 405, 538 (BC93)
3. Bruzual A., G., Magris C., G., & Calvet, N. 1988, ApJ, 333, 673 (BMC88)
4. Salpeter, E.E. 1955, ApJ, 121, 161
5. Witt, A. N. *et al.* 1994, ApJ, in press

The bus

INCLINATION-DEPENDENCE OF SPIRAL GALAXY PHYSICAL PROPERTIES: HISTORY AND TESTS

DAVID BURSTEIN
Arizona State University
Department of Physics and Astronomy, Code 1504
Tempe, AZ 85287-1504 U.S.A.

JEFFREY A. WILLICK
Carnegie Observatories
813 Santa Barbara St.
Pasadena, CA 91101-1292 U.S.A.

AND

STÉPHANE COURTEAU
National Optical Astronomy Observatory
P.O. Box 26732
Tucson, AZ 85726-6732 U.S.A.

1. Introduction

For the past 40 years, studies of the inclination-dependence of galaxy physical properties have relied primarily on statistical tests of photographically-selected galaxy samples (e.g., Holmberg 1958; de Vaucouleurs 1959; Tully 1967; Heidmann, Heidmann & de Vaucouleurs 1972 [H^2dV]; Burstein & Lebofsky 1986; Valentijn 1990; Burstein, Haynes & Faber 1991 [BHF]). The question of interest is deceptively simple: Given isophotal magnitudes, diameters and axial ratios, how do the magnitude and diameter change with inclination? The two seemingly contradictory answers that can be given to this question were, ironically, given by the first two papers that seriously addressed this question with modern data. While Holmberg (1958) found that spiral galaxy optical magnitudes become fainter with greater inclination to the line-of-sight, de Vaucouleurs (1959) found that galaxy diameters become larger with increasing inclination.

73

J. I. Davies and D. Burstein (eds.), The Opacity of Spiral Disks, 73–84.
© *1995 Kluwer Academic Publishers.*

Until the discovery of the Tully-Fisher relation (1977; TF) for estimating the distances to spiral galaxies, statistical studies were made for the general purpose of understanding how dust affects the appearance of a galaxy. The discovery of the TF relation was a tremendous spur for assemblage of much larger samples of spiral galaxies with accurately measured magnitudes and diameters, with the ultimate aim of understanding the large-scale structure of the universe. Ironically, to get to that ultimate goal, workers have to correct their galaxy magnitudes and diameters for inclination-dependent effects.

The scientific goals of our present investigation are twofold: First, to try to understand the reasons why early investigations could analyze essentially similar galaxy samples for inclination-dependent trends and come to opposite conclusions. Second, to inclination-correct, in an internally self-consistent manner, the TF data from various published sources that we are assembling into a Mark III Catalog of Galaxy Peculiar Velocities.

In the present paper we will concentrate more on discussion of the history of this subject. This paper will also give a summary of our test results, with details to be published in a companion paper (Willick, Burstein, & Courteau 1995). §2 discusses the history; §3 summarizes our new results and §4 briefly discusses the implications of our new results in light of history lessons learned.

2. Nomenclature and History

2.1. NOMENCLATURE

As the actual parameters used here are not necessarily the ones used by the original authors, the following are useful definitions:

1. Mean surface brightness at a given wavelength is defined as $SB_\lambda = -2.5 \log[L_\lambda/(\pi A_\lambda B_\lambda)]$ mag pc^{-2}, where L_λ is the luminosity at wavelength λ within a given isophote, A_λ and B_λ are the major and minor axis radii of that isophote, given in units of parsecs. Since SB_λ (at low z) is independent of distance, we can also define $SB_\lambda = m_\lambda + 2.5 \log(a_\lambda b_\lambda)+$ constant mag pc^{-2}; or $m_\lambda + 5.0 \log a_\lambda - 2.5 \log(a_\lambda/b_\lambda) + 1.243$ mag $arcsec^{-2}$, where a_λ and b_λ are the observed major and minor axis radii in units of arcseconds.

2. In general, one can assume that mean surface brightness is some function of axial ratio, $SB_\lambda = f(a_\lambda/b_\lambda)$. With this definition, $m_\lambda + 5 \log a_\lambda - 2.5 \log(a_\lambda/b_\lambda) = f(a_\lambda/b_\lambda)+$ constant. From this, one sees that: a) if a_λ is assumed independent of axial ratio, $m_\lambda \propto 2.5 \log(a_\lambda/b_\lambda)+f(a_\lambda/b_\lambda)$; or b) if m_λ is assumed to be independent of axial ratio, $5 \log a_\lambda \propto 0.5 \log(a_\lambda/b_\lambda)+ 0.2f(a_\lambda/b_\lambda)$. More generally, we can say that $m_\lambda + 5 \log(a_\lambda/b_\lambda) = f(a_\lambda/b_\lambda)+ 2.5 \log(a_\lambda/b_\lambda)$. In other words, the inclination dependence of either absolute

integrated magnitude or absolute diameter depends both on the actual form of $f(a_\lambda/b_\lambda)$ *and* any initial assumptions made.

From these definitions, one notes that the quantities SB_λ, a_λ/b_λ and $f(a_\lambda/b_\lambda)$ are distance-*independent*, while absolute magnitudes and diameters lead to analogous apparent diameters and magnitudes, m_λ, a_λ and b_λ which are obviously distance-*dependent*. BHF and Choloniewski (1991) have shown how the choice of galaxies based on one of the distance-dependent quantities can lead to a wrong conclusion of inclination-dependence for the *other* distance-dependent parameter.

2.2. HISTORY: HOLMBERG 1958

The classic "Holmberg Test" is one in the mean surface brightness (defined by Holmberg to be $L_\lambda/\pi A_\lambda^2$) is plotted against axial ratio. In terms of the present definitions, this test is a plot of SB_λ versus A_λ/B_λ. Plotting SB_λ versus axial ratio instead of $-2.5\log[L_\lambda/\pi A_\lambda^2]$ versus axial ratio simply removes the correlation of b_λ/a_λ with a_λ/b_λ from the diagram, but otherwise the test is identical.

Figure 1 plots SB_λ versus a_λ/b_λ for the Holmberg data set, using Holmberg's photovisual (PV) magnitudes. Holmberg's sample is divided into bins according to morphological classification of the Second Reference Catalog of Bright Galaxies (de Vaucouleurs, de Vaucouleurs & Corwin 1976; RC2). Within the scatter in these plots, the best functional form is $f(a_\lambda/b_\lambda) =$ constant for each subclass of spiral or irregular galaxy; i.e, the mean surface brightness of the galaxies with Holmberg PV data is independent of axial ratio.

From the discussion in §2.1, if one has $f(a_\lambda/b_\lambda) =$ constant, any combination of a_λ or m_λ changing with inclination is possible, as long as the relation $m_\lambda + 5\log a_\lambda - 2.5\log(a_\lambda/b_\lambda) =$ constant is satisfied. If one *assumes* that diameter is independent of axial ratio, as did Holmberg, one concludes that $m_\lambda \propto 2.5\log(a_\lambda/b_\lambda)$. Figure 2 shows that Holmberg's formula for inclination-dependence of magnitude, $\alpha(\sec i - 1)$, is very similar to $2.5\log a/b$ when $\alpha = 0.43$, as Holmberg derived for early-type spiral galaxies. Incidentally, his estimate of a lower value of $\alpha = 0.28$ for late-type spirals appears at odds with the apparent non-dependence of mean surface brightness with axial ratio for ALL spirals, but closer examination shows that this apparent discrepancy owes its existence to a handful of highly edge-on late-type spirals and therefore is not significant.

Holmberg's derivation of magnitude change with inclination was clearly based on his assumption that diameter does not change with inclination. Still, he could have just as easily assumed that magnitude does not change with inclination and derived a diameter dependence on inclination.

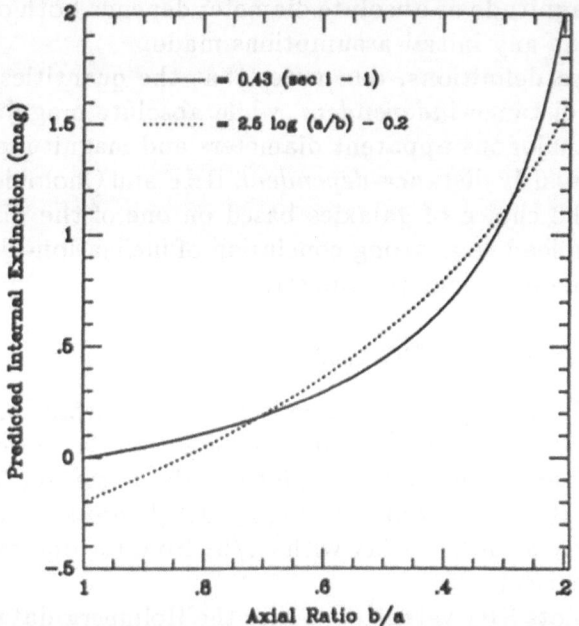

Fig. 1. A comparison of the sec i law of Holmberg (1958) with a log (a/b) law as used by RC3, with arbitrary normalization. Within typical observational error, the two laws are similar from b/a = 1 to b/a = 0.25.

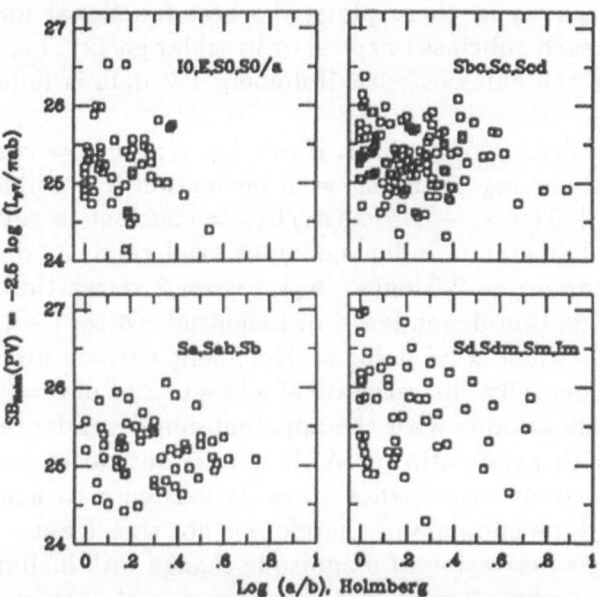

Figure 2. Modfied Holmberg surface brightness test, using original Holmberg PV data, divided by Hubble types. No significant dependence of mean surface brightness on axial ratio is seen for any Hubble type.

Holmberg's conclusion is not necessarily wrong, but it cannot be proven from his own data set.

2.3. HISTORY: DE VAUCOULEURS 1959

de Vaucouleurs (1959) and, subsequently, de Vaucouleurs and de Vaucouleurs (1964, First Reference Catalog of Bright Galaxies; RC1), made plots of apparent diameter versus axial ratio for galaxies *of the same apparent magnitude*. It can be shown that, for both the RC1 and RC2 data, SB_λ is independent of axial ratio, just as for the Holmberg data. As such, if one requires m_λ to be constant, from §2.1 we find $\log a_\lambda \propto 0.5 \log(a_\lambda/b_\lambda)$, whereas de Vaucouleurs (1959) and the RC1 have $\log a_B \propto 0.4 \log(a_B/b_B)$.

Again, it is clear that if one limits a sample of galaxies with similar mean surface brightnesses to have the same apparent magnitude independent of axial ratio, what one is really preserving is the total surface area of the galaxies, $\pi ab = \pi a^2 \times b/a$. For an edge-on galaxy to have the same surface area as a face-on galaxy, the edge-on galaxy must have a larger diameter.

In contrast to Holmberg, the de Vaucouleurs analysis assumed that magnitude does not change with inclination. If de Vaucouleurs had instead chosen a sample of galaxies all with the same apparent diameter, he would have found magnitude to become much fainter with increasing inclination.

2.4. HISTORY: H^2DV 1972

In a series of three papers, H^2dV studied the inclination-dependence of spiral galaxy diameters and magnitudes both observationally and theoretically. The methodology developed in these papers was later used to derive inclination dependencies of m_B and a_B for the data in the RC2. We will use these data to illustrate our points in this subsection.

H^2dV made two explicit assumptions: HI 21 cm flux is inclination-independent and HI 21 cm flux and optical flux are co-extant. Based on these assumptions, H^2dV reasoned that a hybrid 21 cm/optical surface brightness, F_{HI}/D_{25}^2, should be independent of axial ratio if the RC2 B band isophotal diameter D_{25} did not change with inclination. (An analogous test, not made, could also have been done for the RC2 B magnitude, B_T.)

Figure 3 plots F_{HI}/D_{25}^2 versus $\log a_B/b_B$ for all RC2 galaxies with listed values of F_{HI}, with the sample divided into three bins in D_{25}: $\log D_{25} \geq 1.8$, $1.3 \leq \log D_{25} < 1.8$ and $\log D_{25} < 1.3$ (D_{25} in units of 0.1 arcmin). The fourth panel in Figure 3 plots all of the data together. Galaxies in each plot are separated into the same four morphological bins as with the Holmberg test of §2.2.

When these data are sorted into three separate diameter ranges, it is evident that there is no statistically significant trend of F_{HI}/D_{25}^2 with $\log(a/b)$ within a given range of apparent diameter. Rather, the mean value of F_{HI}/D_{25}^2 increases as the mean value of $\log D_{25}$ decreases: galaxies that are apparently smaller have higher HI flux for their size than do galaxies that are apparently larger. In addition, it is also evident that the most edge-on spirals are most plentiful among the largest galaxies, and least plentiful among the smallest galaxies. Only when one puts all of these data together, as in the fourth panel in Figure 3, does an apparent trend of F_{HI}/D_{25}^2 with $\log(a/b)$ appear. However, one can see that this apparent trend is an artifact of combining galaxies from the different diameter bins.

What is happening is that two different selection effects conspire to produce a spurious result. The first selection effect derives from the same source as the problems with the Holmberg and de Vaucouleurs results: **SB_B is independent of a/b for spiral galaxies.** As shown by BHF, if one selects spiral galaxies on the basis of their apparent magnitudes, one is effectively selecting galaxies based on surface area presented to the line-of-sight. As a result, edge-on galaxies can only have enough surface area if they have large major axis diameters, while face-on galaxies can have the smallest possible diameters if seen nearly circular in form. This selection effect can be seen directly in Figure 4, which plots axial ratio versus apparent size for the 458 RC2 galaxies used for this F_{HI}/D_{25}^2 study. The Holmberg data (not plotted) shows the same effect. In both the RC2 and the original Holmberg samples of spiral galaxies, the only edge-on spirals that are included have large diameters, while face-on galaxies can have the full range of possible diameters.

The second selection effect is the classic flux-limited Malmquist bias: In the RC2, the HI flux measurements are signal-to-noise limited, which means an effective apparent flux limit. The flux-limited Malmquist bias effect predicts that, at the flux limit, the galaxies chosen on the basis of apparent flux will have absolute fluxes much higher than average. However, the only galaxies that can be seen at great distances are the face-on galaxies, so that face-on galaxies will preferentially have higher HI fluxes than edge-on galaxies. Hence, smaller-appearing galaxies will be both preferentially face-on *and* have higher HI fluxes than larger galaxies, as we see in Figure 3.

The lesson to be drawn from the H^2dV analysis is that every sample of galaxy is selected in some specific, often complicated manner. The degree to which one ignores selection effects in a galaxy sample is the degree to which any statistical interpretation is subject to selection-caused biases. In the case of the H^2dV study and subsequent RC2 result, the bias resulting from their selection effects led them to conclude that diameters were

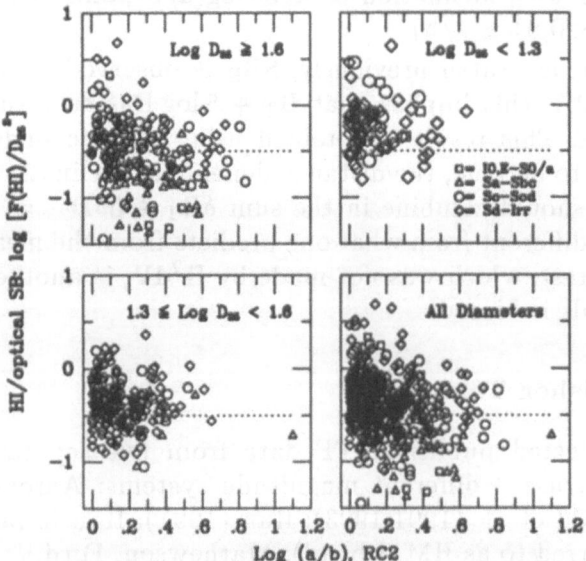

Figure 3. The RC2 HI/optical surface brightness test for the axial ratio dependence of diameter, divided into 3 ranges of diameter. Note lack of correlation within each diameter range, but higher HI/opt SB as galaxies become smaller, producing apparent correlation when all diameters are combined (lower right graph). A line of constant HI/opt SB is plotted in each graph.

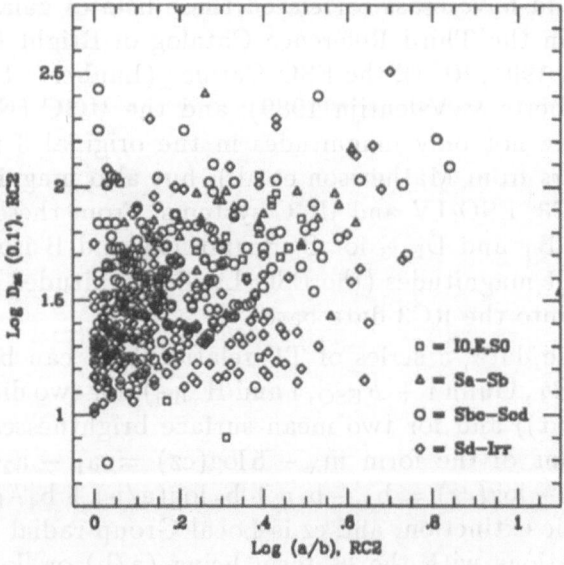

Figure 4. Apparent diameter versus axial ratio for 456 RC2 spiral galaxies with RC2 21 cm HI flux values and Hubble types. Note the lack of edge-on galaxies at low HI flux levels, which gives the distribution of points a triangular shape.

getting larger with inclination $\propto 0.235 \log(a/b)$, and magnitudes were getting fainter $\propto 0.8 \log(a/b)$.

Note that, as stated previously, SB_B is observed to be independent of a/b. From §2.1, this implies that $B_T + 5 \log D_{25} \propto 2.5 \log(a_{25}/b_{25})$, with $a_{25} = 0.5 D_{25}$. This result is obtained using distance-independent quantities. For this to be true, the distance-dependent inclination corrections δ to B_T and D_{25} should combine in the sum $\delta B_T + 5 \delta D_{25} = 1.98 \log(a_{25}/b_{25})$, significantly different from what one predicts from the mean surface brightnesses. This test, which was not made by $H^2 dV$, is another indication that the RC2 result is biased.

3. Tully-Fisher Tests

We have collected published TF data from five separate studies which, among them, use 3 different magnitude systems: Aaronson et al. (1982) ($H_{-0.5}$); Mould et al. (1991,1993); Han (1991); Han & Mould (1992) (collectively referred to as HM, I-band); Mathewson, Ford, & Buchhorn (1992; I-band); Willick (1991, Gunn r-band) and Courteau (1992; Gunn r-band). These data were homogenized into a self-consistent catalog of galaxy peculiar velocities, which will become part of the Mark III Catalog of Galaxy Peculiar Velocities.

As part of our analysis, we have taken the original data, over 3000 galaxies in total, and have cross-correlated these lists of galaxies with existing catalog data in the Third Reference Catalog of Bright Galaxies (de Vaucouleurs et al. 1991; RC3), the ESO Catalog (Lauberts 1982); the ESO-LV Catalog (Lauberts & Valentijn 1989); and the UGC (Nilson 1973). As a result, we have not only magnitudes in the original 3 passbands (plus I band diameters from Mathewson et al.), but also magnitudes and diameters in the RC3, ESO-LV and UGC systems. From these catalog data, we have selected B_T and D_{RC3} for B magnitudes and B isophotal diameters, and ESO-LV R magnitudes (the ESO-LV B magnitudes and diameters are incorporated into the RC3 data base).

Using these data, a series of TF relationships can be formed for four magnitudes (B_T, Gunn r + R_{ESO}, I and $H_{-0.5}$), for two diameters ($\log D_{RC3}$ and $\log D_I(Mat)$) and for two mean surface brightnesses (SB_B and SB_I). We fit relations of the form $m_\lambda - 5 \log(cz) = a_1 - a_2 \eta + a_3 \log(a/b) + a_4 A_B$ and $D_\lambda - \log(cz) = b_1 - b_2 \eta + b_3 \log(a/b) + b_4 A_B$ (where A_B here means Galactic extinction, and cz is Local Group radial velocity). We also looked at relations with the b_3 term being (a/b) or $[\log(a/b)]^2$, but saw little difference from the $\log(a/b)$ results. The coefficients $a_1, \ldots, a_4$ and $b_1, \ldots, b_4$ were determined by means of multiparameter least-squares fits. These fits were iteratively corrected for selection bias following the precepts of Willick (1994).

For each TF fit, we then examined the fit without any axial ratio dependence of magnitude or diameter, versus the fit in which the axial ratio dependence was explicitly determined. In the process we found that not all of the galaxy samples were of equal utility in this exercise. In particular, the Courteau (1992) sample is restricted to galaxies of similar axial ratio, hence, it yields little information on axial dependence of the magnitudes and diameters of the galaxies in that sample.

In each case we examined the best-fit solutions in terms of a plot of TF-residuals versus $\log(a/b)$. If the fit has correctly removed axial ratio dependence from the TF relation, these residuals should be independent of axial ratio. Two examples of these kinds of "null" plots are shown in Figure 5 for RC3 $\log D_{RC3}$ blue diameters (upper) and RC3 B_T magnitudes (lower) versus $\log(a/b)_{RC3}$ for ~1000 galaxies in the Mathewson et al. (MAT). The mean values and 1σ error bars of the mean are plotted for TF residuals in bins of 0.1 dex in $\log(a/b)$. The predicted slopes of axial ratio dependence that produced these null results are those quoted in Table 1 for these data.

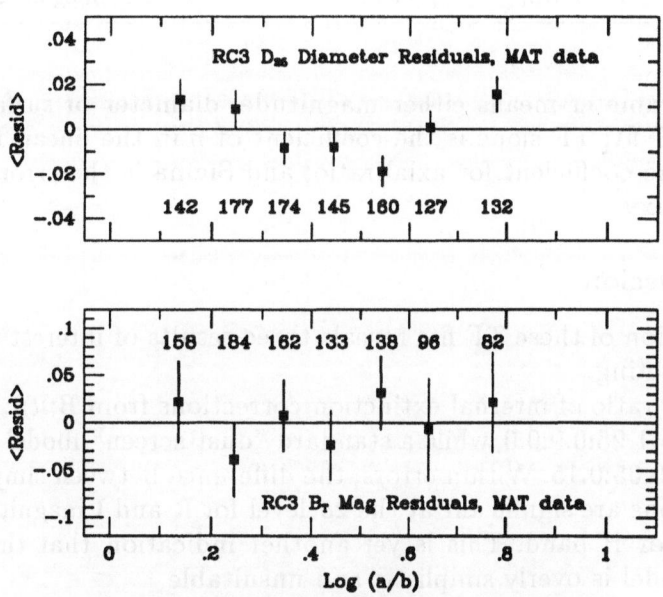

Figure 5. TF test of RC3 diameter (top) and RC3 B mag (bottom) using Mathewson et al. velocity widths. If fit is successful, no trend with axial ratio should be seen. Such is the case for B mag, but RC3 diameter shows a slightly non-linear relation with log (a/b). Number of galaxies in each bin of axial ratio is given, and zero line is plotted. The slope predicted for RC3 diameter is -0.20 and for B mag is 1.45, as quoted in Table 1. TF test must be done this way to avoid selection biases.

While we note that the relation of RC3 Diameter and $\log a/b$ is not quite linear, the effect is relatively smaller ($< 10\%$). Space limitations preclude showing more such fits here; others will be shown in Willick, Burstein & Courteau (1995). Table 1 gives the mean results from these fits for magnitudes, diameters and mean surface brightnesses:

TABLE 1. Results of TF Fits

parameter	TF slope	$\log a/b$ slope	Sigma
RC3 B_T	-5.5	1.45	0.55 mag
R & Gunn r	-6.3	1.25	0.50 mag
MAT + HaM I	-6.8	0.90	0.45 mag
$H_{-0.5}$	-10.3	0.00	0.49 mag
d_{RC3}	0.92	0.20	0.12 dex
MAT d_I	1.03	0.30	0.11 dex
RC3 SB_B	-0.91	-0.20	0.43 mag arcsec^{-2}
MAT SB_I	-1.75	-0.30	0.29 mag arcsec^{-2}

where parameter means either magnitude, diameter or surface brightness for the TF fit; TF slope is the coefficient of η in the linear TF fit; $\log a/b$ slope is the coefficient for axial ratio; and Sigma is the error of the fit per single galaxy.

4. Discussion

Examination of these TF fits reveals three results of interest to the subject of this meeting.

1. The ratio of internal extinction corrections from B:R:I:H is, from Table 1, 1.45:1.25:0.9:0.0, while a standard "dust-screen" model would predict 1.45:0.85:0.65:0.15. Within errors, the difference between simple model and observations are significant at the 2σ level for R and I magnitudes, and less than 1σ for H band. This is yet another indication that the simple dust screen model is overly simplistic and unsuitable.

2. The SB tests use a distance-independent parameter to form a TF-like relation, while the magnitude and diameter TF tests separately use distance-dependent parameters. Yet, one can form a separate surface brightness TF test by combining the magnitude and diameter tests in the usual manner: $\delta_\eta SB = \delta_\eta mag + 5\delta_\eta \log D$, where δ_η indicates the change of this

parameter with the velocity width parameter. For B magnitudes, $\delta_\eta SB_B = -0.91$ while $\delta_\eta B_T + 5\delta_\eta \log D_{RC3} = -5.5 + 5 \times 0.92 = -0.90$. For Mathewson et al. I band magnitudes, $\delta_\eta SB_I = -1.75$ while $\delta_\eta I + 5\delta_\eta \log D_I = -6.8 + 5 \times 1.03 = -1.65$. In both cases, the agreement is within 11%, consistent with the separately determined errors in these fits.

In addition, the combined magnitude and diameter coefficients on axial ratio must also be consistent with the coefficient on surface brightness, as previously pointed out in the case of $H^2 dV$. Here we must use the original Holmberg definition of surface brightness ($L/\pi a^2$), as we have explicitly removed the axial ratio dependence. To do this, we add back the value of 2.5 into the surface brightness dependence on axial ratio. From this, we have $\delta_{a/b} SB_B = 2.5 - 0.20 = 2.3$, while $\delta_{a/b} B_T + 5\delta_{a/b} \log D_{RC3} = 1.45 + 1.0 = 2.45$, while $\delta_{a/b} SB_I = 2.2$ and $\delta_{a/b} I + 5\delta_{a/b} \log D_I = 0.9 + 1.5 = 2.4$. In each case, the distance-dependent test is about 8% larger than the distance-independent test, still less than a 1σ result.

From these tests we conclude that our solutions for axial ratio dependence of B and I magnitude and diameter are self-consistent with each other, and with their mean surface brightnesses.

3. We also note that the change in surface brightness with axial ratio for both B and I band are very similar, even though the galaxy is more transparent in I than in B. We see two possible interpretations for this apparent agreement: a) It is a coincidence owing to the details of how the edge of galaxies change with inclination in different passbands; or b) The change of diameter and magnitude in B and I are being similarly affected by a combination of dust absorption and scattering in a complex geometry of dust, gas and stars in each galaxy.

To differentiate and solve between these hypotheses will require work. The former is difficult to address as the isophotal diameters of edge-on systems, which comprise a large part of most TF samples, are highly sensitive to the outer scale heights of dust and stars which are at present poorly known quantities. One of us (S.C.) has recently embarked on a large multi-color digital survey of late-type spirals which should provide further insight into this question (see Courteau and Holtzman, this conference). The latter requires more sophisticated galaxy models than presented at this meeting, together with Monte-Carlo simulations to reproduce the trends observed here.

5. Acknowledgements

DB wishes to acknowledge support from NSF Grant 90-16930.

84

6. References

Aaronson et al. 1982, ApJS 50, 241

Burstein, D. & Lebofsky, M.J. 1986, ApJ 300, 683

Burstein, D., Haynes, M.P. & Faber, S.M. 1991, Nature 353, 515 (BHF)

Choloniewski, J. 1991, MNRAS, 254, 486

Courteau, S. 1992, Ph.D. Thesis, U.C. Santa Cruz

de Vaucouleurs, G. 1959, AJ 64, 397

de Vaucouleurs, G. & de Vaucouleurs, A. 1964, *Reference Catalog of Bright Galaxies*, Austin: Univ. of Texas Press (RC1)

de Vaucouleurs, G., de Vaucouleurs, A. & Corwin, H.G. Jr. 1976, *Second Reference Catalog of Bright Galaxies*, Austin: Univ. of Texas Press (RC2)

de Vaucouleurs, G., de Vaucouleurs, A., Corwin, H.G. Jr., Buta, R.J., Paturel, G., & Fouqué, P. 1991, *Third Reference Catalog of Bright Galaxies*, New York: Springer-Verlag (RC3)

Han, M.-S. 1991, Ph.D. Thesis, California Institute of Technology

Han, M. & Mould, J. R. 1992, ApJ, 396, 453

Heidmann, J., Heidemann, N. & de Vaucouleurs, G. 1972, Mem RAS 75, Parts 4–6, 85 (H^2dV)

Holmberg, E. 1958, Medd Lund Obs II, No. 136

Lauberts, A. 1982, *The ESO/Uppsala Survey of the ESO (B) Atlas*, Garching-bei-München: European Southern Observatory (ESO)

Lauberts, A. & Valentijn, E.A. 1989, *The Surface Photometry Catalog of the ESO-Uppsala Galaxies*, Garching-bei-München: European Southern Observatory (ESO-LV)

Mould, J.R., Stavely-Smith, L., Schommer, R.A., Bothun, G.D., & Jall, P.J. 1991, ApJ, 383, 467

Mould, J.R., Akeson, R.L., Bothun, G.D., Han, M-S., Huchra, J.P., Roth, J., & Schommer, R.A. 1993, ApJ, 409, 14

Mathewson, D.S., Ford, V.L, and Buchhorn, M. 1992, ApJS, 81, 413

Nilson, P. 1973, *Uppsala General Catalogue of Galaxies*, Uppsala: Roy. Soc. Sci. Uppsala (UGC)

Tully, R.B. 1968, Maryland AJ, Nos. 1 and 4 (also AJ 73, S205)

Tully, R.B. & Fisher, R.A. 1977, A&A 54, 661 (TF)

Valentijn, E.A. 1990, Nature 346, 153

Willick, J. 1991, Ph.D. Thesis, U.C. Berkeley

Willick, J. 1994, ApJS 94, 1

Willick, J., Burstein, D. & Courteau, S. 1995, ApJ, in preparation

WHY A DISTANCE SELECTION EFFECT INVALIDATES THE BURSTEIN, HAYNES AND FABER OPACITY TEST

J. I. Davies, H. Jones and M. Trewhella,
Department of Physics and Astronomy,
University of Wales College of Cardiff,
PO Box 913, Cardiff CF2 3YB

1. Introduction

If you could rise high above the Galactic plane and view our Galaxy from various viewing angles what would you see ? Would, for example, you reach the same conclusions as to the optical luminosity and isophotal size as your viewing angle changed or would your conclusions change along with your angle of view ?

It is of course impossible to carry out this experiment in the way it is described above, but there are many thousands of easily observable galaxies in the sky that present themselves at various viewing angles. Can we use these galaxies to test for a change in galaxy properties with inclination, in the same way as we would ideally like to do the test on a single galaxy ? In this paper we discuss one proposed method (Burstein et al. 1991, henceforth BHF) of determining the inclination dependent properties of galaxies. An understanding of how the observed properties (specifically for this paper the optical isophotal diameter and luminosity) of galaxies change with inclination gives us an insight into the opacity of galaxies. That is the probability of a `typical' photon (in the B band for example) being able to escape from the galactic disk.

If the galaxy is optically thin then we would expect its isophotal diameter to increase and its magnitude to remain constant with increasing viewing angle. If optically thick then its size should remain constant and its luminosity decrease. Collectively we can refer to these tests as surface brightness inclination tests because it is an increase in surface brightness with inclination that causes the isophotal diameter to increase with inclination in the optically thin case. In the optically thick case it is the constancy of surface brightness with inclination that causes the galaxy to appear less luminous as it is inclined (see Burstein et al. and Disney this volume for a historical perspective). The above picture, although relatively straight forward to describe, is much more complicated in practise. This is not only because we have to carry out the test on a sample of galaxies that might be individually quite different from each other, but also because differences in the relative geometry of stars and dust lead to changes in the way surface brightness

J. I. Davies and D. Burstein (eds.), The Opacity of Spiral Disks, 85–87.
© *1995 Kluwer Academic Publishers.*

changes with inclination that do not match the simple predictions of the past (see Disney et al. 1989, henceforth DDP, Witt and Kylafis this volume).

2. The BHF method

BHF, quite rightly in our view, highlight the problems of observational selection (see also Davies et al. 1993 and Jones et al. this volume) and how they have probably been a major influence in our mis-understanding of the opacity problem. They conclude that the result you obtain depends critically on the way in which the sample is selected and more importantly that you need more information than just the size, magnitude and inclination of the galaxy to decide on the opacity issue. BHF show (see their fig 1) that galaxies appear to change their position in a magnitude diameter plot in a way that is related to their inclination (their inclination trajectories). The important point is exactly how they change their position. Is it along a line of constant size (optically thick), a line of constant magnitude (optically thin) or something between the two ? The critical assumption in their paper is that you can decide on the path that galaxies follow in the magnitude size plane if you have a sample with measured distances to all of the galaxies.

There are rather subtle selection effects that BHF say are avoided by using their test (selection effects Valentijn 1990 did not consider). Can we be sure that we detect galaxies over the same average distance independent of their inclination (see Jones et al. this volume) ? For example if galaxies are optically thick then they become intrinsically fainter as they are inclined and so we might then only include relatively nearby inclined galaxies in a magnitude selected sample. If galaxies were optically thin then they would grow in size as they are inclined so we would include inclined galaxies at larger distances in a diameter limited sample. In both cases the average distances sampled will change with inclination and so the BHF test would show that some form of selection effect was operating, we agree that a test of this nature is essential, but are there other selection effects ? For example what if the sample was artificially cut-off at some distance which is independent of a galaxies inclination. BHF state that `Redshifts are available for 77% of UGC spirals (their sample) with morphological classes Sa - Sc and 80% for Sc alone'. They also state that there is a velocity cut-off (due to technical reasons) of about 12,000 km s^{-1} for those galaxies observable by Arecibo and about 7,000 km s^{-1} for galaxies further north. It is this potential selection effect that we model in this paper.

4. A brief description of the results of the numerical simulations of the BHF sample

The numerical simulations are described in Jones et al. (this volume). From our simulated samples of galaxies we select magnitude and diameter limited samples for both the optically thin and optically thick cases. The only difference between these samples and those described in Jones et al. is that we have a maximum distance cut-off that corresponds to the limit placed by the 21cm observations described by BHF. From these simulations we find that the distance dependent test that BHF propose does not lead to any better understanding than previous opacity tests, although in principle we agree it should. The reason is that the 21cm instrumental distance cut-off leads to galaxies being sampled to the same distance as a function of inclination irrespective of their opacity.

We conclude that although the BHF test has been specifically designed to overcome selection effects the necessity of requiring redshifts has introduced yet another selection effect. By having a relatively shallow distance cut-off it is not possible to include in the sample those distant highly inclined galaxies that are needed to explore the possibility that galaxies are optically thin. We are not saying that galaxies are optically thin, but just that the BHF test does not prove that they are optically thick.

4. Summary

There is an enormous discrepancy between the derived opacity obtained using different methods. Rowan-Robinson 1992 has concluded that the optical depth is as low as 0.1 over most of the galactic disk. The surface brightness inclination tests described above have inferred much higher values. In this paper we do not offer support for or argue against either point of view. We merely point out that one line of evidence, derived from a surface brightness inclination test, does not, contrary to the authors strongly made arguments, show that galaxies are optically thick at the 25 Bμ isophote.

5. References

1. Burstein D., Haynes M. and Faber S., 1991, *Nat*, **353**, 515

2. Davies J., Phillipps S., Boyce P. and Disney M., 1993, *MNRAS*, **260**, 491

3. Disney M., Davies J. and Phillipps S., 1989, *MNRAS*, **239**, 939

4. Rowan - Robinson M., 1992, *MNRAS*, **258**, 787

5. Schechter P., 1976, *ApJ*, **203**, 297
6. Valentijn E., 1990, *Nature*, **346**,153

The table

STATISTICAL TESTS FOR OPACITY

EDWIN A. VALENTIJN
SRON, Space Research Groningen
PO Box 800
NL 9700 AV The Netherlands

1. Introduction

Here, I will discuss recent statistical tests related to opacity studies. In essence, most of these tests are "inclination tests", *i.e.* studies how particular photometric parameters depend on the inclination angle of the galactic disks with respect to the line of sight of the observer.

My main conclusion is that the current literature on this subject is less controversial than often stated. At various places of the analysis the conclusion is drawn that observational data can be best understood when, next to a cirrus-like dust component which follows the radial distribution of the stars, a second component exists with an exponential scalelength larger than that of the stars. This second component has a typical optical depth[1] $\tau^B_{face} \sim 1$ at the effective radii of galaxies (and beyond), as suggested in my 1990 paper (hereafter V90). Overall face-on values of $\tau^B \sim 1$ will result in galaxies that appear optically thick and an overall optical depth > 1 will be common for disks that are inclined with respect to the external observer.

Obviously, overall values of the optical depth around unity will affect the determination of intrinsic luminosities, but possibly even more important related astrophysical questions include:

- what is the nature of the extended component?
- does the existence of extended dust components lead to mass models, which solve the flat rotation curves of spiral galaxies, without invoking the existence of dark haloes? (González-Serrano and Valentijn, 1991, = GV91, and Valentijn 1991, = V91).

---

[1] optical depth defined as the ratio between the projected thickness of the disk and the mean free path of photons

89

J. I. Davies and D. Burstein (eds.), The Opacity of Spiral Disks, 89–102.
© *1995 Kluwer Academic Publishers.*

2. The Opacity Study Ladder

The analysis of inclination tests in relation to the opacity problem involves a large amount of steps, which are summarized in Figure 1.

2.1. SAMPLES, PARAMETERS

After obtaining a sample of galaxies with detailed surface photometry (ESO-LV in our work) both a parameter and a sub-sample have to be selected. Today's detailed photometric catalogues actually allow the usage of at least 8 different photometric parameters and 8 different ways to select a sample (e.g. magnitude selected, diameter selected, surface brightness selected, redshift selected, V/V_{max} volume selected). Each of these parameter-sample combinations has its own problems regarding selection effects and possible biases. The unique logistics of selection effects for each of these combinations implies that one can not simply transfer a selection effect noted for one parameter-sample combination to another combination. It is this wealth of ways to perform the inclination test, and the variety of effects for different parameter-sample combinations that has lead to the diversity of the modern literature on this subject (see V91b and V94 for a more extensive discussion). There is no way to discuss all the different combinations here, (an attempt is made in V94), but instead, I will mention a few of the most remarkable results.

The Holmberg approach involved a plot of the *projected* surface brightness ($B + 2.5 log a^2$, with a the length of the observed major axis) versus the apparent axial ratio. Both B and a^2 depend on the value of the optical depth, but in a reverse way and with an approximately equal amplitude for most models. Thus, the projected surface brightness is not a very critical parameter for measuring the optical depth, and the Holmberg graph is not providing much statistical information on the opacity problem.

More powerful are parameters which involve the *local* surface brightness of galaxy disks. The local surface brightness is a distance independent parameter, which avoids a number of distant dependent selection effects. One of the strongest inclination tests results is, perhaps, the constancy of the local surface brightness at the effective radius[2]. In Fig. 2 data are plotted for a sample of Sb galaxies, for which detailed radial surface brightness profile fits indicate that these systems have a perfect exponential light profile ($N \sim 1$ when fitting $I(r) \sim exp(-r/r_\alpha)^N$). For pure spheroidal galaxies with a $r^{1/4}$ profile the fit would return a value $N \sim 0.25$. Bulges are expected to follow a $r^{1/4}$ profile and their presence in disks results in $N < 1$. Thus, the $N = 1$ galaxies in Fig. 2 are expected not to contain any significant bulge, which

---

[2]radius from within half of the total light is originating

The Opacity Study Ladder

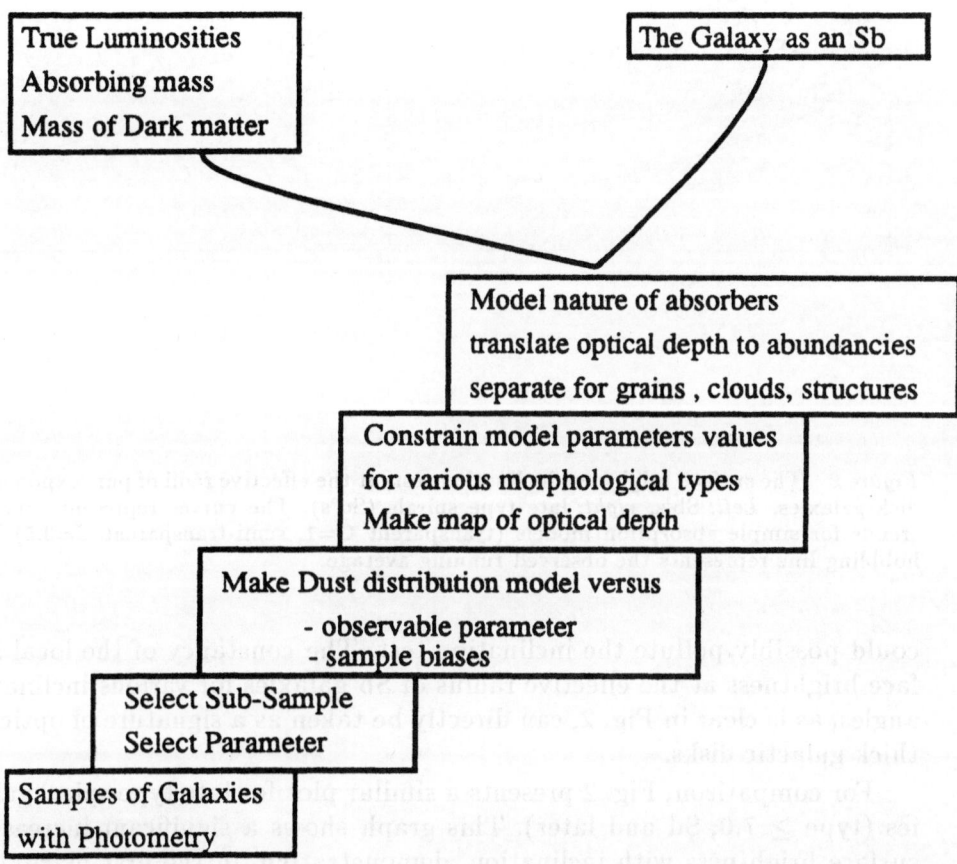

True Luminosities
Absorbing mass
Mass of Dark matter

The Galaxy as an Sb

Model nature of absorbers
translate optical depth to abundancies
separate for grains , clouds, structures

Constrain model parameters values
for various morphological types
Make map of optical depth

Make Dust distribution model versus
- observable parameter
- sample biases

Select Sub-Sample
Select Parameter

Samples of Galaxies
with Photometry

Figure 1. The various steps involved in opacity studies

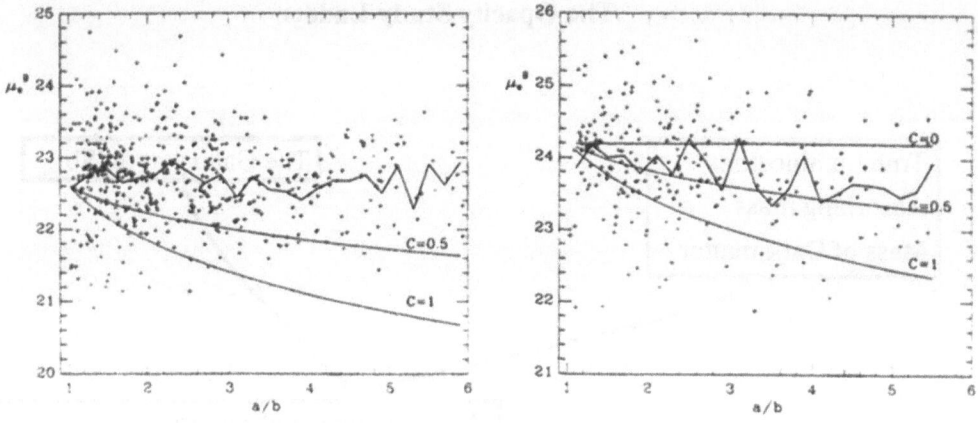

Figure 2. The surface brightness inclination test, at the effective radii of pure exponential disk galaxies. *Left:* Sb's, *right:* late type spirals (Sd's). The curves represent expected trends for simple absorption models (transparent C=1, semi-transparent C=0.5). The hobbling line represents the observed running average.

could possibly pollute the inclination test, The constancy of the local surface brightness at the effective radius of Sb galaxies for various inclination angles, as is clear in Fig. 2, can directly be taken as a signature of optically thick galactic disks.

For comparison, Fig. 2 presents a similar plot for late type spiral galaxies (type $\geq$ 7.0, Sd and later). This graph shows a significant increase of surface brightness with inclination, demonstrating, in the first place, that the inclination test is capable to trace this effect and that late type spirals behave rather different when compared to earlier type (Sb-c) spirals. A detailed map of regressions, C, as function of galaxy type and central surface brightness (the second critical parameter which determines the opacity of spirals) is given in V90 and V94.

The problems with various sample/parameter combinations, the second step on the ladder, have been extensively discussed elsewhere (Burstein *et al.* 1991, V94). Burstein *et al.* rightfully warn against combinations in

which both distant dependent *parameters*, such as the apparent diameter or
the apparent magnitude, and distant dependent *selection criteria* are used.
However, these objections do not relate to the V90 results, for which the,
distant independent, local surface brightness was used as the parameter.
This is further illustrated in V94, in which for a variety of sample selections,
the same result is obtained: a very modest inclination dependency of the
local surface brightness of Sb and Sc galaxies, at least at their effective
radius.

When redshifts of the samples are known, a number of distant dependent selection effects can be circumvented. By applying Tully-Fisher type of
relations the intrinsic magnitudes or diameters can be predicted, which can
be compared with observed, inclined, magnitudes or diameters. Burstein
et al. (1991) claim that such an analysis leads to an optical thick result,
which view is however revised by Burstein's presentation at this conference.
Gouguenheim *et al.* (this conference and Bottinelli *et al.*, 1994) apply the
same technique and conclude that isophotal diameters are not inclination
dependent, as expected in the optical thick case.
It is important to note, that also redshift selected samples can be subject
to biases. Gouguenheim *et al.* circumvented these biases by selecting galaxies from the 'plateau' in the diagram of the Hubble constant versus the
kinematic distance. This way, an unbiased TF relation is obtained from an
unbiased sample. At this point, it is not clear whether such precautions
were undertaken in Burstein's recent result.

In V90, the opacity at the outer regions was studied by evaluating the
distant independent ratio of the photometric diameters D_{26}/D_e. The absence of an inclination dependency of this ratio was interpreted as evidence
of a similar opacity at both locations.

Another way to assess distance dependent selection effects is to compute
V/V_{max} values of the sample, while varying the cutoff limit of the sample.
This way, V94 demonstrated that incomplete samples, with $V/V_{max} < 0.5$,
exhibit an increase of isophotal diameters with increasing inclination angle, a property of transparent outer regions. On the other hand, complete
samples with $V/V_{max} = 0.5$, which have been cut at a higher limit, do not
show any significant increase of isophotal diameters with inclination angle
(see Fig. 3).

The issue of the opacity of spiral disks is most important in the regions
at and beyond the effective radii of the disks. When a value of the optical
depth around unity would be deduced in these regions, then this will have
important consequences for several astrophysical issues. A face-on optical
depth around unity is generally believed to result into the absence of a
diameter increase with inclination, but transparent disks would lead to
a significant diameter increase. Fig. 3 show that no diameter increase is

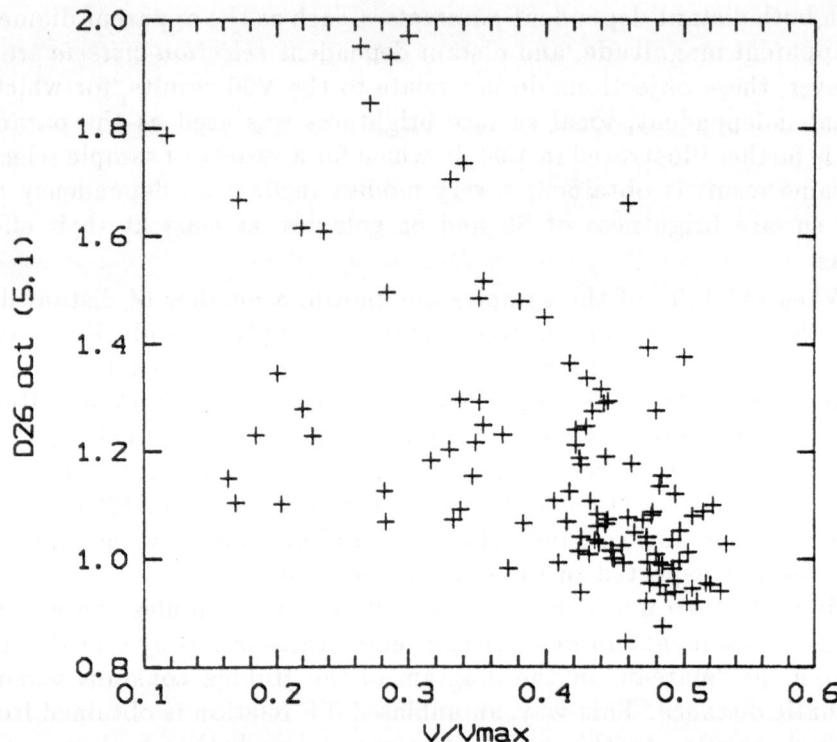

Figure 3. The apparent isophotal diameter increase (D_{26}^{oct}, between a/b=1 - nearly face-on- and a/b=5 - nearly edge-on-) versus V/V_{max} values. Complete samples with $V/V_{max} \sim 0.5$ do not show a significant diameter increase.

observed in unbiased samples, a result which agrees with earlier V90 results, with that of Gouguenheim *et al.* and with Choloniewski (1990), but not with Burstein's presentation given at this conference.

Huizinga and van Albada (1992, see also Huizinga's thesis, 1994) deduce transparent outer regions by assuming that the original visual diameter estimate of Lauberts (D_{org}) are incorrect and that the ESO-LV value of the diameter at the 26th isophote (D_{26}) are correct, in spite of the notion by the authors of the ESO-LV catalogue that, due to a technicality, an artificial inclination dependency was introduced in the D_{26} listed in ESO-LV [3].

V94 announces a recipe for a small correction to D_{26} leading to D_{26oct} which uses the computerized determinations of axial ratios (a/b_{oct}) rather

[3] D_{org} = visual, UGC, Lauberts eye, ESO-Uppsala
D_{26} = ESO-LV, computerized using $(a/b)_{org}$, obtained visually
D_{26oct} = ESO-LV, computerized using $(a/b)_{oct}$ (also computerized)

than the visual estimates of a/b. This recipe is confirmed photometrically by Peletier *et al.* (1994b) and numerically by Rönback and Bergvall (1994). The validity of using the octant axial ratios for computing diameters, rather than the original visual axial ratios is also evident from a graph which is presented by Huizinga. When comparing with Buyn's axial ratios, the dispersion of ESO-LV's (a/b_{oct}) is much smaller than that of $(a/b)_{org}$.

After applying this recipe, the corrected ESO-LV D_{26oct} values are in a very good agreement with the original visual estimates for all inclinations. Both diameters do not show any significant correlation with inclination and thus conform the expectations for optically thick outer regions. Unfortunately, the conclusions of Huizinga and van Albada are caused by neither using the original visual diameters, nor the corrected ESO-LV values.

I conclude, that current studies still strongly suggest the absence of a diameter increase with inclination in the B band of Sb, Sbc galaxies.

2.2. MODELS VERSUS DATA

The regression coefficient C of the observed local surface brightness versus $2.5 \log(a/b)$ has often been used to interface observed trends to predictions from detailed models. Analytical solutions for the absorption in disks with various layers of homogeneously distributed stars and dust are pretty straightforward and provide in general a good first order guideline for the interpretation of observational data, like C. However, both the effects of scatterings and inhomogeneous dust distributions are complicated and have to be taken into account. Good progress on the modeling side has been reported at this conference (*e.g.* Bruzual, Buyn, Kylafis, Witt). Particularly, Witt's (also Witt *et al.* 1992) results on brightening of face-on galaxies and possible de-reddening (bluing) should be taken as a warning against the over-interpretation of simple models. On the other hand, many of the basic notions of simple multiple layer absorption models are reproduced in the more advanced evaluations of the radiation transfer problem.

For a simple slab model with a homogeneous mix of stars and absorbers, or for a sandwich model with transparent outer layers, the regressions (values of C) reported in V90 and V94 correspond to optical depth listed in Table 1. The observed absence of an increase of isophotal diameters corresponds in these models to an optical depth around unity at the outer radii. As outlined in V90, the tabulated radial variation of τ is approximately a factor 10 less than the radial variation of the emitted star light. Thus, the data suggest an absorbing component with an exponential scale length larger than that of the stars (this was actual the motivation for the alternative two-disk rotation curve solutions, GV91).

Note, that at this state of the process (fourth box on the ladder, Fig. 1)

TABLE 1. Very global face-on B band optical depths of Sb galaxies

Center:	$C \sim 0$	$\tau \sim 5\text{-}30$
effective radius:	$C \sim .2$	$\tau \sim 1\text{-}2$
D_{25}, D_{26}:	$C \sim .2\text{-}.4$	$\tau \sim 1$

we need not yet to define the nature of the absorbers, they are only assumed to be well mixed with the stars. The next step involves the identification of the nature of the absorbing bodies. With this I mean, that several models allow to *a postiori* discriminate between the relative contributions of compact opaque clouds and cirrus-type of dust grains. Fundamentally, a measured value of τ relates to value of the product of the disk geometrical thickness, the obscurers cross sections and their spatial density ($T\sigma\rho_0$). This propagates to a total mass of absorbers in the disk, $M_{disk} \sim \tau(m/\sigma)R^2$, with m the mass per absorbing body and R the disk radius. Obviously, for solid bodies $m/\sigma \sim r$, with r the radius of the body. Then, a value of $\tau \sim 1$ immediately rules out any significant contribution to the opacity of disks by means of solid bodies with radii larger than a fraction of a micrometer, thus, for example, excluding asteroids, planets etc., as these will lead to unacceptable total disk masses. However, virialized clouds have the property m/σ independent of r and thus they can contribute significantly to the opacity without trespassing total disk masses. In the models, compact clouds can be taken as black balls, optically thick at the shorter wavelength, which act as pure absorbers without scattering. In this way, I originally introduced molecular clouds in the simple models and identified them as possible candidates for the obscuring component with a scalelength larger than that of the stars. It is noteworthy, that Witt's findings on the effect of clumpiness of the interstellar medium in fact corresponds closely to what one would expect from the 'black ball' models. He reports:

- in clouds scattering can be ignored,
- absorption is relatively independent of cloud structure
- transparency goes up with clumpiness
- extinction is grey, once clumps are optically thick.

3. (Near) Infrared studies

An absorbing component which extends to the outer regions of disks influences the exponential radial scalelength in the passbands that are affected by the absorption. As the near-infrared K passband is believed to be hardly affected, a scalelength observed at K would closely correspond to that of the intrinsic stellar distribution. Thus, a comparison of scalelength at short wavelength with that at the K band will provide valuable information on the radial distribution of the absorbers, when the effect of radial population gradients (colour gradients) can be controlled. For this reason, Peletier *et al.* have mapped at K a representative sample of 38 Sbc galaxies and compared the observed scalelength ratios in B and K to models with dust radial scalelength as a free parameter (Peletier *et al.*1994a,b and this Volume). They find the scalelength ratio (B/K) in the range 1.3-1.9, with a trend for that ratio to increase with inclination angle. Analytical solutions of a radial version of the equation of radiation transfer were made for a variety of dust spatial distributions. In its most simple form, the model predicts a relation between central optical depth at the observed passband (*e.g.* B band), the scalelength at the observed wavelength, α_{obs}, the scalelength of the dust component (α_d), the intrinsic scalelength of the stars ($\alpha_\star$, as mapped at the K band images if colour terms in stellar populations can be handled) of the form:

$$1 = \frac{\alpha_{obs}}{\alpha_\star} - \frac{\alpha_{obs}}{\alpha_d} - 0.5\, ln\, \frac{(1 - e^{(-\tau_0 exp(-3\alpha_{obs}/\alpha_d))})}{(1 - e^{(-\tau_0 exp(-\alpha_{obs}/\alpha_d))})},$$

when the scalelength fits are made in the radial range between 1 and 3 scalelength. τ_0 is the central opacity in the observed pass band, $\tau(r) = \tau_0\, e^{(-r/\alpha_d)}$.

Thus, a particular observed B/K scalelength ratio ($\frac{\alpha_{obs}}{\alpha_\star}$) corresponds to a set of solutions of τ_0 and α_d. We find that observed B/K values can not be satisfactorily reproduced with this model when we require $\alpha_d = \alpha_\star$ and an optical depth in the B band at the outer radii of about 1, as suggested by the inclination tests summarized in the previous Section. But, acceptable solutions are found when the requirement of semi-transparent outer regions is dropped, which solutions, however, do not reproduce the observed trend in the B/K scalelength ratio inclination test.

Again, this implies that we have to consider models in which at least one dust component has a scalelength larger than that of the stars. Two-component models were considered: one dust component with a scalelength following that of the stars, the other component with a larger scalelength. We evaluated such models with various scaleheights of the dust distribution, compared to that of the stars. A reasonable description of all available observational data can be made when the first component, which follows the radial distribution of the stars, has a central optical depth in the range 0.6 - 3.0 and is fully transparent at the outer regions, while the second component has a scalelength in the range 2-4 times that of the stars, and an optical depth around unity at the outer radii (see Table 3).

The dust component that follows the stars, is nearly transparent at the effective radius and its spatial distribution matches to the dust as detected by IRAS. It can, most likely, be associated with the warm dust as seen by IRAS.

Conversely, the second component, with a scalelength larger than that of the stars, could be associated with a much cooler dust component $< 20K$, such as molecular clouds and would not have been detected by IRAS.

Van Driel *et al.*(1994 and this Volume) report on a similar type of scalelength ratio study, but now involving the FIR bands at 50 and 100 μm. The mean observed scalelength ratios $\alpha_{100\mu m}/\alpha_{50\mu m}=1.21$ and $\alpha_B/\alpha_{100\mu m}=1.12$ are most difficult to describe with single component dust models, especially when the B/K scalelength ratios as observed by Peletier *et al.* would also apply to van Driel *et al.*'s sample. Accordingly, van Driel *et al.* conclude that multiple component dust models, such as suggested above, are required to understand the FIR data.

4. Concluding Remarks [4]

I have been asked to compare the V90, V94 results with presentations given at this conference. My main conclusion is: 'that the modern literature is considerably less controversial than is often stated, but that opacity studies of the outer regions of Sb and high surface brightness Sc galaxies require an absorbing component with an exponential scalelength which is larger than that of the stars'.

On Tuesday afternoon we had a discussion session during which we compared the results of several studies. When comparing these with my 1990,1994 results, it appears that there is a **reasonable agreement** about the following items:

●$<$ 1989 studies needed a critical review in the light of more recent results.

In many presentations substantial column densities of dust were reported.

Different views on the overall optical thickness of spiral disks were well presented at this conference, but actually, it was interesting to hear that Bosma and Freeman, who arrive at the lowest optical depth values after analysing rotational velocities of edge-on galaxies, acknowledged that they had to put there slit next to the ·dust lane, otherwise they 'could equally well go observing the night sky'.

●Relatively high optical depth ($\tau^B_{face\_on}$) in the centers of spirals (3-10),

e.g. Pravda's steepening of rotation curves with wavelength

e.g. Peletier's almost 2 times larger dispersion of the central surface

---

[4]Summary of Panel presentation, given at the conference review session

brightness in K compared to that in the Blue band.

e.g. Peletier's high B/K scale length ratios.

•A $\tau^B_{face\_on} \sim 1$ at the effective radius (1.7 scalelength) in many Sb,Sbc's

•The variation of optical depth along morphological type:

low for early types, higher for types 3-5 and again lower for the later types (e.g. Table 2).

Also Bosma listed this variation, but perhaps with a smaller amplitude.

•IRAS seen dust does not account for high optical thickness.

•Cold, $T_{dust} < 20$ K, dust is very important.

•Galaxies are nearly transparent at the K-band.

Table 2 contains a compressed summary of my results, when observed trends are transformed to face-on optical depth by means of a sandwich model, ignoring the effects of various types of scatterings.

TABLE 2. Global values of the optical depth at various positions

Galaxy type			τ^B_{face} values		
type	type	central sb	center	D_{eff}	$D_{25} - D_{26}$
Sa	0-2.5	all (20.5)	<0.5	<0.4	<0.4
Sb,Sbc	3-5	all (21.0)	>3	1	1 *
Scd	5.5-6.5	21.0-21.5	>2	.7	.6 *
		21.5-22.0	1	.4	.2
Sd	7-9	>22.0	.5	.2	.1

Most values are in reasonable agreement with the conference results with the exception of the two values marked with *, which correspond to the optical depth around unity in the outer regions of Sb, Sbc and high surface brightness Sc galaxies. On this issue there is **no uniformity** of results. This is an important issue, as it relates to about 50% of the spiral galaxies as seen by the ESO-LV survey. Note also, that the regions beyond the effective radii contain by definition half of the observed total light.

Table 3 lists the basic properties of my currently favorite model that results in an optical depth around unity at the outer regions. This is a two component dust model with one component following the scalelength of that of the stars, the other one having a larger scalelength.

This model is consistent with and has been discussed here by:

- the V90,V94 results,

TABLE 3. Range of parameters of acceptable two-component models

	α_{dust}/α_*	range of τ^B_{face} values		
		center	D_{eff}	D_{25}
first component :	1	0.6 - 3.0	0.1 - 0.3	0 - 0.05
second component :	3-4	4 - 2	2 - 1	1 - 0.8

- the B/K scalelength ratios presented by Peletier *et al.*.
- the FIR/optical/K scalelength ratios presented by van Driel.

Unfortunately, none of the other detailed model calculations presented at this conference compute dust configurations with scalelengths larger than that of the stars (or models which show no increase of diameters with inclination) and the above table remains a challenge to 'the model makers'. Models including scattering and clumpiness might help to solve the problem of the opacity in the outer regions.

The issue of the opacity at the outer regions is important as it relates directly to two important astrophysical items:
- what is the nature of the absorbing regions that contribute to the optical depth: cold dust, molecular clouds?
- the basic rationale of the maximum disk analysis (which let us believe that we are forced to consider dark haloes to solve flat rotation curves) does not apply when a dust component with a scalelength larger than that of the stars exists. Silk's proposal at this conference for a large number of compact molecular clouds in the halo might be related to this issue, but I still prefer a geometry with compact clouds in the disk.

TABLE 4. The controversy of inclination test results at the outer regions.

	< 1.1	~1.4
Diameter increase ($a/b = 1- > a/b = 5$):		
τ^B_{face} at D_{26}:	~ 1	<.2
Tully-Fisher technique:	Gouguenheim	Burstein
direct diameter-inclination test:	V94	Huizinga
D_{26}/D_{eff} ratio technique:	V90	

Papers relating to the opacity at the outer regions, as presented at this conference, are categorized in Table 4. In a perhaps slightly oversimplified way these papers can be divided into two groups: those who report on the absence of a diameter increase with inclination and a corresponding relatively high optical depth (column 2) and those who report on a significant diameter increase and a low optical depth (column 3).

Gouguenheim et al. demonstrated possible selections effects also in the T-F tests, and they have carefully avoided these by selecting objects from the 'plateau' of their diagram. Such precautions were not evident in Burstein's presentation. An easy verification would be to run the V/V_{max} test both on Burstein's sample and on Gouguenheim's sample.

The differences between Huizinga's and my 1994 result are caused by a technicality as explained in Section 2.1, which I believe is resolved by a small update of ESO-LV parameters, as proposed in V94.

I think it is great we have achieved so much understanding at this conference, and a rather surprising agreement about quite a number of issues, but the remaining issues will keep this field alive for some time to come.

References

1. Bosma A., Byun Y., Freeman F.C., Athanassoula E., 1992, ApJ 400, L21
2. Bothun G.D., Rogers C., 1992, AJ 103, 1484
3. Bottinelli L., Gouguenheim L., Paturel G., and Teerikorpi, P., 1994, *A&A* in press
4. Bruzual G.A., Magris G.C., Calvet N., 1988, ApJ 333, 673
5. Burstein D., Haynes M.P, Faber S.M., 1991, Nature 353, 515
6. Buyn Y.I., Freeman K.C., 1991, Proc. ASA 9, 86
7. Choloniewski J., 1991, MNRAS 250, 486
8. Disney M., Davies J.I., Phillipps S., 1989, MNRAS 239, 939
9. González-Serrano J.I., Valentijn E.A., 1991, (=GV91) Peletier, R.F., A&A 242, 334
10. Huizinga J.E., van Albada T.S., 1992, MNRAS 254, 677
11. Huizinga, J.E., 1994, Ph.D Thesis, University Groningen
12. Lauberts A., Valentijn E.A., 1989, The Surface Photometry Catalogue of the ESO-Uppsala Galaxies (= ESO-LV), ESO, München
13. Peletier, R.F., Valentijn, E.A., Moorwood, A.F.M. and Freudling, W. 1994a, A&A Suppl., 108, 621
14. Peletier, R.F., *et al.* 1994b, A&A (Letters), submitted
15. Rönnback J., Bergvall, N., 1994, A&A Suppl., 108, 193
16. Valentijn E.A., 1990, (= V90) Nature 346, 153
17. Valentijn E.A., 1991a, (= V91) Proc. IAU Symp. No 144, ed. H.Bloemen, Kluwer, p 245
18. Valentijn E.A., 1991b, ESO Messenger 63, 45
19. Valentijn E.A., 1994, (=V94) MNRAS 266, 614
20. van Driel, W., Valentijn, E.A., Kussendrager, D. and Wesselius, P.R., 1994, *A&A (Letters)*, submitted
21. Witt A.N., Thronson, H.A., Capuano, J.M., 1992, ApJ 393, 611

Question
Davies

How can you bin galaxies by surface brightness and then do a SB inc test?

Answer
Valentijn

The question is relevant indeed. In my view I can justified this as we have now convincingly demonstrated that the central s.b. does not depend on inclination. In the original papers I avoided this binning as I did not want the conclusions to depend on it. However, the constancy of central s.b. justifies the binning, as there is consensus about the constant central s.b. (e.g. Peletier, Kodaiza et al).

Comment
Witt

You suggested, based on your statistical tests, that the dust scale length in spirals is larger than that of stars, with $\tau \sim 1$ near the edge of the disk. Given that the attenuation by dust is much more dependent on the distribution of dust relative to the stars than on the actual amount of dust, you could possibly explain your results equally well with an assumption that the ratio of dust scale height to star scale height increases with galactocentric radius, while the scale lengths of dust and stars might actually be the same. Such a variation of scale height ratio with distance form the centre is apparently present in our galaxy and is plausible for disk systems in general.

STATISTICAL MEASURES OF INTERNAL ABSORPTION IN SPIRAL GALAXIES

BARBARA CUNOW

Astronomisches Institut, Universität Münster,
Wilhelm–Klemm–Str. 10, D-48149 Münster, Germany

AND

Department of Mathematics and Astronomy, University of
South Africa, PO Box 392, 0001 Pretoria, South Africa

Abstract. A sample of 2300 spiral galaxies is used for the statistical determination of internal absorption by investigating projected surface brightness and colour. The absorption values for face-on view measured in the photographic b_J band are $A_J(0) = 0^{\text{m}}.80$ for the sandwich model and $A_J(0) = 0^{\text{m}}.53$ for a more realistic three-component model.

1. Introduction

The first measurements of internal absorption in spiral galaxies were statistical investigations made by Holmberg (1958), who introduced the basic method. He defined a projected surface brightness S, with $S = m + 5\log a$, where m is the apparent magnitude and a the apparent semimajor axis. The change of S with inclination angle i is related to the variation of the absorption A with i: $S(i) - S(0) = A(i) - A(0)$. For the determination of the face-on absorption $A(0)$, a model of the galaxy is needed. The change of colour $C = m_{\lambda_1} - m_{\lambda_2}$ with inclination is $C(i) - C(0) = \Delta A(i) - \Delta A(0)$ where $\Delta A = A_{\lambda_1} - A_{\lambda_2}$ is the difference in absorption of the two passbands.

The classical investigations by Holmberg (1958) and Heidmann et al. (1972) led to the conclusion that galaxy discs are optically thin. This was widely accepted until the IRAS data became available. Disney et al. (1989) pointed out that the FIR fluxes of spirals measured by IRAS are significantly larger than predicted by absorption and re-emission of starlight in

103

J. I. Davies and D. Burstein (eds.), The Opacity of Spiral Disks, 103–113.
© 1995 *Kluwer Academic Publishers.*

104

a model with an optically thin disc. A galaxy with an optically thick disc, however, may contain enough dust to explain the large FIR flux via thermal reradiation from the dust. Disney et al. (1989) showed that optically thick discs can be similar to optically thin discs with respect to the variation of surface brightness with inclination. The assumption of galaxy discs being necessarily optically thin was therefore no longer convincing.

The work of Valentijn (1990) was the first statistical investigation suggesting that galaxy discs are optically thick. Kodaira et al. (1992) also found that spiral galaxies are optically thick. Huizinga & van Albada (1992) investigated Sc galaxies. These galaxies show strong absorption in the central parts, but they are optically thin in the outer regions. This is in agreement with the measurements of Cunow (1992). Valentijn (1994) considered in his sample selection effects and biases very carefully and found strong absorption for spiral galaxies with high surface brightness, and low absorption for spiral galaxies with low surface brightness.

This paper presents an investigation of internal absorption in the b_J (green) and r_F (red) bands for 2300 spiral galaxies with $b_J \leq 18$. Three disc models and a more realistic three-component model, including a bulge as well as emitting and absorbing disc components, are chosen and the results are intercompared.

2. Data Catalogue

The data are taken from filmcopies of the ESO/SERC(J) and ESO/SERC(R) surveys. Six adjacent fields near the South Galactic Pole were used. The filmcopies were digitised with the PDS 2020 GM^{plus} in Münster. The catalogue contains 2300 spiral galaxies with $b_J \leq 18$. For each galaxy, total magnitudes in b_J and r_F are determined from aperture magnitudes. Semimajor and semiminor axes, a and b, respectively, are measured from intensity-weighted moments of the J-band images. The apparent ellipticity ϵ, which is used as a measure of the inclination angle i is defined as $\epsilon = 1 - \frac{b}{a}$. Details are found in Cunow (1992, 1993).

For each galaxy, the projected surface brightness S for the J-band (computed from b_J and a), the colour $b_J - r_F$ and the apparent ellipticity ϵ are measured. The random errors of S, $b_J - r_F$ and ϵ are determined from the overlap regions of neighbouring fields. They are $\sigma(S) = 0.17 \, \text{mag arcsec}^{-2}$, $\sigma(b_J - r_F) = 0^{\text{m}}.14$ and $\sigma(\epsilon) = 0.06$. The morphological classification of the galaxies is an automated process where prototypes are selected visually. Five groups of Hubble-equivalent types are used: E, S0, Sa, Sb and Sc. The random classification error is about 1 Hubble-equivalent type. Details are given in Spiekermann (1992).

For galaxies with $b_J > 16$, the apparent ellipticity ϵ is affected by seeing which makes the objects appear rounder. Corrections are obtained by comparing the distributions $N(\epsilon)$ for different magnitude intervals with the adopted undisturbed reference relation $N(\epsilon)$ for $15 < b_J \leq 16$. The correction for ϵ thus obtained is ≤ 0.1

An important aspect is the shape of the PSF. Most faint stars show an apparent ellipticity between 0.05 and 0.1. This may be due to noise because an intrinsically round object will become elongated by a disturbance. For $\epsilon < 0.1$, the differences between apparent semimajor and semiminor axes of stellar images are smaller than 1 pixel. On these small scales, systematic image distortions cannot be excluded. Because the observed shapes of round galaxies are affected in the same way as the PSF for stars, galaxies with $\epsilon < 0.1$ are not used.

To minimize problems due to misclassification of morphological types a division into the subgroups Sa, Sb and Sc is not made. Therefore, the results are valid for a "mean spiral galaxy".

3. Models

3.1. DISC MODELS

The data were used to test three disc models: screen, slab and sandwich model. The screen model assumes a layer of stars obscured by dust lying in front of it. In case of the slab model, stars and dust form a homogeneously mixed disc. The sandwich model assumes two star layers which sandwich a slab (mixture of stars and dust). Detailed descriptions are given in Disney et al. (1989). The following equations represent the absorption A as function of the inclination angle i, (1) for the screen model, (2) for the slab model and (3) for the sandwich model. τ_0 is the face-on optical depth, ζ (for the sandwich model) with $0 < \zeta \leq 1$ gives the thickness of the dust layer relative to the star layer.

$$A(i) \;=\; 1.086\,\frac{\tau_0}{\cos i} \tag{1}$$

$$A(i) \;=\; -2.5\log\left[\frac{\cos i}{\tau_0}\left(1 - \exp\left(-\frac{\tau_0}{\cos i}\right)\right)\right] \tag{2}$$

$$A(i) \;=\; -2.5\log\left[\frac{1-\zeta}{2\cos i}\left(1 + \exp\left(-\frac{\tau_0}{\cos i}\right)\right)\right.$$
$$\left. + \frac{\zeta}{\tau_0}\left(1 - \exp\left(-\frac{\tau_0}{\cos i}\right)\right)\right] - 2.5\log(\cos i). \tag{3}$$

τ_0 and ζ are the parameters determined from the data. Due to the fact that the data are available in b_J and r_F, the relevant parameters are τ_0^J, τ_0^F, ζ_J and ζ_F. It is assumed that ζ does not vary with wavelength. τ_0^J and τ_0^F are related to each other by the extinction law. The investigation by

Knapen et al. (1991) indicates that the mean extinction law of the Galaxy $(\tau_0^F = 0.629\,\tau_0^J)$ is equally valid for other galaxies.

3.2. THREE-COMPONENT MODEL

A more realistic model for spiral galaxies is described by Christensen (1990). The model galaxy is specified by the following components: luminous bulge, luminous disc and absorbing disc. Only absorption is considered, while the more complicated effects, like scattering, are neglected. The emitting components are described by their volume emissivity distributions. The bulge emissivity is given by

$$I_b(r) = I_e \left(\frac{r}{r_e}\right)^{\frac{7}{8}} \exp\left(-7.668 \left(\frac{r}{r_e}\right)^{\frac{1}{4}}\right) \tag{4}$$

with radius r defined as

$$r^2 = \left(x^2 + y^2 + \left(\frac{z}{q}\right)^2\right) q^{\frac{2}{3}}, \tag{5}$$

where (x, y, z) are the coordinates of a given point in the galaxy and q is the true axial ratio of the bulge. The disc emissivity is given by

$$I_d(r) = \quad I_0 \exp\left(-\frac{R}{r_0}\right) \exp\left(-\frac{|z|}{z_0}\right) \quad R \leq R_{\max} \tag{6}$$
$$I_d(r) = \qquad\qquad 0 \qquad\qquad R > R_{\max}$$

with

$$R = \sqrt{x^2 + y^2}. \tag{7}$$

The overall emissivity I at a given point (x, y, z) is

$$I(x, y, z) = I_b(x, y, z) + I_d(x, y, z). \tag{8}$$

The dust distribution, specified by the absorption coefficient κ, is assumed to have the same form as the emitting disc:

$$\kappa(r) = \quad \kappa_0 \exp(-\frac{R}{r_0'}) \exp(-\frac{|z|}{z_0'}) \quad R \leq R'_{\max} \tag{9}$$
$$\kappa(r) = \qquad\qquad 0 \qquad\qquad R > R'_{\max}.$$

The face-on optical depth at the center is given by $\tau_0 = 2z_0'\kappa_0$.

The optical depth τ at (x, y, z) is determined by integrating κ along the line of sight from the observer to this point (x, y, z). The intensity contribution at (x, y, z) seen from outside the galaxy is then

$$I_{\text{obs}}(x, y, z) = I(x, y, z) \exp(-\tau). \tag{10}$$

TABLE 1. Parameter values for the three-component model.

$\dfrac{r_e}{r_0}$	q	$\dfrac{z_0}{r_0}$	$\dfrac{R_{\max}}{r_0}$	$\dfrac{r_0'}{r_0}$	$\dfrac{z_0'}{r_0}$	$\dfrac{R_{\max}'}{r_0}$	$\left(\dfrac{E_e}{E_0}\right)_J$	$\left(\dfrac{E_e}{E_0}\right)_F$	$\dfrac{\tau_0^F}{\tau_0^J}$
0.27	0.75	0.14	4.0	1.0	0.05	4.0	1150	1520	0.629

The intensity of a given image pixel is determined by integrating $I_{\text{obs}}(x, y, z)$ along the line of sight. Hence, the galaxy image as seen by the observer can be calculated.

The model is specified by 11 independent parameters per wavelength band. With the exception of τ_0^J, all parameters are taken from the literature. Values for the geometrical parameters, which are assumed not to vary with wavelength, are taken from Christensen (1990). The non-geometrical parameters are calculated from the bulge-to-disc ratio B/D for given geometrical parameters. B/D values from Simien & de Vaucouleurs (1986) are used, which are valid for the B-band. They are transformed to b_J and r_F using $B - V$ for bulge and disc (van der Kruit & Searle 1981a, 1981b, 1982a, 1982b) and colour relations from King et al. (1981), Shanks et al. (1984) and Lauberts & Valentijn (1989). Table 1 gives the values for the model parameters. To represent a "mean spiral galaxy", values for Sb galaxies are used throughout.

4. Model fits

This study investigates $S(\epsilon)$ for the J-band and $(b_J - r_F)(\epsilon)$. For the disc models, $S(i) - S(0) = A(i) - A(0)$ and $(b_J - r_F)(i) - (b_J - r_F)(0) = (A_J(i) - A_F(i)) - (A_J(0) - A_F(0))$ (see Section 1) are adopted. For the three-component model, S, $b_J - r_F$ and ϵ are obtained from simulated images with known A. As discussed above, galaxies with $\epsilon < 0.1$ are not considered.

4.1. DISC MODELS

The assumption $S(i) - S(0) = A(i) - A(0)$ requires that the apparent semimajor axis a is free from inclination effects. This means that for a given Hubble type the derived value for a must be the same for all directions of view. To estimate the influence on a, galaxies of the same type must be compared for various inclinations. It is important that the mean distance of the galaxies in a sample is the same for different intervals of inclination. In this study, the effects on a are not corrected, because accurate redshift measurements are not available.

TABLE 2. Results for the models.

Model	τ_0^J	ζ	$A_J(0)$	$A_F(0)$	χ^2
Screen	0.22 ± 0.04	–	$0\overset{m}{.}24 \pm 0\overset{m}{.}04$	$0\overset{m}{.}15 \pm 0\overset{m}{.}03$	6.12
Slab	0.67 ± 0.15	–	$0\overset{m}{.}34 \pm 0\overset{m}{.}07$	$0\overset{m}{.}22 \pm 0\overset{m}{.}05$	4.08
Sandwich	1.9 ± 0.9	0.77 ± 0.09	$0\overset{m}{.}80 \pm 0\overset{m}{.}32$	$0\overset{m}{.}56 \pm 0\overset{m}{.}25$	0.96
3-comp.	5 ± 2	–	$0\overset{m}{.}53 \pm 0\overset{m}{.}19$	$0\overset{m}{.}34 \pm 0\overset{m}{.}11$	1.91

To measure $S(i) - S(0)$, ϵ must be transformed to i. For the determination of this relation, the mean true axial ratio $< q >$ is needed (Holmberg 1958), which is measured by deprojecting $N(\epsilon)$ in order to obtain the distribution of the true axial ratios $n(q)$. For more detail about this method see, e.g., Sandage et al. (1970). For simplicity, $n(q)$ is adopted to be Gaussian, specified by $< q >$ and σ_q. To avoid biases, the sampling volumes must be the same for different inclinations. Due to internal absorption, inclined galaxies appear systematically fainter than face-on galaxies. If a magnitude-limited sample is used and no magnitude correction is applied, the sampling volume decreases systematically with increasing inclination. Therefore, the magnitudes which would limit the sample are corrected to give face-on magnitudes by using $S(\epsilon)$ without applying a model. Since the faintest galaxies on the photographic plates for which morphological types are available have $b_J = 19$, and the magnitude correction is $\Delta b_J \leq 2\overset{m}{.}0$, only galaxies with face-on magnitudes $b_J^0 \leq 17$ are used for $N(\epsilon)$.

$< q > = 0.103 \pm 0.009$ and $\sigma_q = 0.033 \pm 0.007$ is obtained. Binney & de Vaucouleurs (1981) found $< q > \approx 0.2$ for spiral galaxies. They used the RC2 catalogue (de Vaucouleurs et al. 1976), where the magnitude correction for inclination effects is smaller than for this investigation. This smaller correction leads to a lower number of galaxies with large apparent ellipticities and thus to a higher value of $< q >$. In the following, $< q > = 0.103$ is adopted.

The results are shown in Table 2 and Fig. 1. Because the disc models assume a disc with an infinite extension in the galaxy plane, no predictions for $i = 90°$ can be made. Therefore, only $\epsilon \leq 0.8$, which corresponds to $i \leq 80°$, is considered. For the screen and slab model, low absorption values are obtained, while for the sandwich model they are relatively high. According to χ^2 (see Table 2) and Fig. 1, the fitted sandwich model represents the data best.

The result of $\zeta = 0.77$ (for sandwich model) indicates an absorbing disc which is geometrically thinner than the luminous disc. This is in agreement with the results from investigations of surface brightness profiles of edge-

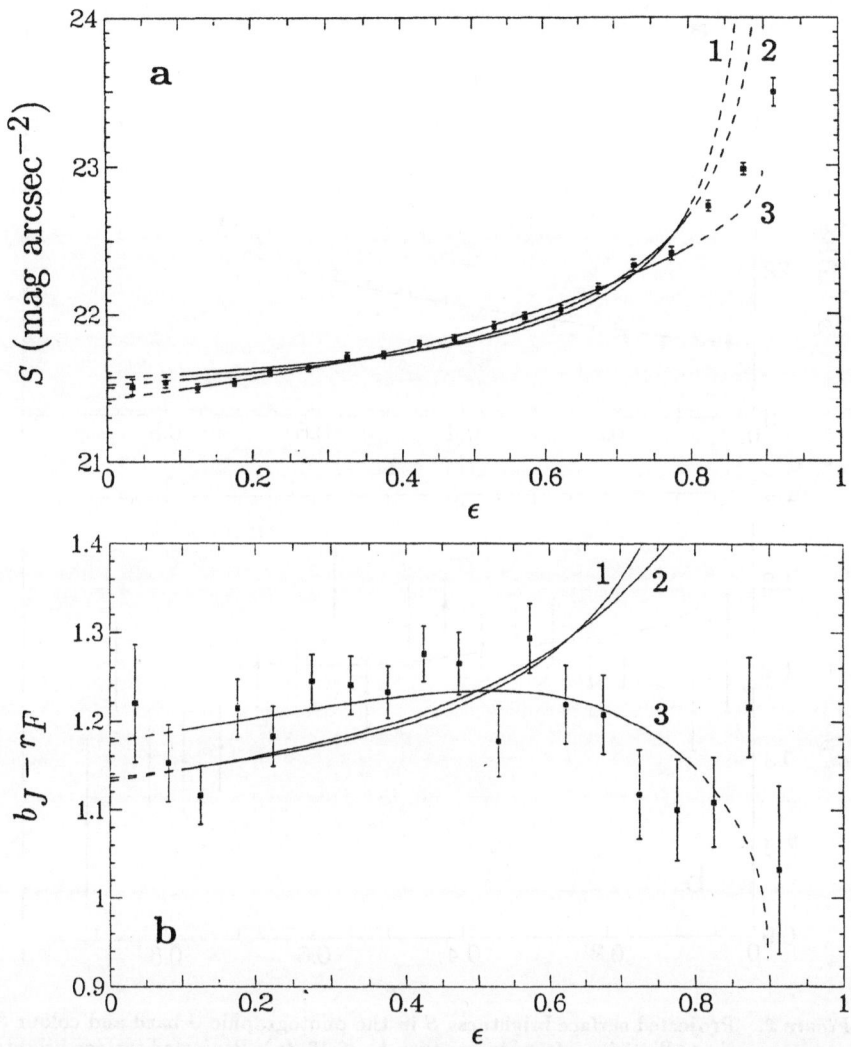

Figure 1. Projected surface brightness S in the photographic J band and colour $b_J - r_F$ versus apparent ellipticity ϵ for galaxies with $b_J \leq 18$. (a): Projected surface brightnesses, (b): colours. The lines give the best-fitting model curves for the disc models: 1 for screen, 2 for slab and 3 for sandwich. The model fits are performed for the interval $0.1 < \epsilon \leq 0.8$ ($\epsilon = 0.8$ corresponds to $i = 80°$, if $< q >= 0.103$).

on galaxies (van der Kruit & Searle 1981a, b, 1982a, b, Kylafis & Bahcall 1987).

4.2. THREE-COMPONENT MODEL

A set of templates is calculated for different inclinations and different optical depths. τ_0^J values of $0, 0.5, 1, 3, 5, 7, 10$ and 15 are chosen. S, $b_J - r_F$ and ϵ are determined by the same algorithms as used for the actual data. This

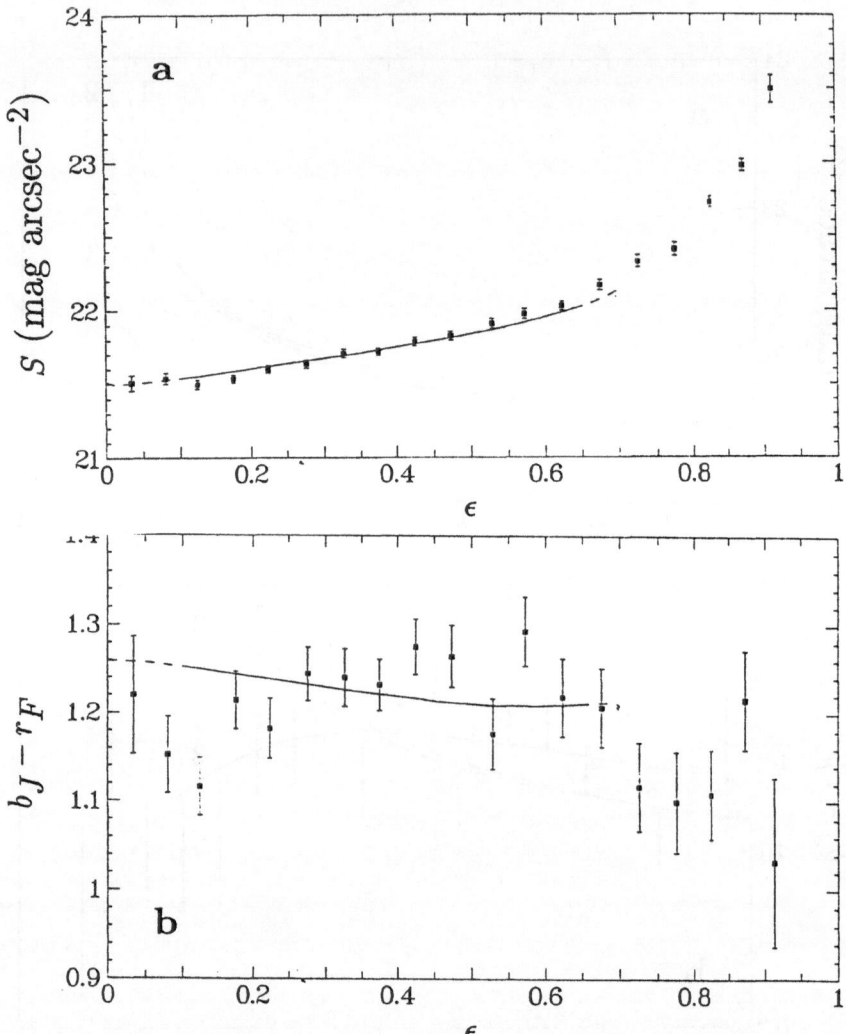

Figure 2. Projected surface brightness S in the photographic J band and colour $b_J - r_F$ versus apparent ellipticity ϵ for galaxies with $b_J \leq 18$. (a): Projected surface brightnesses, (b): colours. The lines give the best-fitting model curves for the three-component model with $\tau_0^J = 5$. The model fit is performed for the interval $0.1 < \epsilon \leq 0.65$.

means a more consistent treatment than for the disc models is possible.

For large values of τ_0^J, the model galaxies seen edge-on appear rounder than for small values of τ_0^J because the discs are becoming fainter. For $\tau_0^J = 15$, $\epsilon = 0.653$ is obtained for $i = 90°$. To be able to use in the fitting procedure the same ϵ-intervall for all simulated optical depths, only galaxies with $\epsilon \leq 0.65$ are considered.

As compared to the data, the best fit is achieved for $\tau_0^J = 5$ with $A_J(0) = 0^m.53$ (Fig. 2). The quality of the fit is comparable to the one for the sandwich model. However, the change of colour with ϵ is represented better by the sandwich model than by the three-component model.

For edge-on view, the chosen model galaxy with $\tau_0^J = 5$ gives $\epsilon = 0.70$. Galaxies with $\epsilon > 0.70$ are not predicted by this model. The present sample contains a significant number of galaxies with $\epsilon > 0.7$. These galaxies are highly inclined Sc galaxies. They show larger apparent ellipticities than edge-on Sb galaxies, because they are intrinsically flatter (Binney & de Vaucouleurs 1981).

5. Discussion

For the three-component and the sandwich model, larger absorption values than for the screen and the slab model are found. The model fits are significantly better for the sandwich and the three-component model than for the screen and the slab model. This, together with the fact that the sandwich and the three-component model give a more realistic description of the structure of a spiral galaxy than the screen and the slab, leads to the conclusion that spiral galaxies show strong internal absorption.

The differences between the absorption values obtained for the four models show that the results certainly depend on the assumed model. This is in agreement with the statements of Disney et al. (1989) and stresses the importance of using a realistic model for the galaxies.

The absorption values obtained are similar to the values found by Huizinga & van Albada (1992) and Valentijn (1994). They are also similar to the values found in an earlier investigation by Cunow (1992), where only surface brightnesses were considered. The χ^2 values obtained for the models were similar. In the present study, surface brightnesses and colours are used, and the χ^2 values are smaller for the more realistic models.

The sample used here is magnitude-limited. In such a sample, galaxies seen at high inclinations are systematically nearer than face-on galaxies, as the former ones appear fainter due to internal absorption. This may lead to biases. Corresponding effects arise in diameter limited samples (Burstein et al. 1991, Choloniewski 1991, Valentijn 1994). To avoid this problem, a space-limited sample of galaxies is needed whose intrinsic properties (e.g. luminosity, diameter, colour) are unaffected by dust and do not change systematically with inclination. As the appearance of a galaxy (e.g. observed magnitude, observed diameter, observed colour) is always influenced by dust absorption, it is extremely difficult to create an unbiased sample.

In a future study it is planned to investigate individually the surface brightness profiles of a number of selected galaxies, which are representative of the statistical sample. This would allow a direct comparison of the absorption values obtained from the different methods. In order to consider the UV and NIR properties of the galaxies, U- and I-data will also be included.

Acknowledgements

I thank Profs. W. Seitter and W.F. Wargau, Drs. P. Schuecker and R. Duemmler for many useful discussions, and Dr. H. Horstmann for help concerning data reduction. G. Spiekermann kindly provided the morphological types of the galaxies used in this study. This work is part of the Muenster Redshift Project MRSP. Financial support of the MRSP under numbers Se 345/14–1,2,3, Se 345/20–1 and Se 345/21–1,2 by the Deutsche Forschungsgemeinschaft (DFG) is gratefully acknowledged.

References

Binney J., de Vaucouleurs G., 1981, MNRAS 194, 679
Burstein D., Haynes M.P., Faber S.M., 1991, Nature 53, 515
Choloniewski J., 1991, MNRAS 250 486
Christensen J.H., 1990, MNRAS 246, 535
Cunow B., 1992, MNRAS 258, 251
Cunow B., 1993, A&A 268, 491
de Vaucouleurs G., de Vaucouleurs A., Corwin H.G., 1976. Second Reference Catalogue of Bright Galaxies, The University of Texas Press, Austin.
Disney M., Davies J., Phillipps S., 1989, MNRAS 239, 939
Heidmann J., Heidmann N., de Vaucouleurs G., 1972, Mem. R. astr. Soc. 75, 85; 76, 105; 76, 121
Holmberg E., 1958, Medd. Lunds Obs., Ser. 2., No. 136
Huizinga J.E., van Albada T.S., 1992, MNRAS 254, 677
Lauberts A., Valentijn E.A., 1989, The Surface Photometry Catalogue of the ESO-Uppsala Galaxies, European Southern Observatory, Garching
King D.J., Birch C.J., Johnson C., Taylor K.N.R., 1981, PASP 93, 385
Knapen J.H., Hes, R., Beckman J.E., Peletier R.F., 1991, A&A 241, 42
Kodaira K., Doi M., Shimasaku K., 1992, AJ 104, 569
Kylafis N.D., Bahcall J.N., 1987, ApJ 317, 637
Sandage A., Freeman K.C., Stokes N.R. 1970, ApJ 160, 831
Shanks T., Stevenson P.R.F., Fong R., MacGillivray H.T., 1984, MNRAS 206, 767
Simien F., de Vaucouleurs G., 1986, ApJ 302, 564
Spiekermann G.,1992, AJ 103, 2102
Valentijn E.A., 1990, Nature 346, 153
Valentijn E.A., 1994, MNRAS 266, 614
van der Kruit P.C., Searle L., 1981a, A&A 95, 105
van der Kruit P.C., Searle L., 1981b, A&A 95, 116
van der Kruit P.C., Searle L., 1982a, A&A 110, 61
van der Kruit P.C., Searle L., 1982b, A&A 110, 79

Question
Giovanelli

You applied the Holmberg test assuming diameters do not change with inclination. That's a daring assumption. Could you comment on how this affects your conclusions on disk opacity (and everything else)?

Answer
Cunow

You are right that there should be effects concerning the results from the disc models. How large the effects are, I do not know. But since the three-component model yields an <u>image</u> of the model galaxy and I apply <u>exactly</u> the same algorithms to the model galaxies and to the real galaxies when calculating the apparent diameter, I do not think that there are significant biases in the resulting values of opacity.

Comment
Davies

The distribution of inclination tells you very little about opacity it has more to do with the selection criteria.

THE DISTRIBUTION OF GALACTIC INCLINATIONS

A CLUE TO OPACITY?

H. JONES, J. DAVIES AND M. TREWHELLA
University of Wales College of Cardiff
Department of Physics and Astronomy
PO Box 913
Cardiff CF2 3YB

Abstract. We have simulated samples of exponential disk galaxies, modelling specific samples with different selection-criteria. We have found that selection criteria is influential in the observed distribution of inclinations. We have compared the distributions of inclinations of the models to those of a real sample taken from the UGC. We find that the UGC sample distribution can be identified with high opacity. This behaviour is in agreement with previous authors. We briefly discuss possible causes of this result.

1. Background

A currently contentious issue - essential to a full understanding of the opacity of spiral galaxies - is that of the effect of selection bias on surface-brightness - inclination tests. Is there some way of determining the opacity of galaxies without resorting to a method requiring a simultaneous study of (what amounts to) the change in magnitude, and the change in diameter of a galaxy? Historically, these two effects have been notoriously difficult to separate. One simple test is merely to consider the distributions of galactic inclinations.

To help us understand this problem we have constructed a computer routine capable of generating different samples of galaxies. Extracting data from these simulations and examining this together with observational data allows us to draw inferences from the comparison.

2. The model

Our model is based on the following:

115

J. I. Davies and D. Burstein (eds.), The Opacity of Spiral Disks, 115–119.
© 1995 *Kluwer Academic Publishers.*

1. We simulate exponential disc galaxies.
2. Magnitudes are chosen from a Schecter function (Schecter 1976), with $\alpha = -1$, characteristic magnitude $M_B = -21.0$.
3. The surface brightness is chosen from a Gaussian distribution with a mean of $21.7 B_\mu$ (Freeman 1970, Van der Kruit 1987).
4. We adopt the following relation between observed surface brightness and inclination:

$$\mu_{obs} = \mu_{face} + 2.5 C log(b/a). \tag{1}$$

   where C is the parameter of extinction. (Disney, Davies and Phillips, 1988)
5. Similarly, the disc magnitude is given by:

$$M(i) = M(0) - 2.5(1 - C)log(b/a). \tag{2}$$

6. Inclinations are selected from a uniform distribution of cos(i), from edge-on (b/a = 0.0) to face-on (b/a = 1.0).
7. Each galaxy has associated with it a distance, calculated from a distribution - uniform over volume. We select isophotal magnitude-limited samples and isophotal diameter limited-samples and we consider the two extreme cases; (i) C = 1.0, (completely transparent) (ii) C = 0.0, (completely opaque).

3. Theoretical background

By considering the galactic content within a finite region of space, we can predict the likely effect on the inclination distribution of a sample.

For a fully transparent galaxy, apparent magnitude is independent of inclination. So, for a magnitude-limited sample there should be a random distribution of orientations. However, for an optically thick sample, magnitude-limited selection will mean that it is the inclined galaxies that are fainter, these faint (inclined) galaxies are selected against. The resulting distribution will, therefore, be dominated by galaxies of low inclination.

Selection, for a diameter-limited sample, is independent of apparent magnitude. To get into the sample a galaxy must have a minimum diameter, and for fully opaque galaxies, diameter is independent of inclination. Orientations therefore, will be randomly distributed. Whereas, for transparent galaxies diameter is dependent upon inclination. Inclined galaxies can be seen to greater distances so they will tend to dominate the distribution.

4. Results

The model was run with all galaxies subject to the following selection criteria: Magnitude limited sample - selected to $15 B_\mu$ within the 25th magnitude

$arcsec^{-2}$ isophote. Diameter limited - selected to a diameter of 1 arcmin at the 25th mag $arcsec^{-2}$ isophote.

The distributions of inclinations for each sample are shown in figures 1(a)-(d).

[For $b/a \leq 0.14$, the distribution for a sample of randomly oriented oblate spheroids is superimposed (dashed-line) on the histogram. Here we adopt a distribution of the form:

$$\phi(b/a)d(b/a) = b/a[((b/a)^2 - q^2)(1 - q^2) \tag{3}$$

with the intrinsic axial ratio, $q = 0.13$ (Giovanelli et al, 1994)]

Figure 1(a) shows the distribution of inclinations for the diameter-selected, optically thick case. The inclinations correspond well to a random distribution. In figure 1(b) the distribution for the diameter-limited, optically thin case is shown. This sample is clearly dominated by highly inclined galaxies. The inclination distribution for a magnitude-limited, optically thick case is shown in figure 1(c). This sample is dominated by face-on galaxies. Figure 1(d) shows the results for the magnitude-selected, optically thin case. The inclinations agree well with the random distribution. In each case the inclinations follow the predicted distributions. As can be seen, the distribution of the optically thick diameter selected sample, and that of the optically thin magnitude limited sample, are close to a random distribution.

Clearly, the selection criteria used has a substantial influence on the observed distributions of inclination.

5. The UGC Data

The Uppsala General Catalogue (UGC), selection criteria is diameter-based. We have selected all 1982 Sc galaxies in the catologue. The distribution of inclinations for this sample is shown in figure 1(e). The distribution equates well with a random distribution. Relating this to our model indicates high opacity. So what can we conclude?

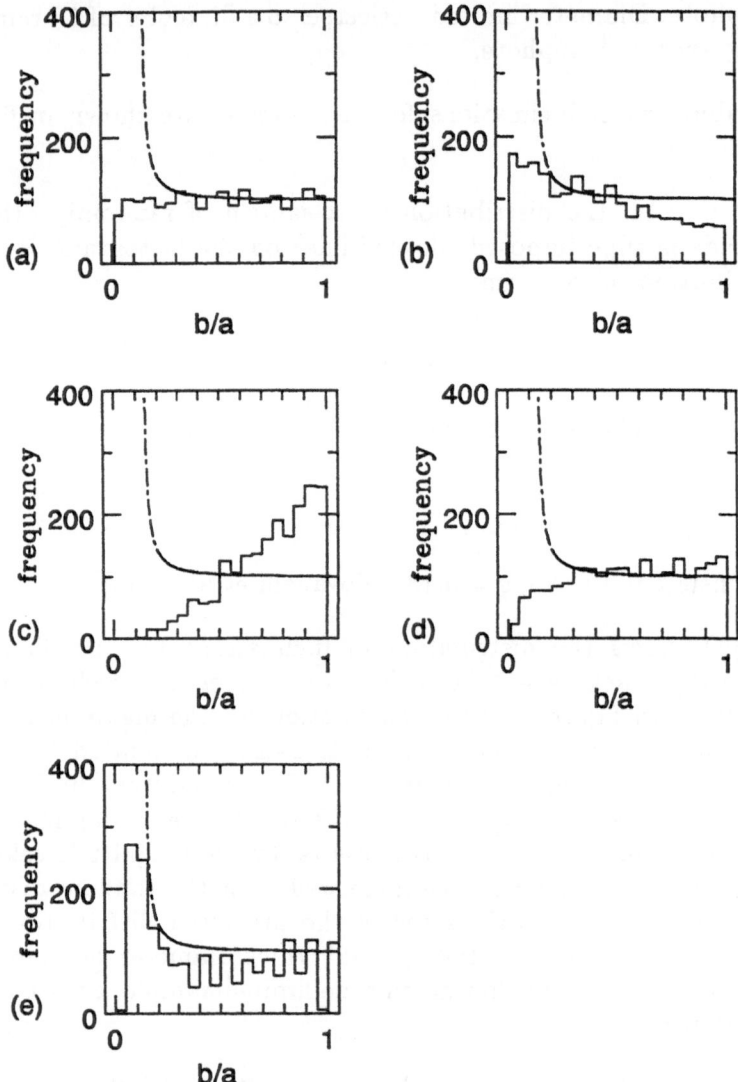

Fig. 1. a The distribution of inclinations for a diameter-selected sample of opaque spirals. **b** Distribution of inclinations for a diameter-limited, optically thin case. **c** The inclination distribution for a magnitude-limited, optically thick case. **d** The magnitude-selected, optically thin case. **e** The distribution of inclinations for the UGC sample.

6. Conclusions

What is evident is that the UGC galaxies behave as though they are optically thick. Two possible interpretations of this result are that:-

1. Galaxies *are* optically thick.
2. Global dust configuration is a significant influencing factor.

Observation, both locally and of other galaxies imply a complex nature in the large-scale structure of the dust. It is worth considering the types of geometry that could allow a galaxy to behave as though optically thick, regardless of its' true opacity.

Observations indicate that the large scale configuration of dust is often concentrated in regions along the inner edges of the spiral arms. The obscuring effect may be negligible when such ringlike structures are face-on to our line of sight. However this effect will increase markedly when inclined - depending on width and scale height - as the dust at large radii obscures the inner regions. Further study of the consequences of such a distribution is required, for until the global dust-distribution is known in much greater detail, the true opacity of spiral galaxies will remain an enigma.

7. References

1. Burstein D., Haynes M. and Faber S., (1991), *Nat*, **353**, 515.
2. Davies J., Jones H., and Trewhella M., (1994), *(this conference)*.
3. Disney M., Davies J. and Phillips S., (1989), *MNRAS*, **260**, 491.
4. Giovanelli R., Haynes M., Salzer J., Wegner G., Da Costa L., and Freidling W., (1994), *Preprint*.

OPTICAL THICKNESS OF Sb-Scd GALAXIES FROM THE TULLY-FISHER RELATION

L. Gouguenheim and L. Bottinelli
Observatoire de Paris, F-92195 Meudon Cedex

G. Paturel
Observatoire de Lyon, F-69230 Saint Genis Laval

P. Teerikorpi
Tuorla Observatory, SF-21500 Piikkiö

Abstract. We point out the implication of our work on the Tully-Fisher distance indicator, using an angular size limited sample of more than 5 000 spiral galaxies, on the opaqueness problem. We conclude that the inclination dependent behaviour of these galaxies is consistent with a high optical thickness at the surface brightness limit 25 magnitude per square second, in any case significantly larger than the classical value of about 0.2.

1. Introduction

The optical thickness of galaxy disks has been discussed in several recent papers with contradictory conclusions ranging from optically thin (1,2) through "moderately optically thick" (3) to "opaque"edges (4,5) of the spiral disks. One reason contributing to the unclear situation, is the problem how to treat correctly the data in statistical studies (3).

Our contact with the present problem comes from a practical goal of how to correct magnitudes and diameters to zero inclination. This knowledge, which is closely related to the question of optical thickness, is needed in our programme (6) for measuring the kinematics of the local universe using the Tully-Fisher (TF) distance indicator together with an angular size limited sample of 5 171 spiral galaxies.

2. Method

From observed log V_M (V_M = maximum rotation velocity), we predict diameter D_{25} and magnitude B for spiral galaxies of different inclinations (log R_{25}) using the TF relations

121

J. I. Davies and D. Burstein (eds.), The Opacity of Spiral Disks, 121–125.
© 1995 *Kluwer Academic Publishers.*

together with kinematical distances derived from a Virgo infall model. In the ideal case, the observed and predicted parameters, log $[D_{obs}/D_{pred}]$ and $B_{obs} - B_{pred}$ vs. log R_{25} diagrams should directly give the corrections. In practice, such an approach may be influenced by inclination dependent Malmquist bias. Hence, we first constructed unbiased samples, using the concept of normalized distance d_n which has proved so useful in studies of H_o (7). We add to the normalization formula an inclination term, in order to guarantee that at all inclinations the samples are free from the Malmquist bias, when one cuts the sample at sufficiently small d_n.

This approach overcomes the problems pointed out by Burstein et al. and by Choloniewski for the methods not utilizing distance information. An important advantage is the possibility to construct sufficiently large unbiased magnitude limited subsamples - apparent magnitudes B_T are known for almost 90% of the sample - and hence to test simultaneously the diameter and magnitude effects.

In order to intercompare the results from diameters and magnitudes, we adopt a simple disk + spherical bulge model for the galaxies, and assume that the bulge is free from inclination dependence, while the surface brightness of the disk may change with the viewing angle.

Writing in a linear approximation (R stands for R_{25}):
$$\log [D_{obs}/D_{pred}] = C \log R \tag{1}$$
and denoting $k = L_{bulge}/L_{tot}$ and $K_{25} = dlogD/d\mu$, the corresponding magnitude change within the 25 magnitude isophote will be:
$$m_{obs} - m_{pred} = -2,5 \log[k + (1-k)R^{2C(0.2/K_{25}+1)-1}] \tag{2}$$
In principle, the bulge fraction k and the slope of the surface brightness profile K_{25} are known for different galaxy types, and the coefficient C **is the only free parameter**. For the model to be realistic, C should be consistently given by both the diameter and magnitude behaviour.

We cut from the sample very face-on and (for the diameter test) very edge-on galaxies (allowing 0.07<logR<0.8), because face-on galaxies have large errors in $logV_M$ and edge-on galaxies have a systematic effect in their diameters (1, 3).

3.   Results

We show first that for Sb - Scd types there is very small, if any, diameter effect up to log R = 0.8. The smaller unbiased sample already gave us an indication for C≈0, which means that we can check the value of C using the whole angular size limited sample in the diameter test: with C≈0, the Malmquist bias at each logR would be quite similar. We find an impressively constant mean value up to logR = 0.8, a formal least-squares solution giving C=0.04±0.02.

The complementary test consists in checking whether such a small value of C is consistent with the magnitude behaviour. Here it is important to use the unbiased subsample of a magnitude limited sample (m<12.5) which is different for each value of C (entering the formula for the normalized distance). The predicted magnitude effect is quite sensitive to C. C=0.1 is clearly too large, while C≈0.04 gives the best fit, in agreement with the result from the diameter effect.

Note that the magnitude effect is less than expected for a pure disk (C=0 would imply the slope =2.5), roughly by the amount required by the known contribution from the bulge. We concluded that the diameters and the magnitudes alike indicate a very small value of C, and a correspondingly high value of the optical thickness τ. But how high? A rough estimate comes from the model where stars and dust are uniformly mixed at the disk edge (8). Ignoring any scattered light, the radiation transfer problem can be solved for each inclination and the coefficient C=dlogD/dlogR becomes

$$C=2.5K_{25}R\tau \ exp(-\tau R)/[1-exp(-\tau)] \qquad (3)$$

which is not constant but may be integrated to give the total change of the size when the galaxy is turned away from the face-on position:

$$logD/D_0=2.5K_{25}log\{[1-exp(\tau R)]/[1-exp(-\tau)]\} \qquad (4)$$

where D_0 denotes the face-on diameter and D the diameter for a galaxy seen with the axis ratio R.

It is not possible to determine τ accurately (for τ >1.0 its value is far too sensible to small changes in the observed C). However, the small values of C suggests typically τ >0.8, or at least "moderately optically thick" Sb - Scd disks. τ <0.5 is clearly excluded, together with the "classical" inclination corrections corresponding to C≈0.2.

4. References

1. Huizinga, G.E., Van Albada, T.S. (1990) *MNRAS* **254**, 677-685
2. Byun, Y.I. (1993) *Publ. Astron. Soc. Pac.* **105**, 993-995
3. Burstein, D., Haynes, M.P., Faber, S.M. (1991) *Nature* **353**, 515-521
4. Valentijn, E.A. (1990) *Nature* **346**, 153-155
5. Choloniewski, J. (1991) *MNRAS* **250**, 486-504
6. Paturel, G., Bottinelli, L., Fouqué, P., Garnier, R., Gouguenheim, L., Teerikorpi, P. 1990, *The Messenger*, n°62, 8
7. Bottinelli, L. Gouguenheim, L., Paturel, G., Teerikorpi, P. (1986) *A&A* **156**, 157-171
8. Disney, M.J., Davies, J.L., Phillips, S. (1989) *MNRAS* **239**, 939-976

Question
Pelletier

A recent paper by Bernstein & Raychaudhury (CFA preprint) shows that for a Coma sample of spirals the I band Tully-Fisher residuals correlate quite strongly with axis ratio. This indicates a very high opacity in the B-band, although their paper maybe does not contain enough galaxies.

Answer
Gouguenheim

OK - No comment.

Question
Greenberg

What do you mean by optical depth? Do you mean the optical depth when looking through the galaxy?

Answer
Gouguenheim

We considered a layer of dust uniformly mixed with stars at the disk edge. Ignoring any scattered light in the radiation transfer problem we obtained:

$$I_o = I_i \frac{\left[1 - e^{-\tau}\right]}{\left[1 - e^{-\frac{\tau}{\cos i}}\right]}$$

In which τ is the conventional optical depth. Using this relation instead of $I_0 = I_i(\cos i)^{\delta}$ leads to the value of C

$$C = 2.5 K r \tau e^{-\tau R} / \left(\lambda - e^{-\tau R}\right)$$

In order to calculate the total change of the diameter when R is changed from $R=1$ to R, we integrate $\dfrac{\partial \log D}{\partial \log R}$ which gives

$$\log \frac{D_i}{D_0} = 2.5 K \log\left(\lambda - e^{-\tau R}\right) / \left(\lambda - e^{-\tau}\right)$$

Now C can be determined for different values of τ - which were given in my table.

I agree that the values of τ corresponding to the observed slope C are LOWER LIMITS, because stars above the dust layer would artificially decrease the derived τ.

For this reason, and also because the corrections are not inconsistent with a lower value of C, we cannot exclude even remarkably opaque disks:-

Question
Giovanelli

Since you derive corrections as offsets from an adopted TF relation, and since the latter is influence by the adopted diameter - and magnitude - inclination connections, could you comment on how you derived such relation and how that procedure affects your results?

Answer
Gouguenheim

Yes the $\log(D_{obs} - D_0)$ (e.g. $M_{obs} - M_0$) formulae contain the unknown inclination connection C. Hence some iterations are needed and were used. For the diameters, we started by assuming $C=0$ (opaque case). If actually the disks are not opaque, we do not in this manner accept biased data (some unbiased ones being of course lost).

Question
Burstein

a)   In you plots of $\log(D_{obs}/D_0)$ vs log a/b, I believe a slope of 0.2 log a/b can be fit as well as your slope as long as the zero point is not fixed. Could you show what such a line looks like on your graph?

b)   How is D_0 defined, as you apparently require it to be known?

Answer
Gouguenheim

a)   For the unbiased range (332 galaxies sample) where no bias is expected, the $\log(D_{obs}/D_0)$ vs log R plot is expected to start from zero, which is confirmed by the distribution of points in the graph. Anyway, the 0.2 slope that you mention is definitely not compatible with our data, even if the zero point is left free.

b)   As I explained D_0 is predicted from the TF diameter relation, which is determined (calibrated) from the unbiased plateau data.

The Giovanelli

EXTINCTION IN SC GALAXIES AT I BAND AND IN THE 21CM LINE

R. GIOVANELLI
Cornell University
Dept. of Astronomy and NAIC
Ithaca, NY 14853, USA

Abstract.
We report significant dependence on luminosity of the disk central surface brightness of spiral galaxies, as well as of the relations that convert isophotal radii, scale lengths and magnitudes to the edge–on perspective. These are tightly wedded to extinction processes that operate in disks. The HI flux is virtually inclination independent.

1. Introduction

The early estimates of the opacity of spiral disks were based on statistical studies, which investigated the mean dependence of photometric parameters on the inclination of galaxies to the line of sight. The work of Holmberg (1958) and de Vaucouleurs (1959) and the fact that we have a mostly unimpeded view of the extragalactic sky influenced the assumption that spiral disks are largely transparent. However, the work of principally Disney, Davies and Phillipps (1989) and Valentijn (1990) has recently fueled debate by arguing that spiral disks may in fact be quite opaque, or at any rate far less transparent than previously thought, and largely provided the motivation for this conference.

Statistical studies, such as those of Holmberg and that which will be presented here, cannot provide unambiguous numerical estimates for the opacity of spiral disks. Alone they will not yield, for example, a mean value of $\tau(0)$, the optical depth at the center of a putatively exponential absorbing disk, in the face–on aspect, nor a mean value of relative thickness of the stellar to the dust disk. What they can yield, however, are reliable

J. I. Davies and D. Burstein (eds.), The Opacity of Spiral Disks, 127–140.
© 1995 *Kluwer Academic Publishers.*

estimates of the mean changes with inclination of the emerging flux, as well as the mean relative sizes of disks, when measured at a fixed isophotal level. The large, homogeneous samples of single band images available for disk systems, allow detailed separation in subclasses, and a fair understanding of the effects of selection biases. These results can then provide a broad guideline to encase more detailed multiband studies of selected systems, and strategies for techniques capable of providing direct estimates of τ.

In a recent contribution (Giovanelli et al. 1994, hereinafter "ScI") we have reported results on extinction in spiral disks at I band and in the 21cm line, based on CCD I–band observations of a very extensive sample of Sbc/Sc galaxies. Here we extend that study by addressing the question of how the photometric properties of disk galaxies depend on their luminosity. Mainly we wish to verify whether the internal extinction correction (for background references to this topic see ScI, Huizinga 1994, Valentijn 1994 and refs. therein), which depends on the disk inclination, can be parametrically related to the galaxy's luminosity. Such an effect would be important for cosmological applications: neglect to account for it could not only add to the scatter, but also alter the slope of the Tully–Fisher relation, introducing potentially severe biases in the reconstruction of the local density and peculiar velocity field.

Here, we utilize the data base of Sbc/Sc galaxy images in the I band (and associated spectroscopy) described in paper I (the "Sc sample"). With the same selection criteria described in ScI, the sample has now been extended to 1714 objects. For applications that do not require availability of scale lengths and central disk surface magnitudes, we also include galaxies in the survey of Mathewson, Ford and Buchhorn (1992, hereinafter MFB), which together with the Sc sample yield an expanded sample of 2816 objects.

In order to conform to common practice and facilitate comparison with other work, we express the inclination of disks with the expression (a/b), the major–to–minor size ratio, which is related to the measured ellipticity of the outer isophotes via $\epsilon = 1 - (b/a)$. A $^\circ$ superscript indicates values of the photometric parameters in the face–on aspect. In the computation of distances and absolute magnitudes, we assume a Hubble constant of $H_\circ = 100$ km s^{-1} Mpc^{-1}.

2. Selection Bias in the Sc Sample and L–Dependence of $\mu_d(0)$

In a series of papers (Disney 1976; Davies et al. 1994 and refs. therein, hereinafter collectively referred to as) the Cardiff group has quantified the treatment of selection biases of a catalog via so–called *visibility functions*. Given an observed quantity, e.g. central disk surface magnitude $\mu_d(0)$, the visibility function gives a measure of the likelihood that a galaxy, character-

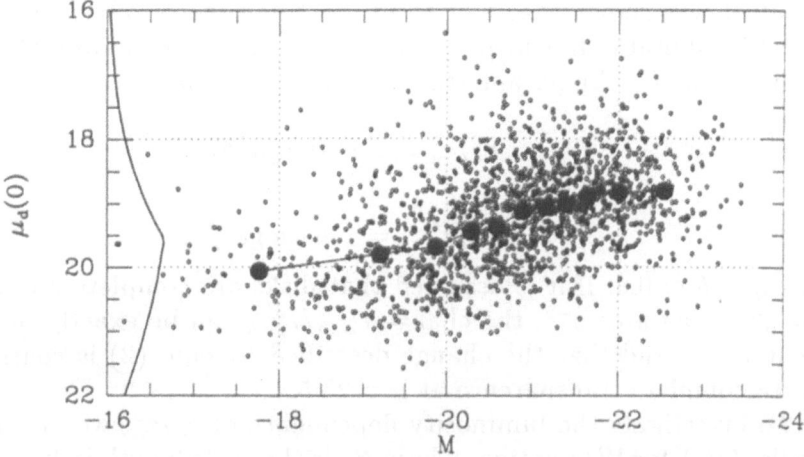

Figure 1. Distribution of $\mu_d(0)$ and absolute magnitude for galaxies in the Sc sample. Mean values are given by large symbols. The sample visibility (thick line) is computed for a limiting $r_{23.5} = 25"$, a threshold magnitude $m_L = 13.5$ and a surf. mag. saturation level of 19.0, using relations derived by the Cardiff group.

ized by some value of the observable, be included in the observed sample.

In fig. 1 we display the bivariate distribution of galaxies in the Sc sample, in the ($\mu_d(0)$, absolute magnitude) plane. Superimposed along the vertical axis is the sample visibility curve as a function of $\mu_d(0)$, computed adopting the Cardiff group treatment. Mean values of $\mu_d(0)$ computed for data selected along vertical strips are denoted by large filled symbols. The Sc sample sits well near the maximum of the visibility function and within the uncertainties, the distribution of observed $\mu_d(0)$ falls roughly as expected given the visibilities. Thus, as pointed out by the Cardiff group, the interpretation of Freeman's (1973) law (constancy of $\mu_d(0)$) must be tempered by the allowance of the effects of selection biases. However, the different distribution of intrinsically faint and bright objects cannot be produced by selection effects. We are unaware of a catalog bias that would select against $M = -22$, $\mu_d(0) = 20$ objects, and not against $M = -19.5$, $\mu_d(0) = 20$ objects. So the former must be relatively rarer in space. When samples are restricted to relatively more luminous galaxies, it is found that Freeman's law applies fairly tightly; however, when the luminosity range is expanded a significant deviation occurs, intrinsically fainter spirals having lower surface brightness. This effect is less marked in the B band (van der Kruit 1987), while it is steeper in the K band (Giovanelli et al. 1994b). We postulate that the progressive blurring of the $[M, \mu_d(0)]$ correlation with decreasing wavelength is partly due to differences in the degree of extinction occurring in galaxies of different luminosity (see section 4).

3. Scale Length, Isophotal Radii and Luminosity

In ScI, it was shown that at I band, both the mean values of scale length and of isophotal radius at $\mu = 23.5$ vary with the inclination of the disk to the line of sight. We parametrized the behavior via

$$r_d/r_d^o = 1 + \eta \log(a/b) \tag{1}$$

$$r_{23.5}/r_{23.5}^o = 1 + \delta \log(a/b) \tag{2}$$

with $\eta \simeq \delta \simeq 0.6$. If it is assumed that disks are completely trasparent at $\mu = 23.5$ mag arcsec^{-2}, the change $r_{23.5}/r_{23.5}^o$ can be exactly predicted. In ScI it was found that the change described by eqn. (2) is consistent with nearly complete transparence at $\mu = 23.5$.

To investigate the luminosity dependence of r_d/r_d^o, we first linearly fit the $[\log(R_d), \log W]$ relation, where R_d is the scale length in kpc and W the edge–on velocity width; then we fit the residuals of that relation with axial ratio $\log(a/b)$, separately for groups of different luminosity. The procedure is repeated twice, fitting first $\log R_d$ on $\log W$, then reversing the order of the fit; the results are similar, and we adopt the relation corresponding to the mean of the two procedures. The same procedure is applied for $[\log R_{23.5}, \log W]$. The relationships obtained from fitfing the residuals are, unlike eqns. (1) and (2), power laws:

$$R_d/R_d^o = r_d/r_d^o = (a/b)^\alpha, \tag{3}$$

$$R_{23.5}/R_{23.5}^o = r_{23.5}/r_{23.5}^o = (a/b)^\beta, \tag{4}$$

The dependences of α and β on absolute magnitude are displayed in figs. 2(a) and (b). *The corrections of both sizes to the face–on aspect are larger for less luminous galaxies.*

Eqn. (4) yields a mean inclination correction $r_{23.5}/r_{23.5}^o$ for galaxies of given a/b and M (via the dependence of β on luminosity). That correction depends on the slope of the photometric profile which, as shown in eqn. (25) of ScI, enters via the ratio $r_d/r_{23.5}$ and is also luminosity dependent. Eqn. (4) thus implies the adoption of a mean value of $r_d/r_{23.5}$, for each value of M. In reality, for each value of M there will be a range of values of the ratio $r_d/r_{23.5}$, and to each corresponds a slightly different correction, scattered around the mean value predicted by eqn. (4). We can then improve on eqn. (4) by introducing a second order term as follows. Let $< r_{23.5}/r_{23.5}^o >$ be the correction obtained from eqn. (4) and $< r_d/r_{23.5} >$ the ratio that corresponds on the average to a galaxy of axial ratio (a/b) and absolute magnitude M. Then, if the galaxy's $r_d/r_{23.5}$ differs from the

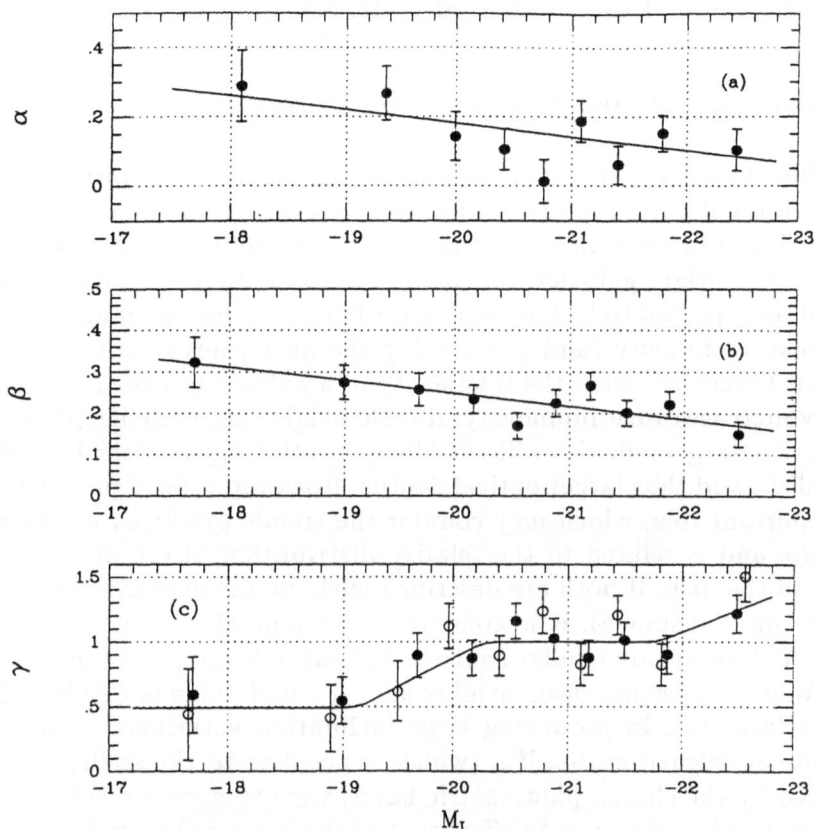

Figure 2. Absolute magnitude dependence of inclination exponents.

mean expectation value, such that $\Delta(r_d/r_{23.5}) = r_d/r_{23.5} - \; < r_d/r_{23.5} >$, the derivative of eqn. (25) in paper I yields

$$r_{23.5}/r^{\circ}_{23.5} = \; < r_{23.5}/r^{\circ}_{23.5} > \; + \; \Delta(r_{23.5}/r^{\circ}_{23.5}) \qquad (5)$$

with

$$\Delta(r_{23.5}/r^{\circ}_{23.5}) = 2.302 \log(a/b)(r^{\circ}_d/r_d) < r_{23.5}/r^{\circ}_{23.5} >^2 \Delta(r_d/r_{23.5})$$

where r°_d/r_d is given by eqn. (3), $< r_{23.5}/r^{\circ}_{23.5} >$ by eqn. (4), with α and β depending on luminosity as shown in figs. 2(a) and 2(b), and the ratio $r_d/r_{23.5}$ is the specific value for the galaxy in question. The use of eqn. (5) is of course only possible if individual values of both r_d and $r_{23.5}$ are available for each galaxy. The adjustment is generally small in comparison with the mean value obtained from eqn. (4), but it can be very important when a galaxy's $r_d/r_{23.5}$ differs significantly from the mean value for objects of its luminosity, and it produces a more accurate estimate of $r^{\circ}_{23.5}$ as we show in fig. 3.

4. Luminosity Dependence of Extinction

The statistical tests carried out in this section utilize the expanded data set, which includes our Sc sample and MFB sample, for a total of 2816 objects.

Why should we expect a luminosity dependence of internal extinction laws? Internal extinction increases with inclination of the disk axis to the line of sight i because the path length through the disk increases roughly as $1/\cos i$. For bright galaxies, the amount of extinction at a given inclination should be expected to be larger than for faint ones for two main reasons: (a) the mean metallicity (and presumably the dust content of the interstellar medium) increases with the luminosity of a galaxy; (b) the linear size of a galaxy increases with luminosity; if scale height and scale length are correlated, then larger galaxies will yield longer path lengths of the l.o.s. through their disks and thus larger optical depths. However, a third parameter plays an important role, which may counter the trends produced by the preceding two, and is related to the relative distribution of the dust and stellar layers in the disk. If both are described with double exponential models (in radial and z–distance), it is common to assume that scale lengths of the two distributions are similar (but see Valentijn, these proceedings) and the ratio of dust to stellar scale heights is ζ. A small value of ζ is less effective than a large one in producing large inclination variations in the amount of internal extinction. So, if ζ (which is peculiar to the stellar population sampled by the chosen photometric band) were to decrease with increasing luminosity, this effect would offset that of the increased metallicity and size. It is also likely that the effects of clumpiness in both the stellar and dust distributions play an important role, especially in less luminous galaxies. We unfortunately have few secure observational handles on the interplay of those parameters, and a better than coarse a priori prediction of the emerging flux as a function of both inclination and luminosity is difficult.

In ScI we inferred an internal extinction law for the whole Sc sample, which was parametrized in the form

$$m^\circ - m = -\gamma \log(a/b) \qquad (6)$$

Several techniques were used to get γ, which yielded values ranging between 0.95 and 1.15. We adopted a value of 1.05 ± 0.08. Two texhniques lend themselves to investigate the luminosity dependence of γ: the inspection of residuals of the M–$\log W$ relation, in the manner described in the preceding section, and that which was referred to as the "modified Holmberg test". The latter consists in monitoring the inclination dependence of the quantity $\Sigma = m(\infty) + 5 \log r^\circ_{23.5}$: the mean slope of Σ on $\log(a/b)$ is γ. Since to the first order Σ is not a distance dependent quantity, it can be profitably used with subsets of the data, e.g. selected in bins ranked by luminosity.

By studying the behavior of a simulated sample endowed with various sets of transformation relations and constructed mimicking the selection criteria that apply to our expanded sample, we verified that no significant biases are introduced in the distribution of Σ or of the TF residuals, with respect to either inclination or luminosity, as a result of the sample selection process (see Giovanelli et al. 1994b for details).

We separate the 2816 galaxies in the expanded sample in 10 groups of 280 objects each, ranked in order of increasing absolute magnitude. Fig. 2(c) displays the values of γ obtained for each group, using the TF residuals technique (filled symbols) and the modified Holmberg test (unfilled symbols). In the latter case, an inclination law needs to be adopted for the isophotal radius, for which we used eqn. (4), with β as given by the solid line fit to points in fig. 2(b). Both methods agree well within the errors, yielding lower values of γ for less luminous than for more luminous galaxies: while for galaxies at the faint end the mean difference in detected flux between edge–on and face–on appearance is about 0.5 mag, for the brightest galaxies in the sample that difference is between 0.9 and 1.2 mag.

In fig. 2(c) the decrease of γ which occurs for galaxies fainter than $M = -20$ bottoms out at about 0.5 mag. Thus, significant extinction still appears to occur even for the smallest, presumably metal and dust poor, galaxies in the sample. Could scattering play an important role in low luminosity galaxies? If a brightnening of the disk occurs in the face–on perspective due to scattering, the amplitude of the difference between face–on and edge–on fluxes will be larger than if no brightening occurred. According to the calculations reported by Witt (1995) in these proceedings, the brightening effect of scattering in face–on systems of moderate central optical depth ($\tau(0) < 5$) amounts to no more than 0.1 mag, for any reasonable assumption for the dust albedo; for nearly transparent or very opaque systems, the face–on brightening effect due to scattering is negligible or nonexistent. We then conclude that the difference of 0.5 mag between face–on and edge–on fluxes is principally due to extinction, and postulate that this arises largely from an increase of ζ or, at any rate, from an increasingly good mixing — spatially and kinematically — of the dust and stellar populations, with decreasing luminosity of galaxies. This circumstance generally increases the amplitude of the inclination dependence of extinction, for a galaxy of given dust content and size. For more luminous galaxies the stellar population sampled by the I band conceivably becomes less tightly confined to the plane of the disk than does the population I material with which the dust is mainly associated, resulting in a decrease of ζ; so, although they are dustier and larger, the increase in γ is partly suppressed.

5. The Magnitude–Size Relation

It is well–known that luminosity and size of disk systems are well correlated quantities. Depending on the band at which they are measured, a varying amount of scatter is contributed to that correlation by the effects of inclination. The relationship can thus be used as a tool to sample the efficacy of the adopted corrections. In this representation, we restrict ourselves to the Sc sample alone, as one of the correction steps involves the use of scale lengths, which are not available to us for the MFB sample. Fig. 3 compares several steps of the correction process: panel (a) shows the observed apparent magnitude and angular size, corrected for extinction within our Galaxy and other cosmological effects (see ScI, section 2), but not for inclination effects; panel (b) shows the effects of correction of $r_{23.5}$ to $r^\circ_{23.5}$ alone, adopting eqn. (4) and β as given in fig. 2(b); panel (c) displays the effects of magnitude corrections alone, i.e. the size uncorrected for inclination and the magnitude corrected as per eqn. (6), with γ as given in fig. 2(c); panel (d) shows the results of correcting both magnitudes and sizes according to the global recipes given in ScI, i.e sizes with a value of $\delta = 0.62$ in eqn. (2) and $\gamma = 1.05$; panel (e) displays magnitudes and sizes jointly corrected for inclination as in panels (b) and (c); panel (f) shows the result when the isophotal radius is corrected using eqn. (4) *and* the adjustment described by eqn. (5). As anticipated in section 3, the adjustement to the isophotal radius correction represented by eqn. (5), which is possible when scale lengths are available for each galaxy, produces a very significant improvement on the quality of the correction, as suggested by the much reduced scatter of the fits. The small arrow in fig. 3(a) represents the mean correction vector to face–on, for a galaxy of $M = -20.5$ and $\log(a/b) = 0.5$.

6. Inclination Dependence of the HI Flux

In ScI (and refs. therein) it was found that the variation of the detected HI flux with inclination is quite mild. Parametrizing the ratio between the observed and the face–on fluxes as $S_{21}/S^\circ_{21} = (a/b)^\kappa$, a value of $\kappa \simeq 0.12$ was found to be consistent with the data. However, if we apply the analog of the modified Holmberg test used to obtain γ in section 4, i.e. we monitor the inclination dependence of the hybrid quantity $\Sigma_{21} = S_{21} + 5 \log r^\circ_{23.5}$ and take in consideration the luminosity dependence of the conversion $r_{23.5}/r^\circ_{23.5}$, the resulting value of κ *is indistinguishable from zero*. We conclude by proposing that no significant dependence of the detected flux on inclination is discernible, and that the solid angle of spiral disks covered by regions of high HI optical depth increases very little between face–on and nearly edge–on aspect.

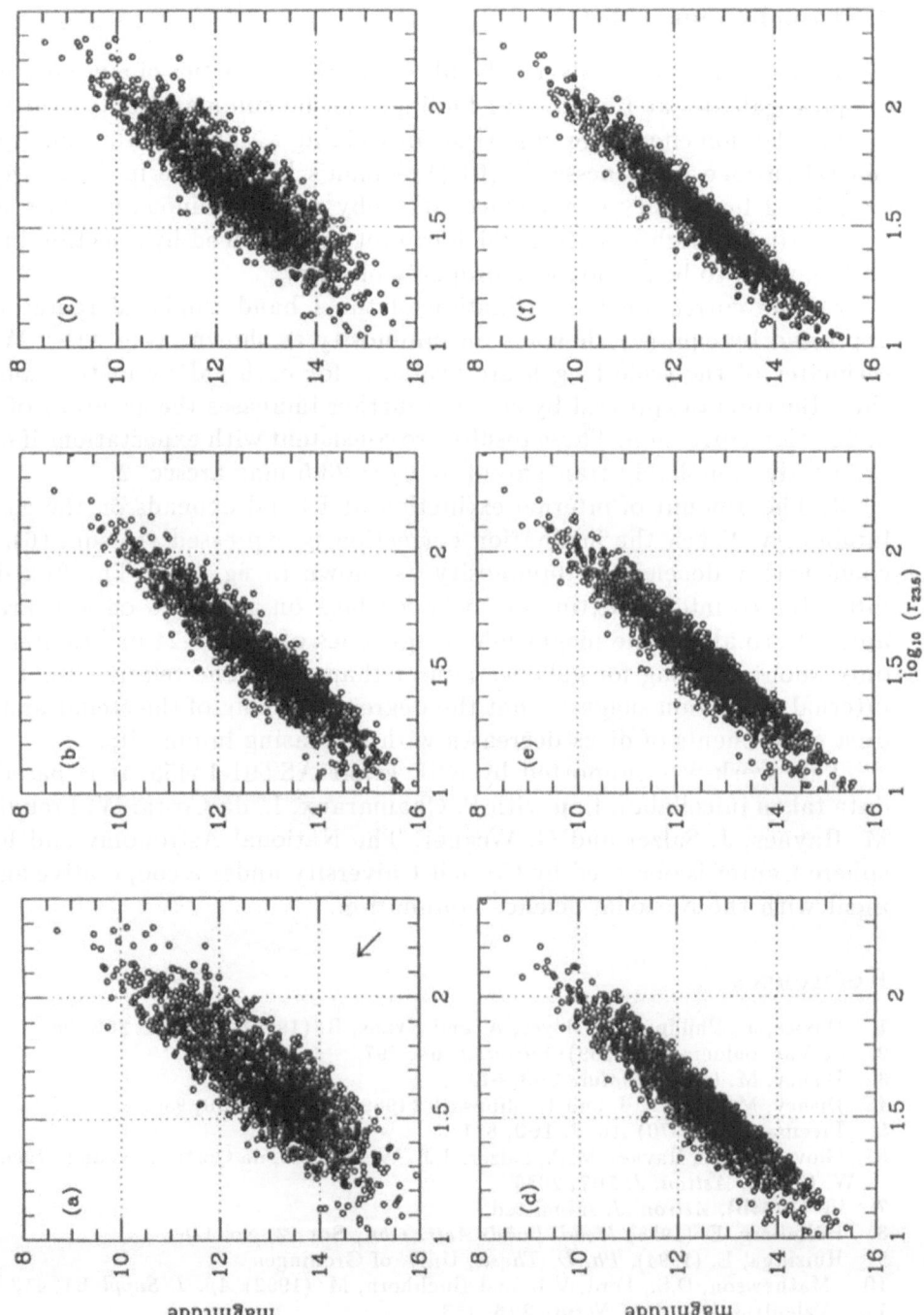

Figure 3. Apparent magnitude and isophotal radius: various correction schemes to face–on perspective are compared (see section 5 for details).

7. Conclusions

We have statistically analyzed the photometric properties of a large sample of spiral galaxies at I band. Our findings can be summarized as follows:

1) Selection effects are important in yielding a tight distribution of disk central surface brightnesses $\mu_d(0)$ (Freeman's law); more in detail, $\mu_d(0)$ exhibits a luminosity dependence, whereby more luminous galaxies have preferentially brighter $\mu_d(0)$, which cannot be produced by selection biases and appears to be an intrinsic property of disks.

2) The correction for inclination of the I band isophotal radius $r_{23.5}$ expressed by eqn. (4), depends on luminosity as shown in fig. 2(b). When estimates of the scale length are available for each galaxy in the sample, the adjustment expressed by eqn. (5) further increases the accuracy of this inclination correction. These results are consistent with expectations if disks are nearly completely transparent at $\mu_I = 23.5$ mag arcsec^{-2}.

3) The amount of internal extinction at I band depends on the galaxy luminosity. When the inclination correction is expressed via eqn. (6), the coefficient γ depends on luminosity as shown in fig. 2(c). The flux dimming due to internal extinction between face–on and edge–on appearance amounts to about one magnitude for galaxies of $M = -21$ or brighter and only about 0.5 mag for galaxies fainter than -18. The relative amount of internal extinction suggests that the degree of mixing of the stellar and the dust components of disks decreases with increasing luminosity.

This work was supported by NSF grant AST91-15459. It is based on data taken in collaboration with P. Chamaraux, L. da Costa, W. Freudling, M. Haynes, J. Salzer and G. Wegner. The National Astronomy and Ionosphere Center is operated by Cornell University under a cooperative agreement with the National Science Foundation.

References

1. Davies, J., Phillipps, S., Boyce, A. and Evans, R. (1994), *MNRAS* **268**, 984
2. de Vaucouleurs, G. (1959)ʻ*Astron. J.* **64**, 397
3. Disney, M. (1976), *Nature* **263**, 573
4. Disney, M., Davies, J. and Phillipps, S. (1989) *MNRAS* **239**, 939
5. Freeman, K. (1970) *Ap. J.* **160**, 811
6. Giovanelli, R., Haynes, M.P., Salzer, J.J., Wegner, G., da Costa, L.N. and Freudling, W. (1994), *Astron. J.* **107**, 2036
7. id. (1994b), *Astron. J.* submitted
8. Holmberg, E. (1958) *Medd. Lunds Astr. Obs.*, **Ser. 2**, No. 136
9. Huizinga, E. (1994), *Ph. D. Thesis*, Univ. of Groningen
10. Mathewson, D.S., Ford, V.L. and Buchhorn, M. (1992) *Ap. J. Suppl.* **81**, 413
11. Valentijn, E. (1990) *Nature* **346**, 153
12. Valentijn, E. (1994), *MNRAS* **266**, 614
13. van der Kruit, P.C. (1987), *Astron. Ap.* **173**, 59
14. Witt, A. (1995), these proceedings

Question

Mike Disney

I would like to believe your results. However, I suspect all you are seeing is surface-brightness (optical) selection. All your galaxies, irrespective of inclination, have the same SB, to within a magnitude or so. The same for Burstein et al's and Valentijn's samples. See Davies et al. MN...

If the samples are dominated by selection they will be useless for determining extinction, using the Holmberg test.

Answer

Riccardo Giovanelli

Inclination dependences have been estimated with a variety of different tests. They are affected in different ways by selection effects, yet they all yield the same answer. I also don't share your scepticism on the application of the surface magnitude - inclination test.

Question

Valentijn

Your results are obtained at the I band and as far as I can see there are no major inconsistencies with my B-band results, given the fact that the bands are far away from each other. A modest or small diameter increase with inclination at B will become more apparent at I. However, at one point, your paper proposes a difference, when you comment that the inclination dependence of the ratio of I diameters and B visual diameters confirm that computerised diameters, such as your I and ESO-LV Blue diameters give a faulty result when they are used for the inclination test.

This is an important issue, but I do not accept that you can resolve this on the basis of a comparison of B and I diameters, as your observed trend can equally well represent the increase of diameters of inclined galaxies at larger wavelength (I) compared to B.

Answer

Riccardo Giovanelli

The effect you mention was first pointed out by Huizinga and Van Albada in B band. We confirmed their results at I band. As you know, we introduced in print your concern in our paper, as it is a legitimate one. On the other hand let me add a footnote to your comment. If the result we observe - which is quite clearly outlined - were due to extinction, then it would be different at B and I band. The fact that it appears similar at both bands suggests that it might be a legitimate visual bias. But, I agree with you, more should be done to understand it.

138

Question
de Vaucouleurs

1. Can we say that $r_{23.5}$ in I is roughly equivalent to r_{25} or $r_{25.5}$ in B?

2. You find $\Delta M / \Delta \log\left(\dfrac{a}{b}\right) \approx 1.1$ in I at Sc. This agrees rather well with the RC3 formula where the slope is 1.5 at Sc (in B), but it does vary with morphological type T.

Answer
Riccardo Giovanelli

1. $r_{23.5}^{I}$ in about 3.5 scale lengths out, which I believe is somewhat further out than r_{25}^{B}.

2. This work only involves Sbc and Sc galaxies, so I cannot comment - on the basis of it - on variations with morphological type.

Question
Byun

Your analysis suggests that the isophotal diameters increase with inclination by about 50%. If so, it is natural to expect that intrinsically small and faint galaxies are included in your sample, more preferably if they are inclined. Would this lead to an overestimated magnitude-inclination relation (and may also be underestimated diameter-inclination dependence)?

Answer
Riccardo Giovanelli

As I described, we have tried to disentangle differential behaviour, in their inclination laws, between galaxies of different intrinsic size, and do find such and effect. The results are in the direction one would expect, but I am personally surprised that the differences are not larger. Perhaps still undisclosed selection effects are hiding in the data, as you suggest.

Question
Gouguenheim

As your sample is not distance limited you should be check that it mimics a distance limited one i.e. that there is no artificial effect in $\log\dfrac{a}{b}$. Your survey deals with a sample of Sbc-sc galaxies (a rather tight type range) which are expected to have a rather restricted $\log V_m$ range. However, their luminosities seem to encompass a rather large range (from ~-17 to -23). I wonder whether the slope of the TF relation which has been used depends strongly on a rather small number of small $\log V_m$ data ?

Did you check whether the distribution of $\log\dfrac{a}{b}$ is the same for the different $\log V_m$?

Answer
Riccardo Giovanelli

Part I:
The slope of the TF relation was not constrained in the exercise I described. We asked the question: "What is the value of δ - in $\dfrac{r_{23.5}}{r_{23.5}^0} = \lambda + \delta\log\dfrac{a}{b}$ - or that of γ in the magnitude relation, that minimise scatter in the TF relation, with offset and slope left as free parameters.

Part II

Yes, we did check the mean distance of each $\left\langle\log\dfrac{a}{b}\right\rangle$ subset

Comment
Davies

It is very puzzling that the change in diameter and magnitude conspire to produce a constant SB irrespective of inclination.

Also to do the magnitude, inclination distance test you need a magnitude limited sample.

140

Question
Dave Burstein

Riccardo - what do you make of the fact that the mean I-band surface brightness is independent of a/b? In particular, we see a dependence of mean SB with a/b for 90 galaxies, why not for 50 galaxies at I-band if their outer parts are transparent?

Answer
Giovanelli

But we do- the outer parts <u>are</u> transparent, and the change in the isophotal diameters with inclination take about half of the change in SB, defined as $m_{23.5} + 5\log r_{23.5}$. The other half is due to internal extinction.

The simplicity of it all

EXTINCTION IN THE GALAXY AND IN GALACTIC DISCS

G. de Vaucouleurs
Department of Astronomy
University of Texas at Austin

ABSTRACT. Three current models (I: Sandage-Tammann = RSA; II: de Vaucouleurs *et al.* = RC2; III: Burstein-Heiles = RC3) for the heliocentric angular distribution of interstellar extinction in the disc of the Galaxy are tested by means of faint galaxy and cluster counts, bright galaxy colours and surface brightnesses, and HI indices.

The derivation of the formulae used in RC3 to evaluate the internal extinction corrections to the total B-band magnitudes and colours of spirals is explained .

Direct tests for the opacity of galactic discs by means of the visibility or invisibility of background galaxies seen through nearly face-on large spirals are indicated. Examples are given. The data suggest that on average spirals are substantially opaque inside the 24-th B-m/ss isophote and practically transparent outside the 26-th isophote.

1. INTRODUCTION

Three models of the heliocentric angular distribution of interstellar extinction, $A(l,b)$ (mag), in the disc of the Galaxy have been widely used during the past 20 years. Model I (RSA) [1] assumes zero extinction at all galactic latitudes $\mid b \mid \geq 50°$, and a B-band extinction, $A_B = 0.13(\mid csc\, b \mid -1)$ at all $\mid b \mid \leq 40°$, with an unspecified smooth transition between 40 and 50 degrees (and no longitude dependence). A meteorological analogy is "broken cumulus cloud in a clear atmosphere" with a cloudless patch near the zenith.

Model II (RC2) assumes a polar extinction, $A_B \simeq 0.2$ (0.19 at the North Galactic pole, 0.21 at the South Galactic pole), and essentially a $\mid csc\, b \mid$ law (as in the plane parallel approximation), but with a dependence on l and b derived from a 4-th order spherical harmonic respresentation of the Lick and Mt Wilson counts of faint galaxies (de Vaucouleurs and Malik 1969, de Vaucouleurs and Buta 1983, Appendix C). A meteorological analogy is "widespread cirrus cover in a hazy atmosphere".

---

[1]First suggested as a "hypothetical model" by McClure and Crawford (1971) was adopted by Sandage (1972, 1973) and Sandage and Tammann (1981, 1987 = RSA).

J. I. Davies and D. Burstein (eds.), The Opacity of Spiral Disks, 143–165.
© 1995 *Kluwer Academic Publishers.*

Model III (RC3) assumes that A = 0 wherever the local HI column density $N_H \leq$ $2.2.10^{20}$cm^{-2}, and A $\propto N_H$ elsewhere, with a correction depending on the local surface density of faint galaxies at $\delta \geq -23°$, the southern limit of the Lick counts (but no correction at $\delta \geq +65°$). The resulting value of A_B (adopted and listed in RC3 for some 23,000 bright galaxies) is not given by an analytical expression, but is read by interpolation on contours maps (Burstein and Heiles 1978, 1982). A meteorological analogy is "scattered cirri in a locally clear atmosphere" .

The following sections summarize the results of various tests of the three models made by R. Buta, H. Corwin, S. Odewahn and the author during the preparation of RC3.

2. SYSTEMATIC DIFFERENCES BETWEEN MODELS

The differences between the three models are subtle, amounting to colour excess differences at the poles, $\Delta E(B - V)(90°) \leq 0.05$ mag. The differential B-band extinction between models II and I, ΔA_g(II−I), varies from 0.2 mag at the poles to 0.3 mag at csc $\mid b \mid$ = 2.5 (de Vaucouleurs and Buta, 1983, Fig. 2). The average difference, $< \Delta A_g >$ = 0.25 mag, is largely responsible for the systematic zero point difference of 0.25 mag between the long and short extragalactic distance modulus scales within the Local Group (de Vaucouleurs 1993, Fig. 2 and 3).

The systematic difference in B-band extinction between models III (adopted in RC3) and II (used in RC2), depends statistically on $\mid b \mid$, as follows:

$$\log A_g(RC3)/A_g(RC2) = -0.0146(\mid b \mid -11°) \tag{1}$$

with a dispersion $\sigma \simeq 0.10$. There is no appreciable systematic difference in this relation between the northern and southern galactic hemispheres. This relation was used in RC3 to calculate the galactic extinction for galaxies within 9.4° from the galactic plane or at $b \leq -63.8°$ for which no reddening or HI data were given by the B-H charts. The Model III B-band extinction is 0.18-0.19 mag less than in the RC2 model at high galactic latitudes, 0.06-0.07 less at $\mid b \mid$ at intermediate latitudes, and (in the mean) equal at $\mid b \mid$ = 11°. The average difference , $< \Delta A_g > \simeq 0.15$ mag (depending on the latitude distribution of the objects) accounts for most of the recent revision of $\simeq 10\%$ in the short scale (de Vaucouleurs 1993). Model III has patchy areas of zero or negligible extinction near both galactic poles, less extensive than Model I and without circular symmetry about the poles.

3. TESTS FOR MODELS I AND II

The main differences between models I (RSA) and II (RC2) are (a) the presence or absence of extinction-free cones of $\simeq 80°$ apertures around each pole and (b) the differ-

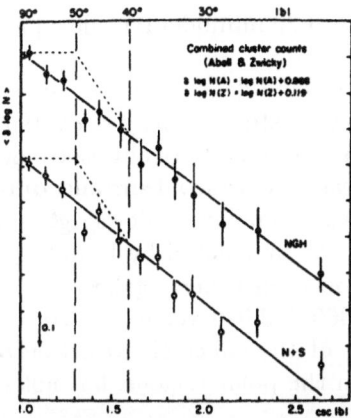

FIG. 1. Cosecant *b* law from combined cluster counts normalized at |*b*| = 40°. Mean points of Abell and Zwicky catalogs in NGH (above) and average of NGH and SGH (N + S). The csc *b* law is precisely obeyed up to the galactic poles (solid lines). The dashed lines show the trend expected if the "polar window" Model I were valid.

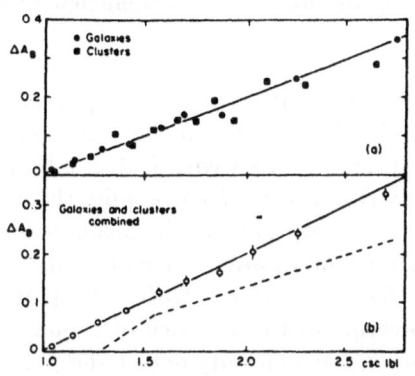

FIG. 2. Cosecant *b* law in combined faint galaxy and cluster counts normalized at poles and scaled to $A_B = 0.20$ mag. Note near perfect agreement with csc *b* law including polar caps (|*b*| > 40°) where mean errors of combined mean points are smaller than symbol size. Dashed line shows expected trend if Model I were valid.

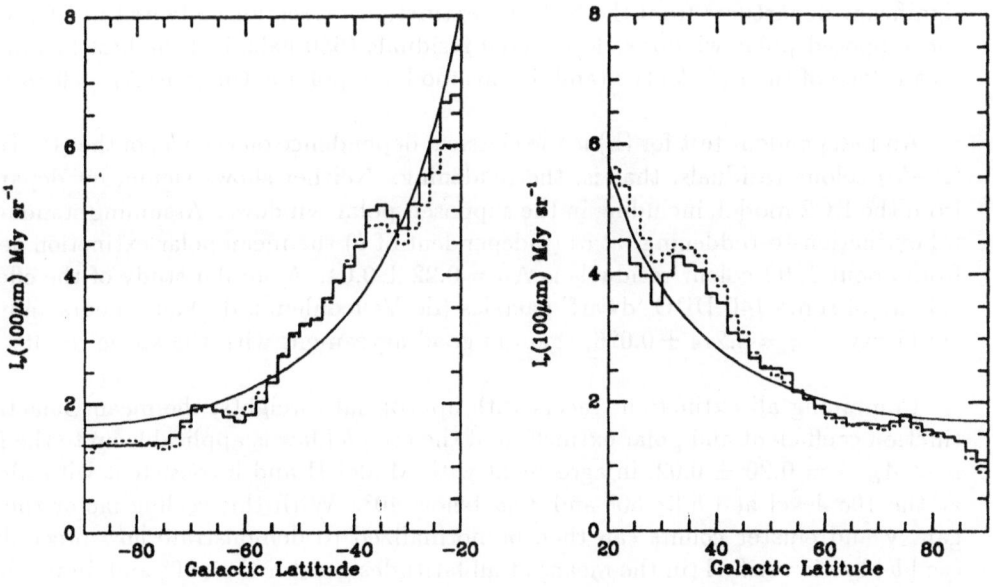

Figure 3. Mean galactic latitude distribution of IRAS 100 μm (solid histogram) and Bell Laboratories HI (dotted line), after Boulanger and Perault (1988). The HI data were scaled to the IRAS flux in each hemisphere. In the SGH the averages refer only to the declination zone covered by the Bell horn survey.

ent extinction laws and coefficients (0.13 in model I, 0.20 in model II) below $| b | = 40°$.

To test (a) one can use the dependence on csc b of faint galaxy counts (Lick = Shane & Wirtanen 1967, Zonn 1968) and of distant galaxy clusters (Zwicky *et al.* 1962-1968, Abell 1958), assuming the true space distributions to be statistically isotropic (or at least uncorrelated with Galactic latitude). Both tests (de Vaucouleurs and Buta 1983) show that the csc b law applies throughout the presumed polar windows right up to the poles in both galactic hemispheres (Figures 1, 2). As first noted by Holmberg (1974) cluster counts are even more sensitive to extinction than general galaxy counts, that is, the slope $d(\log N)/d(\text{csc} | b |)$ is greater (-0.360 ± 0.020 versus -0.319 ± 0.010). However, in the absence of a quantitative theory of the effect of extinction on such counts, one can only detect the presence of dust in the polar regions, but not find the actual value of the extinction coefficient.

To estimate the latter via test (b) one can use the dependence on csc b of a number of other extinction tracers, such as the mean surface brightness of nearby galaxies within a metric diameter (*e.g.* the *effective* diameter), or the neutral atomic hydrogen index, being a measure of the HI to B-band flux ratio (the former is not affected by Galactic extinction and the latter is directly dimmed by A_B mag). Neither shows significant departures from the Galactic extinction corrections adopted in RC2 within the supposed polar windows. From 1100 residuals (550 galaxies) the best fit implies a mean slope of the csc $| b |$ law (and, in this model, a polar extinction) $A_B = 0.18 \pm 0.02$.

An independent test for (b) is the classical dependence on csc $| b |$ of the (U−B) and (B−V) colour residuals, that is, the reddening. Neither shows significant departures from the RC2 model, including in the supposed polar windows. Assuming standard total extinction-to-reddening ratios (independent of b) the mean polar extinction derived from about 1200 colour residuals is $A_B = 0.22 \pm 0.04$. A similar study of the effective colours of some 140 DDO 'dwarf' galaxies (de Vaucouleurs, de Vaucouleurs and Buta 1981) gave $A_B = 0.214 \pm 0.026$, again in good agreement with the above results.

Combining all extinction tracers with appropriate weights, the mean Galactic extinction coefficient and polar extinction, if the csc $| b |$ law is applicable up to the poles, is $< A_B > = 0.20 \pm 0.02$, in agreement with Model II and inconsistent with Model I at the 10σ level at $| b | \geq 50°$ and 3.5σ below 40°. With this scaling factor the faint galaxy and cluster counts can then be normalized to demonstrate how precisely the csc $| b |$ law is verified (in the mean) at all latitudes greater than 20° and, in particular, in the polar windows postulated by Model I (Figure 5). Strictly speaking the polar extinction of Model II is applicable to magnitude-limited samples of galaxies which tend to be more easily seen in regions of lower-than-average extinction. The true average extinction for an unbiased sample (i.e. randomly selected or uniformly distributed in l, b) must be larger, perhaps by factors of 1.5 to 2, according to some models of the clumpiness in the dust distribution (see review and references in de Vaucouleurs and Buta 1983, section VII).

4. TESTS FOR MODELS II AND III

The same six extinction tracers were used to compare the predictions of models II and III. The parameters were first corrected for inclination and redshift (although there is no evidence of a correlation with galactic latitude) and for galactic extinction according to each model. Then residuals δy from their median values for each morphological type were calculated and analysed for possible dependence on either csc $| b |$ for both models, or A(B) for each model. Solutions were made for the coefficient c in $\delta y = c \mid \csc b \mid$, or c' in $\delta y = c'A(B)$, where the residuals were mutiplied by appropriate factors to correspond to magnitude residuals in the B-band extinction (that is x4.33 for the colour residuals, 1 for the surface brightness residuals, and -1 for the the HI residuals).

Starting with 3826 galaxies with colour data in an early version of RC3 (RC3.0), we first rejected all those with extreme values of the parameters (log $R_{25} > 1.3$, csc $| b |>$ 2.5, cz $> 5,000$ km s^{-1}), then excluded Local Group members, peculiar galaxies, those with strong emission lines or uncertain total magnitudes, and all galaxies within 6° from the centre of the Virgo cluster (HI deficiency). The solutions (after two cycles of 2σ rejection) for about 2500 remaining objects are collected in Tables 1 and 2. Both show that in general the dispersion σ of the residuals is very slightly less for model III than model II (except in U−B where it is the same), but the improvement is hardly significant. The weighted means of the coefficients in Table 1, $< c > = + 0.007 \pm$ 0.005 for model II, versus $+ 0.020 \pm 0.010$ for model III suggest a possible tendency of both models to underestimate the extinction coefficient (by $\simeq 0.01$ mag), but at a non-significant level. The agreement between the different extinction tracers is significantly better in model II (s = 0.014) than in model III (s = 0.020).

The solutions in Table 2 suggest that the relative extinction, A(b) − A(90), is overestimated by 11.9 ± 5.7 % in model II (the negative signs indicates that the corrected colours get bluer and the corrected surface brightness brighter as A increases) and , possibly, underestimated by a non-sgnificant 6.9 ± 7.6 % in model III. The agreement between the different extinction tracers is again sgnificantly better in model II (s = 0.113) than in model III (s = 0.152).

These tests check only the dependence on csc $| b |$ or on A(B) of the *relative (i.e. differential)* extinction, but tell us nothing on the zero point, that is the polar (model II) or minimum (model III) extinction. On the whole they provide little evidence to choose between the two models. The slightly smaller dispersion of the residuals in model III, may be attributed mainly to its better angular resolution, but this is not a verification of its zero point A(min) = 0.

Among the photometric extinction indicators not used previously, is the Balmer decrement (restricted to the Hα/Hβ ratio) in extragalactic HII regions (Sparks et al. 1986). Departures from the Case B typical recombination value of 2.86 (for $T_e = 10^4 K°$, $N_e = 10^2$) are attributed to differential extinction by dust in our Galaxy or in the other. Then for the standard galactic extinction curve

$$E(B - V) = 2.10 \log(2.86 H\alpha/H\beta)$$

is an upper limit to the Galactic extinction alone. The dependence on csc | b | of the *lower* bound to the E(B−V) distribution for 80 HII galaxies suggests that E(B−V)(90°) ≤ 0.066 at the 95% confidence level, and ≤ 0.057 at the 90% level. The favored value is E(B−V) ≃ 0.03 | csc b | which within the rather large errors is generally consistent with Model II, but marginally inconsistent (at the 6% level) with Model III. Because of the small number of objects and the dominant contribution of the HII regions to the measured extinction (compared to that of our Galaxy), this is not a critical test.

Finally, in a few instances it is possible to compare for individual galaxies the galactic reddenings predicted respectively by Model II (RC2) and Model III (RC3) with the actual values derived from the colour excesses of field stars. Some examples are given in table 6. For what it is worth this very small sample of three objects, all at intermediate latitudes (21° <| b |< 41°), tends to support Model II rather than Model III. This is merely an illustration. Many more precise determinations of foreground reddening from the colours of field stars will be needed to provide a convincing test of extinction models.

5. SPECIFIC TEST FOR MODEL III

Our final choice of Model III for RC3 was determined by a specific test (c) we made for the absence of detectable extinction in the fields where this model predicts A = 0. These fields are scattered across a rather large range of galactic latitudes (1.0 ≤ csc | b |≤ 1.5). If, then, the residuals of each of the extinction tracers described in section 3, are plotted versus csc | b |, their weighted average slope should be zero or negligible if Model III is correct, but ≃ 0.2 if Model II is correct.

The extinction tracers are of two types: (a). Photometric: total and effective colours, mean effective surface brightness, and hydrogen index. All galaxies for which these parameters were known and model III predicted A = 0 were extracted from a preliminary version of RC3 (RC3.0) and the residuals from the *median* values (for the A = 0 set) were plotted versus csc | b | for each model. The coefficients of linear solutions of the form

$$y- <y>= a(\csc \mid b \mid -1) + b \qquad (2)$$

are collected in table 3. All, except HI (which has the smallest number of test objects and the largest mean error), give a negative slope — indicative of over-correction for A_B — in the case of Model II, but a negligible slope for Model III. The weighted means of all indicators,

$$<a>= -0.165 \pm 0.098 \text{ for Model II, and } <a>= -0.015 \pm 0.092 \text{ for Model III,}$$

show that the latter gives a significantly better representation of galactic extinction (i.e. substantially independent of $\csc b$ where the model predicts $A = 0$), although the departure of Model II is only at the 1.7σ level. Note again that the comparison tests only the slope $\delta A_B/\delta(\csc \mid b \mid)$, not the zero point. If the B-H model had predicted $A(\text{minimum}) = \text{const.} > 0$ in the low-extinction areas, the test would have given the same result. However, in view of the various arguments given by B-H in favour of their assumption that $A(\text{minimum}) = 0$, this test was accepted as adequate supporting evidence in favour of Model III which was adopted in RC3.

(b). Galaxy and Cluster Counts: a similar test was made for the surface density of faint galaxies and distant clusters in fields centered within 3 ° from the galaxies used in test (a). The residual of the $\log N$ in each field from the relevant median (in the $A = 0$ fields) was plotted versus $\csc \mid b \mid$ and linear solutions of the form (2) were made. The results are collected in Table 4 for the four sources of counts used (Lick and Zonn galaxy counts, Abell and Zwicky cluster counts). All four give a negative slope whose weighted average, $<a>= -0.24 \pm 0.06$, is different from zero at the 4σ level. The negative sign indicates that the counts decrease with increasing $\csc \mid b \mid$ as would be expected if the cosecant law applied right up to the poles even in the $A = 0$ areas of Model III. [2] The surface density tests, therefore, disagree with the photometric tests (except HI). They tend to support a model intermediate between II and III, as if some residual low-density dust were present in the intercloud regions. This would be the case if the HI cutoff density were less than $2.2.10^{20}$ cm^{-2} (Model III).

6. HIGH-LATITUDE NEBULOSITY AND DIFFUSE EMISSION

Photometric tests and galaxy or cluster counts are poor tracers of extinction by dust at high-galactic latitudes. Light scattering by the dust may be a more sensitive indicator of its presence. This was first observed at low galactic latitudes more than 50 years ago (Henyey and Greenstein 1941). The photographic detection of faint diffuse nebulosities at intermediate galactic latitudes, in particular in the vicinity of the

---

[2] Note, however, that the dependence on $\csc \mid b \mid$ is less in the $A = 0$ set than in the unrestricted sample (− 0.24 versus − 0.33).

Magellanic Clouds (de Vaucouleurs 1954, 1955a; de Vaucouleurs and Freeman 1972, Fig. 8, 9), over the South celestial polar cap (de Vaucouleurs 1955b, 1960), and later in other areas (Lynds 1965, Sandage 1976), demonstrated the widespread distribution of gas and dust in many directions as far as 40 and 50° from the galactic plane. Their visibility on both red-, green- and blue-light photographs (i.e. Hα and continuum) and the detection of concentrations of neutral hydrogen in some of these nebulosities (McGee et al. 1963; Heiles and Cleary 1979), indicated a strong correlation between dust and gas, both neutral and ionized, leading to the conclusion that HI emission is, at least in a first approximation, a good tracer of dust. This fact was used by Burstein and Heiles (1978, 1982) to produce Model III. However, their assumption that optical extinction drops to zero wherever $N(HI) \leq 2.2.10^{20} cm^{-2}$ remains questionable (because the HI maps used by Heiles were obtained with parabolic reflectors which suffered from side lobe contamination when pointing toward regions of low emission, in particular the galactic polar caps. [3]

The all-sky maps of the so-called "infra-red cirri" detected at 60 and 100 μm by IRAS demonstrated the widespread distribution of thermal emission by dust heated by the general radiation field from the galactic plane. The strongest coincide neatly with the diffuse nebulosities previously found by photography in the optical range at least up to $| b | = 45°$. This led to the conclusion (Boulanger and Perault 1988) that 'The far-infrared emission of the Galaxy above 10° is characterized by large scale filamentary structures, resembling terrestrial cirrus clouds (Low et al. 1984), superposed on a continuous background (Hauser et al. 1984)' which is entirely consistent with Model II, if the continuous background originates in the inter-cloud medium and covers the polar regions.

That the dust distribution extends to still higher latitudes, including the polar regions, is indicated by a number of observations:

(a) the photographic detection of reflection nebulosities in the immediate vicinity of the South galactic pole by D. Malin, in particular in four fields with $-79° \geq b \geq -87°$ (reproduced in Figure 16 of de Vaucouleurs and Buta 1983). As noted by Sandage (1976), photography can only detect the denser, brighter regions above some mean background level because 'A uniform dust sheet would add a uniform surface brightness to the night sky, and hence would not be detected'. The brighter regions have surface brightnesses in the range $25 - 27$ (B,V) mag arcsec^{-2}. A fainter, uniform intercloud medium would not be observable. [4]

---

[3]The Bell Laboratories HI survey with a horn collector (2° beam FWHM) was much less affected by this spillover problem (Stark *et al.* 1990). The HI maps produced by this survey do not agree well with those used by Burstein and Heiles, particularly at high latitudes, nor do they correlate closely with the faint galaxy counts. It would be useful to recalculate the Model III extinctions with the Bell Labs HI map.

[4]According to Tyson (private communication) the integrated B-band surface brightness of all the

Because the relation between extinction and scattering depends strongly on the geometry of the illumination and the unknown albedo and phase function of the dust particles, it is not possible to derive the opacity of the clouds from their surface brightness. For a model assuming isotropic scattering by a high-latitude cloud illuminated by the galactic plane Sandage (1976, equation 8) quotes a V-band extinction $A_V \simeq$ 0.08 mag for a SB = 27 mag arcsec^{-2}. To this must be added the unknown base level of the invisible inter-cloud regions (possibly 0.03-0.04 mag if $< SB > = 28$) and many of the detected clouds are brighter than 27. A total minimum B-band extinction of $\simeq$ 0.2 mag is, therefore, not implausible.

(b) the detection by IRAS of thermal emission at 100 μm covering both galactic polar caps (Figure 3). There is a good correlation with the HI column density. Unfortunately, remaining uncertainties in the absolute calibration (zero point) of the IRAS fluxes and in the correction for the thermal emission by the zodiacal light prevents a secure translation of the residual emission into dust column density and optical opacity. According to Boulanger and Perault (1988) a plausible median visual extinction in the polar caps ($| b | \geq 50°$) is $A_V = 0.10 \, ^{+0.06}_{-0.03}$.

(c) the distribution of the far UV (FUV = 1400 - 1850 Å) diffuse background measured in the UVX experiment (Hurwitz, Bowyer and Martin 1990) has led to direct evidence that 'as the HI column density approaches zero, the [non-zero] FUV continuum background arises primarily from scattering by dust, *which implies that dust may be present in virtually all directions*'. Of particular interest is the conclusion that a clumped model of the ISM made up of dense clouds in a transparent inter-cloud medium gives a poorer fit to the I(λ1580) $-$ N(HI) relation than a diffuse model with a continuous density distribution of dust. The continuous model, corrected for emisssion lines, has an optical depth

$$\tau(\lambda) = (R_\lambda/1.086)[E(B - V)_0 + N(HI)/5.0.10^{21}] \tag{3}$$

The best fit solution gives at the 95% confidence level, $-0.016 < E(B - V)_0 < +0.032$ near the poles, or $< E(B - V) > = + 0.015$, corresponding to N(HI) = 8.10^{19} cm^{-2}. This is about one-third of the polar reddening postulated by Model II, but uncertainties in the contributions of the extra-galactic flux and of the two-photon recombination of ionized H, still leave room for a large margin of error in this estimate. Nevertheless, the conclusion that the warm interstellar medium 'is not (strongly) depleted in large grains' is of interest, in view of the argument that, perhaps, evaporation of the icy mantle of interstellar grains may be the reason for the vanishing optical extinction where N(HI) is non-zero in Model III.

---

galaxies is $< SB > = 28.8$ B-m/ss. Within their 29th B-m/ss isophotal level such galaxies cover 15% of the sky near the galactic poles.

In conclusion it would seem that the ideal model of galactic extinction would be one combining (with appropriate weights) all the previous tracers, including the FUV background, visible (emission and reflection) nebulosities and the IR cirri, as well as the HI, HII and molecular clouds. Not all have yet been mapped in sufficient detail, but with currently available surveys an improved extinction map could result from a suitable combination of existing HI and IRAS maps. [5] At 2−3° resolution it will be preferable not to use the Lick galaxy counts as a correction because of the large intrinsic fluctuations in the galaxy density distribution.

7. POLAR EXTINCTION FROM STELLAR PHOTOMETRY AND POLARIMETRY

In the simplest model of galactic extinction by a uniform, plane parallel layer of dust, leading to the traditional csc $\mid b \mid$ law, the zero point (polar extinction, A_1) is equal to the slope $dA/\sin \mid b \mid$ of the relation. However, if the distribution is clumpy and the scattering properties of the dust particles depend on their environment (cloud vs inter-cloud, neutral vs ionized, atomic vs molecular), it is entirely possible that this expected equality will not hold, even if a linear relation with csc $\mid b \mid$ emerges as a large scale statistical average for an inhomogeneous ISM. Then the polar extinction (averaged over a small area) may be smaller (or larger) than the mean extinction at lower latitudes, $< A/\sin \mid b \mid >$ (averaged over all longitudes). The poles may not even be at the extinction minima and centres of symmetry of the extinction pattern. It is, therefore, of interest to examine polar extinction values derived from surveys restricted to the vicinity of the poles.

Multicolour photometry of B stars has been used to test Models II and III with conflicting results. While Nicolet (1982), working in the Geneva system, derived from 129 B- and early A stars at $\mid b \mid \geq 30°$ a mean polar reddening, $< E(B - V) > = 0.04$,[6] Tobin (1985) discounted this result by excluding many stars as 'peculiar' with respect to a variety of more stringent rejection criteria. From Strömgren photometry of 110 B stars (selected out of ≈ 1100) Tobin found much better agreement with Model III and, after further rejections of aberrant points, concluded that '72 selected B stars more than 250 pc from the galactic plane support the [Model III] zero-point of galactic reddening' and that the use of faint galaxy counts 'to provide a first-order correction for variations in the dust-to-gas ratio' seems supported by a few stars.

Similar studies of about 1000 A and F stars at $\mid b \mid \geq +75°$ by Hilditch, Hill and Barnes (1983) also lead to low values of the polar extinction, but illustrate its

---

[5] the Bell Labs horn survey needs to be extended to the South pole.

[6] in agreement with the older values of 0.054 (Holmberg 1974) and 0.057 toward the SGP (Knude 1977), and all consistent with the value 0.046 adopted in Model II.

patchiness. Thus, in the second and third longitude quadrants, $90° < l < 270°$, reddening is negligible and independent of distance beyond 100 pc and out to 400 pc, $< E(b-y) > \simeq +0.001 - 0.002$, while it is small, but significant, $< E(b\text{-}y) > = +0.008$, remaining constant beyond 100 pc, in the fourth and first quadrants, $270° < l < 90°$, and there is a patch of higher extinction, $E(b-y) = +0.024$, in the fourth quadrant apparently associated with an HI (and dust) cloud stretching across the polar cap (Figure 1, *loc.cit.*).

Another indication of significant extinction at high latitudes is derived from interstellar polarisation studies which provide at least *lower limits* to the polar extinction of $A_V \geq 0.05$ at the SGP (Appenzeller 1975) and ≥ 0.03 at the NGP (Markannen 1979). Because of the loose correlation between extinction and polarisation these estimates are at best suggestive.

Various estimates of the polar extinction derived from the slope of the csc b dependence, mainly for extragalactic objects (Table 7A), have a mean value of $A_B = 0.21 \pm 0.01$ (n = 15), while those derived from the zero point of the relation in the polar caps, based on stars and galactic sources (Table 7B), have a mean value of $A_B = 0.11 \pm 0.02$ (n = 13). The systematic difference is significant and could be caused by the presence of very low density dust in the galactic corona extending out to $\simeq$ 50 kpc in the plane and $\simeq$ 10 kpc perpendicular to it (cf. Savage & de Boer 1981). A dust-less gas corona does not appear to be physically plausible (cf. Pecker 1972). Faint galaxy counts around edge-on nearby spirals could be used to test this hypothesis.

8. INTERNAL EXTINCTION CORRECTIONS FOR DISC SYSTEMS IN RC3

The effect of internal extinction on the integrated magnitudes and colours of disc systems and its dependence on inclination has been the subject of much study (Holmberg 1958, de Vaucouleurs 1961, Heidmann, Heidmann and de Vaucouleurs 1972 = H^2V) and it is still actively debated.[7] It is clear, however, that the effect depends greatly on the details of the dust distribution and that it acts differently on the spheroidal (bulge) and disc components of composite systems such as spirals. The contribution of diffraction and forward scattering is difficult to assess (Lindblad 1942, Elvius 1956, Bruzual, Magris and Calvet 1988), but should not be ignored and whether the optical properties of the dust differ or not in different galaxy types (or even among galaxies of the same type or in different parts of a given galaxy) is uncertain. The

---

[7]The question of the dependence of apparent diameter on inclination is discussed by others elsewhere in this volume. In RC3 we adopted Burstein's point of view according to which D_{25} is independent of inclination (see also Choloniewski 1991), but this conclusion is sensitive to selection effects and the inference that the discs of spirals are opaque (optically thick) at the 25th B-m/ss isophote does not necessarily apply to lower isophote levels. Different views have recently been given new support by Huizinga (1994) and Valentijn (1994).

only practical approach is empirical and statistical. Different recipes have been used in various catalogues, in particular for RC1, RC2 and RC3. The tests made during the preparation of RC3 are summarized below.

(a) Integrated colours: using the selected sample of about 2500 galaxies from RC3.0 mean and median total and effective colours (corrected for galactic reddening and redshift as in RC2) were calculated for each morphological type. The residuals from the median, δC, were represented by a linear function of log R_{25} which, as shown by previous studies (H^2V), is a close approximation to $\log(\sec i)$ and avoids the divergence as $i \to 90°$. Solutions were made for the coefficients α, β in

$$\delta C = \alpha + \beta \log R_{25} \qquad (4)$$

where C is in turn the B−V, U−B total or effective colours. Plots of the slopes β versus morphological type T (on the RC2, RC3 system) along the Hubble sequence show a characteristic dependence well represented by

$$\beta(T) = a + b(T - 3)^2 \qquad (5)$$

The values of a, b are collected in Table 5 for the several colour indices. It appears that on the whole $< \beta > \simeq 0$ at all $T \leq -1$ (E and L types) and $T \geq 7$ (Sd-Im types), that is their colours are essentially independent of axis ratio (inclination), but for most spirals ($-1 \leq T \leq 7$) the internal reddening in B−V is given by

$$E_i(B−V) = 0.35 - 0.022(T - 3)^2$$

for both the total and effective B-band magnitudes. For the the U−B colours the slopes and zero points imply X = E(U−B)/E(B−V) = 1.1 ± 0.2, confirming the empirical average used in RC2 after de Vaucouleurs (1961).

(b) Integrated magnitudes: several tests were used to derive the effect of inclination, $A_i(B)$, on the total magnitudes of galaxies.
1. Absolute magnitudes. Let $B_T^1 = B_T - A_g(B) - K_B$, be the integrated magnitude corrected only for Galactic extinction (Model II) and redshift, but not for inclination. The absolute magnitude, M_T^1, is calculated assuming Hubble's law (the choice of H_o is immaterial) from the radial velocity corrected for solar motion and Galactic rotation.

To avoid the catalogue bias against low luminosity objects at low galactic latitudes the analysis was restricted to $| b | \geq 50°$ ($| \csc b | \leq 1.305$). Residuals, $\delta(M_T^1)$, from the median absolute magnitude for each type T were fit by

$$\delta(M_T^1) = const. + \alpha_1(\log R_{25} - 0.2) \qquad (6)$$

A plot of α_1 versus T shows that for the spirals $(0 \le T \le 10)$, $<\alpha_1> = 1.04 \pm 0.25$, in fair agreement with the value adopted in RC2, $<\alpha_1> = 0.8$.

2. Hydrogen Index. The index $HI^1 = m_{21}^1 - B_T^1$ (without inclination corrections) depends mainly on type, but will be affected for each given type by the difference between the 21-cm self-absorption and the B-band extinction. If the former is negligible (as was assumed in RC2, but not in RC3), the residuals should reflect the latter. It has the advantage of not requiring an estimate of distance. The residuals from the medians for each type were fit by

$$- \delta(HI^1) = const. + \alpha_2(\log R_{25} - 0.2) \tag{7}$$

For $3 \le T \le 10$, $<\alpha_2> = 1.23 \pm 0.17(|\csc b| \le 1.305)$ or $1.30 \pm 0.14(|\csc b| \le 2.5)$, suggesting that the B-band extinction might be larger than the RC2 value.

3. Surface brightness. Here we followed Holmberg's (1958) approach, assuming that galaxies are optically thick, i.e. that D_{25} is independent of inclination as in RC3. Magnitudes were corrected for $A_g(B)$ as in RC3 (Model III), but the test is insensitive to the choice of galactic extinction model. Holmberg calculated the "surface magnitude" as

$$S = m + 5\log D = S^* + \alpha_3(\sec i - 1) \tag{8}$$

where i is given by $\cos^2 i = (q^2 - 0.04)/(1 - 0.04)$, for all types. For a sample selected as in sect. 4 and restricted to $\sigma(B_T) \le 0.10$, $A_g(RC3) \le 1.0$, and $V \ge 2500$ km s^{-1} in the Virgo cluster region, we calculated

$$S = B_T - A_g(RC3) - K_B + 5\log D_{25}^0, \tag{9}$$

where, for $T \ge 0$,

$$log D_{25}^0 = log D_{25} - log(1 - A_g/3.35) \tag{10}$$

and for $T < 0$,

$$log D_{25}^0 = log D_{25} + A_g(B)/G_{25} \tag{11}$$

with the intrinsic flattening depending on type as in (Bottinelli et al. 1983). A plot of α_3 versus T shows that $\alpha_3 \simeq 0$ for $T < 0$ and $<\alpha_3> = 0.37 \pm 0.02$ for $2 \le T \le 10$, in fair agreement with Holmberg's values for his m_{pg}, $D_{26.5}$ systems, $\alpha_3 = 0.28 \pm 0.07$ for his types Sc$^-$, Sc$^+$ and 0.43 ± 0.06 for his types Sa, Sb$^-$, Sb$^+$. Whether α_3 is discontinuous at $T = 0$ is not clear.

The $\sec i$ dependence would apply if the absorbing layer were all in front of the galaxy. It can be shown (H^2V) that for an infinitely thin layer in the galactic plane a dependence on $\log(\sec i)$ is more plausible and that it is well approximated by $\log R_{25}$. We considered two cases:

1. Optically thick case (D_{25} independent of i), then S is calculated as in equation (8) and

$$S = S^* + \alpha_4 \log(\sec i) \tag{12}$$

then for $2 \leq T \leq 10$, $< \alpha_4 >= 1.88 \pm 0.11$.

2. Optically thin case, D_{25} depends on i and the corrected face on diameter is given, as in RC2, by

$$\log D_{25}^{\circ} = \log D_{25} - 0.235 \log R_{25} \tag{13}$$

then

$$S_1 = S - 1.175 \log R_{25} = S_1^* + \alpha_5 \log(\sec i) \tag{14}$$

For $2 \leq T \leq 10$, $< \alpha_5 >= 0.86 \pm 0.11$, in fair agreement with the value (0.72 ± 0.16) found by H^2V.

If we knew for certain whether spiral discs are opaque or transparent at the 25-th isophote level we could make a fairly definite choice of correction to the total B-band magnitudes of spiral galaxies. In RC2 we used the optically thin case, where $\alpha \simeq 0.8$. In RC3 we chose the optically thick case, and a value of the inclination coefficient for all $T \geq 0$

$$\alpha(T) = 1.5 - 0.03(T - 5)^5 \tag{15}$$

which is larger than that advocated by Haynes and Giovanelli (1984), but less than $< \alpha_4 > = 1.88$ found above for the opaque disc case. In the absence of scattering by dust, and if the mean surface brightness of galaxies were exactly constant and independent of inclination, the coefficient should be 2.5. In the presence of forward scattering the coefficient will be decreased, with current estimates predicting it to be ≈ 1.8 (Bruzual et al. 1988). The exact relation between extinction and scattering will depend in a complicated fashion on details of the dust distribution (clumped or diffuse, location w.r.t. spiral arms, thickness) and its scattering properties (crystal shape, refraction index, diameter function), and their dependence on morphological type which to a large extent are still unknown.

9. GALAXIES AS PROBES OF OPTICAL THICKNESS OF SPIRAL DISCS

The tests described so far do not provide direct information on the opacity of the discs of spirals, either because they refer to a single location in our Galaxy, or give only the integral of a hopelessly complicated radiation transfer problem. To measure directly the transmission coefficient in disc systems we need to observe light probes through the disc system itself seen close to face-on. This is not usually possible near the centres of galaxies which are often too bright to permit viewing objects of similar or lower intrinsic brightness (although in symmetric systems, digital background subtraction could help, as in the case of globular clusters). Through most of the discs, however, the surface brightness of the (exponential) component is no longer overwhelming and occasionally allows distant background galaxies to shine through the foreground disc.

There are two ways to perform this test. (a) by counting faint galaxies to the plate limit over a large area including the foreground galaxy, correcting for any general gradient across the field, and converting the count reduction factor to a magnitude loss via the slope of the apparent luminosity function in the surrounding field. This method has been applied only to the Magellanic Clouds (Wesselink 1961a,b, Hodge 1969, 1974, 1975). In the Large Cloud measurable extinction has been detected out to $6° = 5.5$ kpc from the centre of the disc and near total opacity prevails within $1° = 0.8$ kpc where the V-band extinction is predicted to be $A_V \simeq 1.5$ mag (Gurwell and Hodge 1990).[8] In the Small Cloud extinction was detected out to $2° = 2$ kpc radius from the center with a maximum of $A_V \simeq 1.3$ mag (Hodge 1974 and references therein). [9]

After the Magelanic Clouds, the next largest galaxies, such as M31, M33, M51, M81, M101, would be among the best targets, but discrimination between background galaxies and diffuse objects (HII regions, globular clusters, and unresolved star clumps) may require spectroscopic observations.

(b) by measuring the dimming and reddening of individual galaxies seen through the disc of large, nearby galaxies. The best and largest example is the Large Magellanic Cloud through which a dozen background galaxies larger than 1' have been identified in the SGC (Corwin, de Vaucouleurs and de Vaucouleurs 1985). The nearest to the centre, NGC 1809, is only $2° = 0.3\ r_e$ from the geometric centre of the bar. It is probably an Sc, 3' in diameter, with an axis ratio of 0.26, but its relatively low surface brightness shows that it is affected by extinction in the LMC, in addition to veiling by the crowded star field on the south-west edge of the bar. The B-band surface brightness of the bar at this location is about 22.5 m/ss (see de Vaucouleurs and Freeman 1972, Fig. 10). This object is known to be moderately reddened (RC3), but it deserves detailed multi-colour photometry. [10]

Numerous other examples of this situation can be found by leafing through photographic atlases or (better) inspecting original large-scale reflector plates. A casual search in the '*Atlas of Galaxies*' (Sandage & Bedke 1988) revealed several examples in the fields of NGC 45, 210, 289, 450, 578, 628, 3642, 3686, 4321, 5885, 7424, 7552, and

---

[8]The centre adopted by Gurwell and Hodge is the geometric centre of the outer loop, which falls slight;y West of the 30 Doradus complex (see de Vaucouleurs and Freeman 1972, Table 2 and Fig. 11), the bar centre is 10 m West, 24' South of it.

[9]Allowance for the inclination, $i \simeq 60°$, of the disc would reduce this to $A_V \simeq 0.65$ in the face-on view, corresponding to a colour excess $E(B-V) \simeq 0.2$ and a transmission factor $t_V \simeq 0.55$.

[10]The total magnitudes $B_T = 13.0$ in RC3 and $m_B = 15.38$ in ESO-LV are highly discrepant and should be checked. The fact that NGC 1809 was discovered by John Herschel (1847) with his 18-inch telescope suggests that the RC3 value is more nearly correct.

IC 5332.

Another option is the (rare) instance of partial occultation of a smooth (i.e. early type) background galaxy by a nearer spiral. One such example, AM1316-241, has recently been discussed by White and Keel (1992) with promising results. Unfortunately, the foreground spiral is rather strongly inclined and the 'background' elliptical has a smaller redshift than the spiral. More examples are needed. [11]

Finally, Hodge (1969) has suggested that observation of interstellar lines (HI, Ca^+, Na I, Fe^+, etc) in absorption in the spectra of background galaxies seen through a foreground, near-face-on spiral could be used to indirectly estimate the extinction in the latter via the statistical relation between column density and colour excess.

In the meantime a provisional conclusion may be that spiral discs seem to be substantially opaque where the (face-on) B-band surface brightness is SB $\leq$ 24 m/ss, and practically transparent where SB $\geq$ 26 m/ss. The standard isophote level of 25 m/ss is in the 'twilight' zone. In addition the evident patchiness of the dust distribution complicates matters at all radii so that extinction will depend on the resolution.

The extragalactic work at the University of Texas and, in particular, the RC3, has, for many years, been supported by the US National Science Foundation.

---

[11]Others are discussed by White and Keel in this volume.

Table 1. Residuals versus csc $|b|$

Tracer	c(II)	σ(II)	N(II)	c(III)	σ(III)	N(III)
$4.33(B-V)_T$	0.015	0.342	1052	0.030	0.333	1053
m.e.	0.018			0.017		
$4.33(B-V)_e$	0.013	0.342	969	0.037	0.182	974
m.e.	0.019			0.018		
$4.33(U-B)_T$	0.016	0.459	900	0.032	0.459	901
m.e.	0.026			0.026		
$4.33(U-B)_e$	−0.022	0.437	841	0.008	0.437	843
m.e.	0.025			0.025		
μ'_e	−0.021	0.690	718	−0.022	0.681	716
m.e.	0.044			0.044		
− HI	0.014	0.624	541	0.009	0.621	541
m.e.	0.048			0.048		
Weighted means	0.005			0.020		
m.e.	0.007			0.010		
s	0.014			0.020		

c = slope of B-band magnitude residual versus csc $|b|$, assuming $A_B/E(B-V)$ = 4.33 and E(U-B)/E(B-V) = 1.0 for the average colours of galaxies. σ = dispersion of residuals from solution. N = number of galaxies. s = dispersion of solutions from weighted mean.

Table 2. Residuals versus $A(B)$

Tracer	c'(II)	σ	N	c'(III)	σ	N
$4.33(B-V)_T$	−0.094	0.342	1052	0.088	0.333	1055
m.e.	0.090			0.068		
$4.33(B-V)_e$	−0.100	0.342	970	0.118	0.333	971
m.e.	0.091			0.070		
$4.33(U-B)_T$	−0.035	0.459	901	0.204	0.463	904
m.e.	0.130			0.101		
$4.33(U-B)_T$	−0.222	0.437	842	0.116	0.437	843
m.e.	0.126			0.098		
$\mu'_{e,0}$	−0.351	0.685	717	−0.2870	0.680	716
m.e.	0.223			0.169		
− HI	0.030	0.625	541	0.081	0.621	541
m.e.	0.259			0.175		
Weighted Means	−0.119			0.069		
m.e.	0.057			0.076		
s'	0.113			0.152		

c' = slope of residuals from median versus A(B). σ = dispersion of residuals from solution. N = number of objects. s' = dispersion of solutions.

Table 3. *Magnitude Residuals in A = 0 areas of Model III*

Tracer	a(II)	b(II)	σ(II)	a(III)	b(III)	σ(III)	N
$4.33(B-V)_T$	−0.258	0.030	0.395	−0.086	0.009	0.385	382
m.e.	0.108	0.026		0.103	0.026		
$4.33(B-V)_e$	−0.232	0.052	0.350	−0.095	0.026	0.345	343
m.e.	0.095	0.026		0.095	0.022		
$4.33(U-B)_T$	−0.076	−0.038	0.520	0.076	−0.057	0.520	329
m.e.	0.151	0.038		0.151	0.038		
$4.33(U-B)_e$	−0.114	−0.005	0.505	0.014	−0.019	0.500	306
m.e.	0.151	0.038		0.147	0.038		
μ'_e	−0.415	0.084	0.679	−0.234	0.062	0.674	240
m.e.	0.261	0.065		0.259	0.065		
−HI	0.141	−0.084	0.601	0.283	−0.103	0.598	176
m.e.	0.267	0.064		0.266	0.063		
Weighted means	−0.165			−0.015			
m.e.	0.098			0.092			

a = slope of B-band magnitude residual versus csc | b |, assuming $A_B/E(B-V)$ = 4.3 and E(U-B)/E(B-V) = 1.0 for the average colours of galaxies. b = zero point. σ = dispersion of residuals. N = number of galaxies.

Table 4. *Log Counts Residuals in A = 0 areas of Model III*

Counts	Source	a(III)	m.e.	b(III)	m.e.	σ(III)	n	Rem.
Galaxies	Lick	−0.214	0.042	0.046	0.007	0.068	190	
Galaxies	Zonn	−0.334	0.095	0.027	0.016	0.126	140	
Clusters	Zwicky	−0.223	0.110	0.028	0.021	0.179	162	1
Clusters	Abell	−0.186	0.227	0.036	0.047	0.286	88	
Weighted	Means	−0.237	0.055	0.040	0.010			

a = slope of log count residual from median versus csc | b |. b = zero point. σ = dispersion of residuals. n = number of fields.
(1) Field Z042 (l = 289.13°, b = + 67.32°), seen through Virgo cluster, is rejected.

Table 5. *Inclination Coefficients*

Tracer	a	m.e.	−b	m.e.	σ
$(B-V)_T$	0.35	0.021	0.022	0.003	0.040
$(B-V)_e$	0.34	0.020	0.021	0.003	0.038
$(U-B)_T$	0.38	0.082	0.024	0.009	0.16
$(U-B)_e$	0.40	0.050	0.025	0.007	0.096
X_T	1.08		1.09		
X_e	1.18		1.19		

a = colour excess at T = 3; b = coefficient of $(T-3)^2$; σ = dispersion of $\alpha(T)$ from fit. X = E(U−B)/E(B−V).

Table 6. *Observed and calculated foreground E(B−V)*

Galaxy	Observed	m.e.	Ref.	Model II	Model III	b°
LMC	0.06	0.005	1	0.10	0.063	−32.9
M 31	0.10	0.02	2	0.095	0.077	−21.6
M 81	0.10	0.02	3	0.07	0.037	+40.9
Means	0.087			0.088	0.059	

(1) from an unpublished compilation (1982) of 14 estimates (1958-1981)
(2) mean of four sources (de Vaucouleurs & Corwin 1986, ApJ, 308, 487)
(3) from U,B,V colours of nine field stars (unpublished)

Table 7A. *Polar extinction from slope of $|$ csc b $|$ relation*

Year	Source	Method	E(B−V)	m.e.	A_B	m.e.	Rem.
1958	Holmberg	P_g−P_v colours of 174 galaxies $\|b\|>11°$	0.06	.01	0.26	.04	1
....	....	P_g surface brightness of 119 gal. $\|b\|>15°$	0.05	.01	0.21	.05	2
1961	de Vaucouleurs	B−V colours of 181 normal gal. $\|b\|>30°$	0.044	.025	0.19	.11	3
1964	Sturch	UBV colours of 108 RR Lyr stars near minimum, $b>+30°$	0.05	.015	0.21	.06	4
1970	Peterson	V−r colours of brightest gal. in 55 groups, $\|b\|>12°$	0.027	.008	0.12	.03	5
1974	Holmberg	V−r colours of brightest gal. in 39 groups, $\|b\|>20°$	0.062	.014	0.26	.06	6
....	....	UBV colours of 108 RR Lyr stars	0.059	.009	0.25	.08	7
....	....	UBV colours of 69 RR Lyr stars	0.047	.016	0.20	.07	8
....	....	Strömgren colours of 67 RR Lyr in S hemisphere	0.057	.011	0.245	.05	9
1976	Rubin *et al.*	HM residuals of 96 faint ScI gal. $3500< V_c <6500$ km s^{-1}	0.036	.007	0.15	.03	10
1981	de Vaucouleurs *et al.*	Effective UBV colours of 140 DDO dwarf gal.	0.050	.006	0.214	.026	11
1983	de Vaucouleurs	SB residuals of ≈ 600 gal.	0.037	.007	0.16	.03	12
....	& Buta	HI residuals of ≈ 400 gal.	0.049	.007	0.21	.03	12
....	....	B−V colour residuals of ≈ 800 gal. $\|b\|>30°$	0.044	.005	0.19	.02	12
....	....	U−B colour residuals of ≈ 450 gal.	0.051	.005	0.22	.04	12
1987	Sparks *et al.*	Balmer decrement in HII regions	<0.066	...	<0.28:	...	13

(1) Medd. Lund Obs. II, No. 136, transformed from 0.062 in P_g−P_v system, $A_B = 4.3E(B-V)$

(2) Medd. Lund Obs. II, No 136, transformed from $A_{pg} = 0.22$ in P_g system

(3) ApJS 48, 233, $|b|<30°$ in paper is a misprint

(4) corrected for period and line blanketing effects and for selection bias, see revision in (7)

(5) AJ 75, 695, corrected for selection bias, see revision in (6)

(6) A&A 35, 361, revision of (5). This is the value given in the text for E(B−V), the value (0.052) listed b Holmberg in his Table 4 is not E(B−V), but E(V−r)

(7) ref. (6), revision of (4)

(8) ref. (6), new analysis of (4), allowing for local departures from mean csc b law

(9) ref. (6), data after Jones 1973

(10) AJ 81, 687, HM = Hubble modulus, residuals of absolute M_{pg} on corrected Zwicky scale

(11) AJ 86, 1429. (12) AJ 88, 939. (13) MN 225, 769

Table 7B. Polar Extinction from Zero Point in Polar Caps

Year	Source	Method	E(B−V)	m.e.	A_B	m.e.	Rem.
1968	Philip	UBV and Strömgren colours of 15 B to A5 stars, $b > +65°$	0.05		0.21		1
1971	Bond et al.	B−V colours of 46 B,A stars near South Galactic Pole	0.029	.005	0.12	0.02	2
1971	Rodgers	UBV colours of 60 stars of types $\leq$ A7, $b < -81°$	0.04	.03	0.17	.13	3
1972	Pfau	Colours of 17 O,B stars at $\mid z \mid > 1$ kpc,$b \mid \geq 60°$	0.033	.008	0.14	.034	4
1972	Feltz	Colours of 182 B,A,F stars within 20° from NGP, $z \leq 1$ kpc	0.000	.002	0.00	.009	5
1975	Appenzeller	Polarisation of stars, $b > +70°$ $z \geq 140$ pc	> 0.011		> 0.05		6
....	....	.... of 13 stars, $b < -70°$ $z \geq 100$ pc	0.016		0.07		6
1976	Hilditch et al.	Colours of $\approx$ A-F stars 1000 $b > 75°$	0.01:		0.04:		7
1977	Knude	Strömgren colours of 158 A-F stars, $b < -52°$	0.057	.004	0.245	.017	8
1979	Markannen	Polarisation of 71 stars, $b \geq 80°$	$\geq$ 0.01:		$\geq$ 0.04		9
1980	Albrecht & Maitzen	Strömgren-Crawford colours of 90 B,A,F stars, $b \leq -79°$	0.025		0.11		10
1982	Nicolet	Geneva colours of 11 B-A stars $\mid b \mid > 70°$	0.04:		0.17		11
1985	Tobin	Strömgren colours of 72 B stars $z > 250$ pc	$\approx$ 0.00:		0.00:		12
1988	Boulanger & Perault	IRAS 100μm cirri $\mid b \mid > 50°$	0.030		0.13		13
1990	Hurwitz et al.	Far UV diffuse background	0.015		0.07		14

(1) AJ, 73, 1000. (2) PASP, 83, 643. (3) ApJ, 165, 581. (4) AN, 293, 195.
(5) PASP, 84, 497, but reddenings are 0.008± 0.008 mag too small by comparison with polarisation measurements, according to Markkanen (A&Ap, 1979, 74, 201)
(6) A&Ap, 1975, 38, 313. (7) MNRAS, 176, 175.
(8) ApL, 18, 115, probably an overestimate, see (10).
(9) A&Ap, 74, 201. General extinction + localized cloud ($A_B \geq 0.13$) at z = 100-200 pc, in fourth quadrant
(10) A&ApS, 42, 9. (11) A&ApS, 47, 199. (12) A&Ap, 142, 189.
(13) ApJ, 330, 964. (14) ApJ, 372, 167.

REFERENCES

Abell, G.O.,1958, ApJS, 3, 211

Appenzeller, I. 1975, A&Ap, 38,313

Bottinelli, L., Gouguenheim, L., Paturel, G. & de Vaucouleurs, G. 1983, A&Ap, 118, 4

Boulanger, F. & Perault, M. 1988, ApJ, 330, 964

Bruzual, G., Magris, G. and Calvet, N. 1988, ApJ, 333, 673

Burstein, D. & Heiles, C. 1978, ApJ, 225, 40

Burstein, D. & Heiles, C. 1982, AJ, 87, 1165

Cholonowski, J. 1991, MNRAS, 250, 486

Corwin, H.G., de Vaucouleurs, A. & de Vaucouleurs, G. 1985, 'Southern Galaxy Catalogue', Univ. of Texas Monograph in Astron. No 4

de Vaucouleurs, G. 1954, Observatory, 74, 23, 157

de Vaucouleurs, G. 1955a, Observatory, 75, 129

de Vaucouleurs, G. 1955b, AJ, 60, 126

de Vaucouleurs, G. 1960, Observatory, 80, 106

de Vaucouleurs, G. 1961, ApJS, 48, 233

de Vaucouleurs, G. 1993, ApJ, 415,10

de Vaucouleurs, G. and Buta, R. 1983, AJ, 88, 939

de Vaucouleurs, G. and Corwin, H.G. 1986, 308, 487

de Vaucouleurs, G. and Freeman, K.C. 1972, in Vistas in Astronomy, ed. A. Beer, Oxford: Pergamon Press, 14, 163

de Vaucouleurs, G., de Vaucouleurs, A. and Buta, R. 1981, 86, 1429

Elvius, A. 1956, Stockholm Obs. Ann., 18, Nr 9

Gurwell, M. & Hodge, P. 1990, PASP, 102, 849

Hauser, M.H. et al. 1984, ApJL, 278, L15

Haynes, M. & Giovanelli, C. 1984, AJ, 89, 758

Heiles, C. & Cleary, M.N. 1979, Austral. J. Phys., ApS, No 47

Heidmann, J., Heidmann, N. & de Vaucouleurs, G. 1971, Mem. RAS, 75, Parts 4-6

Henyey, L.G. & Greenstein, J.L. 1941, ApJ, 93, 70

Herschel, J. 1847, 'Results of Astronomical Observations ...', London: Smith, Elder & Co., 70

Hilditch, R.W., Hill, HG. & Barnes, J.V. 1983, MNRAS, 204, 241

Hodge, P.W. 1969, SAO Special Report 306

Hodge, P.W. 1974, ApJ, 192, 21

Hodge, P.W. & Snow, T.P. 1975, AJ, 80, 9

Holmberg, E. 1958, Medd. Lund Obs., II, Nr 136

Holmberg, E. 1974, A&A, 35, 121

Huizinga, J.E. 1994, Extinction Studies of Spiral Galaxies, Ph.D. thesis, Groningen Univ.

Hurwitz, M., Bowyer, S. & Martin, C. 1990, ApJ, 372, 167

Knude, J.K. 1977, ApL, 18, 115

Lindblad, B. 1942, Stockholm Obs. Ann., 14, Nr. 3

Low, F.J. et al. 1984, ApJL, 218, L19

Lynds, B.T. 1965, ApJS, 12, 163

Markannen, T. 1979, A&Ap, 74,201

McClure, R.D. & Crawford, D. 1971, AJ, 76, 31

McGee, R.X., Murray, J.D. & Milton, J.A. 1963, Austral. J. Phys., 16, 136

Nicolet, B. 1982, A&ApS, 47, 199

Pecker, J.-C., 1972, A&Ap, 18, 253

Sandage, A. 1973, ApJ, 183, 711

Sandage, A. 1976, AJ, 81, 954

Sandage, A. & Bedke, J. 1988, *Atlas of Galaxies*, NASA SP-496

Sandage, A. & Tammann, G.A. 1981, *A Revised Shapley-Ames Catalog of Bright Galaxies*, Carnegie Institution Publ. 635, Washington (2nd ed. 1987)

Savage, B.D. & de Boer, K.S. 1981, ApJ, 243, 460

Shane, C.D. & Wirtanen, C. A. 1967, Publ. Lick Obs., 22, Part I

Sparks, W.B. et al. 1987, MNRAS, 225, 769

Stark, A.A. et al. 1990, ApJS, ..., ...

Tobin, W. 1985, A&Ap, 142, 189

Valentijn, E.A. 1994, MNRAS, 266, 614

Wesselink, A.J., 1961a, MNRAS, 122, 503

Wesselink, A.J., 1961b, MNRAS, 122, 509

Zonn, W. 1968, Acta Astron., 18, 273

Zwicky, F., Herzog, E., Wild, P.m Karpowicz, M. & Kowal, C.T. 1962-68, *Catalogue of Galaxies and Clusters of Galaxies*, California Inst. of Technology, Pasadena, vols. 1-6

The pose I

PROPERTIES OF DUST IN BACKLIT GALAXIES

WILLIAM C. KEEL AND RAYMOND E. WHITE, III
Department of Physics and Astronomy
University of Alabama, Box 870324,
Tuscaloosa, AL 35487, U.S.A.

1. Introduction

Several recent studies have brought renewed interest in the dust content of spiral disks, and especially in how much optical obscuration is produced by the dust in a "typical" galaxy. Various aspects of this problem have been clarified by a variety of approaches: comparison of radiative-transfer models with observed color and surface-brightness data [1,2,3]; kinematics of edge-on galaxies as observed at various wavelengths [4]; comparison of images at widely disparate wavelengths such as B and K [5]; and the by-now notorious inclination-surface brightness test [6,7].

The authors' current interest in the opacity of spiral disks was inspired by Valentijn's [6] statistical reinvestigation of the classical inclination-surface brightness test. That Valentijn [6] found spirals to be largely opaque (since their surface brightnesses were independent of inclination) seemed counter-intuitive to us for two rather primitive reasons: 1) if spirals were opaque, then the galaxy survey used to deduce this would have been difficult to obtain, since we live in a spiral galaxy; 2) there are well-known examples of seeing more distant objects (galaxies, quasars, *etc.*) through foreground spiral galaxies (other than the Milky Way!). Thus, we set out to determine the opacity of spiral disks directly, rather than statistically, by searching for foreground spirals projected against background galaxies. The non-overlapping regions of a partially overlapping galaxy pair can be used to reconstruct, using purely differential photometry, how much light from the background galaxy is lost in passing through the foreground galaxy in the region of overlap.

Meanwhile, in a statistical reassessment of Valentijn's [6] work, Burstein *et al.* [7] concluded that Valentijn got the right answer for the wrong reason,

J. I. Davies and D. Burstein (eds.), The Opacity of Spiral Disks, 167–179.
© 1995 *Kluwer Academic Publishers.*

maintaining that the result was a product of sample selection effects. Using a sample claimed to be less subject to such selection effects, Burstein *et al.* [7] nonetheless found that spirals are optically thick (although not *opaque*, per se). Most recently, however, Burstein [8] now finds in an expanded sample that spirals are *not* so optically thick after all, since the new surface brightness sample exhibits a mild inclination dependence. Huizinga [9] has suggested that the Valentijn [6] result was confounded by the presence of bulge systems in the sample, the surface brightnesses of bulges being inherently more inclination-independent than those of spiral disks. In White and Keel [10], we found absorption to be strongly concentrated in the spiral arms of AM1316-241, which is also where most of the disk light is coming from in the *B* band; we therefore suggested that a spatial correlation between emission and intrinsic absorption in spiral disks may moderate any inclination-dependence in apparent surface brightness, thereby accounting for Valentijn's [1] result.

In light of the ongoing controversy over selection effects in statistical samples, there are several benefits to the direct, differential photometric approach we have adopted:

- it is not subject to the selection effects influencing the statistical studies cited above
- there is no selection against high opacity regions, as there is in some spectroscopic studies of background objects shining through foreground disks (*e.g.* quasars or HII regions in a background galaxy — see James and Puxley [11])
- the imaging technique involves only differential photometry (apart from very mild wavelength-dependent scattering effects)
- large, contiguous areas can be analyzed, allowing average values of the opacity to be estimated (whereas spectroscopic studies of background HII regions or quasars probe relatively few points in a foreground disk)
- there is no need to correct for the internal extinction of the background galaxy or the Milky Way (which is required in some spectroscopic studies of background objects shining through foreground disks)

This technique also has some disadvantages relative to others:

- there are rather few tractable objects nearby enough for spatially well-resolved analysis
- the success of the technique hinges on the degree of symmetry in both the foreground and background galaxies

Our extinction values differ in a significant way from those derived from internal galaxy properties. We measure the dust content in an area-weighted sense, whereas any technique relying on a galaxy's own radiation measures the dust content weighted by the distribution of starlight and dust in the galaxy itself. Our values are directly relevant, for example, to calculations of the cumulative effect of disks on optical quasar counts, but may not be the appropriate ones for calculating Tully-Fisher corrections, depending on the relative distributions of stars and dust.

It is useful to distinguish several regimes of galaxy backlighting, depending on the apparent sizes of the galaxies and the impact parameter of the background light. Quasars in the fields of galaxies, and the many background galaxies accessible at faint magnitudes, represent one limiting case, where the background source is much smaller than the foreground galaxy. These give perfectly respectable measures of extinction, and ones in which scattering into the beam is negligible, but we do not always know how many small background objects that we don't see — that is, use of such small probes is biased toward the clearest lines of sight. Much better are overlaps of galaxies with similar angular size, so that we are not vulnerable to this particular selection effect. Best of all are edge overlaps, with the opposing parts of the galaxies seen free of mutual interference and useful in measuring a mean brightness profile and evaluating symmetry. Central superpositions (such as NGC 3314, [12]) are good for probing the central regions of disks, but there is little constraint on the disk brightness profiles so that results are necessarily limited in accuracy and radial extent. Finally, spirals seen nearly edge-on may have their disks backlit by their own outer bulges (as analyzed by van Houten [13] and recently by Simien et al. [14]). These cases have very well-understood geometry, but effects of scattering can be more important than in galaxy pairs.

2. Constructing opacity maps

Figure 1 depicts the ideal case for constructing maps of opacity using differential photometry: a foreground disk (spiral) galaxy is half-projected against half of a similarly-sized background elliptical galaxy. For the sake of illustration, the (unobscured) surface brightness of each galaxy is taken to be constant, with S and E being the actual surface brightness values of the foreground disk and background elliptical in the overlap region, and τ is the optical depth in the disk. The observed surface brightness in the overlap region is then $\langle S + Ee^{-\tau} \rangle$, where brackets are used to emphasize that this whole quantity is the observable in the overlap region and cannot be directly decomposed into its constituent components. We use symmetric counterparts from the non-overlapping regions of the two galaxies to *estimate* S and E, and denote the estimates as S' and E'. We can then construct an estimate of the optical depth, denoted τ':

$$e^{-\tau'} = \frac{\langle S + Ee^{-\tau} \rangle - S'}{E'}. \tag{1}$$

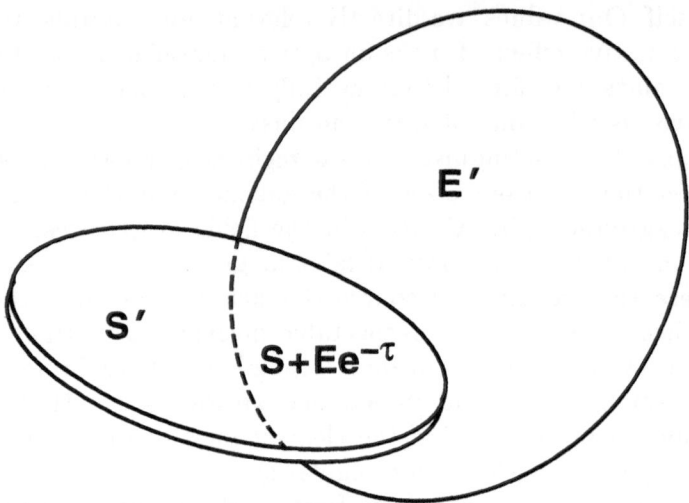

Figure 1. Schematic of the ideal overlapping galaxy pair for the direct measurement of opacity in the foreground spiral.

Here the estimate of the foreground spiral's surface brightness, S', is first subtracted from the surface brightness of the overlap region, $\langle S + Ee^{-\tau} \rangle$; this result is then divided by the estimate of the background elliptical's surface brightness, E'. This creates a map of $e^{-\tau'}$ in the overlap region.

Although it is impossible to actually do so, it is *formally* useful to "break" $\langle S + Ee^{-\tau} \rangle$, the observed surface brightness in the overlap region, into its constituent parts to assess the systematic errors of the above construction:

$$e^{-\tau'} = \frac{(S - S')}{E'} + \frac{E}{E'}e^{-\tau}. \tag{2}$$

The systematic errors induced by departures from symmetry can be estimated from the non-overlapping parts of the galaxies. Note that when the background galaxy has substantially higher surface brightness than the foreground galaxy ($S, S' \ll E'$), the estimate of τ is particularly insensitive to systematic errors induced by deviations from symmetry in the foreground spiral. In this case,

$$e^{-\tau'} \approx \frac{E}{E'}e^{-\tau}, \tag{3}$$

with E/E' being especially close to unity for most ellipticals (and S0s). Also, a lower limit to τ' is provided by simply dividing the overlap region by the symmetric counterpart of the background galaxy and neglecting to scrape off the emission from the foreground galaxy:

$$e^{-\tau'} < \frac{\langle S + Ee^{-\tau} \rangle}{E'} \qquad \Rightarrow \qquad \tau' > -\ln\frac{\langle S + Ee^{-\tau} \rangle}{E'} \qquad (4)$$

Depending on the inclination of the foreground galaxy, different symmetries are used for scraping off the emission due to the foreground spiral in the overlap region: if the spiral is nearly face-on, rotation symmetry is used to swing the unprojected portion around for subtraction; if the foreground spiral is instead more edge-on, its finite disk thickness may require reflection symmetry to be used to flip the unprojected portion of the spiral over for subtraction. Incidentally, a color map allows the orientation of the spiral to be determined since the near-side edges are bluer. Corrections for forward-scattered light from the background galaxy are discussed in [10] and below and are usually found to be minimal.

We explicitly model the effects of scattering on our deduced opacities. Scattering effects may require a significant correction to our optical depth estimates if the amount of light from the background galaxy scattered into the beam is non-negligible in comparison with the fraction reaching us directly; for example, if the background object has a strong intensity gradient within the peak of the scattering function. Our estimate is an upper limit, derived under the assumption that the dust is uniformly spread (giving the largest scattering-to-extinction ratio). We take the major-axis profile of the background galaxy and assume it to be spherical, thus again giving a maximal background level and scattering correction whatever the intrinsic galaxy shape. For various assumed values of the line-of-sight separation, we numerically integrate the scattered-light intensity over all scattering angles; our calculations so far adopt a Henyey-Greenstein phase function $(1 - g^2)/(1 + g^2 - 2g\cos\theta)^{1.5}$, with grain parameters as used by Bruzual et al. [15]. Since we remove the symmetric part of scattered light in a galaxy along with its overall light profile, we are concerned only with the differential scattering correction from one side of the galaxy to the other. This drops dramatically as the line-of-sight galaxy separation increases, by a factor 300 from one to ten times the projected separation in the specific case of AM0500-620. We often have only indirect arguments for the third dimension of separation between pair members; if the isophotes are not measurably disturbed, we can generally assume that the galaxies are far enough apart that they do not physically overlap at the relevant isophotal level, so that they are at least several times their projected separation apart.

3. Galaxy Sample

Suitable galaxy pairs for such analysis are rare; were it not for the galaxy-galaxy correlation function, we would expect virtually no useful nearby candidates. Our observing sample is drawn from a variety of sources, including: numerical searches for overlapping neighbors in the RC3, UGC, and RSA catalogues, the Karachentsev catalogue of galaxy pairs [16], and the Chinese catalogue of double galaxies [17]; visual inspection of the Arp-Madore catalogue [18], the Arp *Atlas of Peculiar Galaxies* [19], and POSS prints of some of the richest galaxy clusters (*e.g.* Shapley 8 in the Shapley Supercluster); and anecdotal lore.

Promising candidates were observed using CCD cameras at Kitt Peak, Cerro Tololo, Lowell, and the European Southern Observatory. We have so far observed 34 galaxy pairs, of which ~ 8 are tractable enough for detailed analysis. We have also observed 3 "peeking bulge" galaxies — nearly edge-on spiral galaxies whose bulges provide backlighting for and can be seen on either side of their intervening disks; such individual systems can be analyzed in a similar, but not identical, fashion as overlapping pairs. We imaged in both B and I bands in order to derive color information and to determine the wavelength-dependence of the opacity.

Finally, we have been able to derive limited information on the spectacular superposition NGC 3314 [12,20,21]. In this case we cannot separate the brightness profiles of the two spirals, as their nuclei coincide to within a few arcseconds. However, by removing a minimum-disk exponential profile from the background galaxy, sharp enough edges are left on its spiral features that simple on/off comparisons for adjacent parts of the arms of the foreground spiral yield useful data; this is our only overlap probing the inner parts of disks (within about 0.3 R_{25}), plus evidence that the background nucleus is seen as a heavily reddened condensation.

4. Individual Systems

4.1. AM1316-241

As reported in White & Keel [10], our best case thus far is AM1316-241, an Arp-Madore catalogue object consisting of a foreground Sbc projected against a background elliptical. Figure 2 shows the B-band image of this pair, which is also interesting because the foreground spiral's recession velocity (10365 km s^{-1}) is 660 km s^{-1} *larger* than that of the background elliptical. The axial ratio of the foreground spiral is 4.42, implying an inclination of 77°.

The symmetry of each of the two galaxies is good enough that we can employ the cut-and-paste technique described above to estimate the opac-

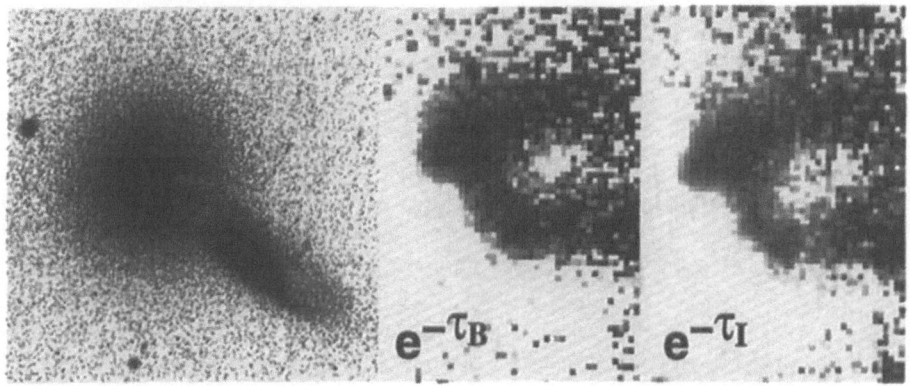

Figure 2. *B*-band image of AM1316-241; maps of $e^{-\tau_B}$ and $e^{-\tau_I}$ shown with the same absolute intensity scale.

ity over a relatively large fraction of the overlap region. Figure 2 also shows maps of $e^{-\tau'}$ in the *B* and *I* bands shown on the same absolute intensity scale, where darker regions are more opaque. The opacity is clearly concentrated in the spiral arm, while the interarm region is nearly transparent. It is also obvious from Figure 2 that the arm is optically thicker in *B* than in *I*. Table 1 lists the face-on-corrected optical depth in the arm and interarm regions, as well as for an average over the disk area seen in the $e^{-\tau'}$ maps of Figure 2. The face-on-corrected optical depths are found by dividing the apparent optical depths by the galaxy's axial ratio (4.42). The resulting optical depths are rather small: $\tau_B = 0.35$ in the arm region and 0.075 in the interarm region, while $\tau_I = 0.15$ and 0.05 in the arm and interarm regions, respectively. The arm is at 0.75 R_{25}^B (where R_{25}^B is the radius at which the blue surface brightness $\mu_B = 25$ mag arcsec^{-2}) while the measurable disk region extends from 0.37-0.75 R_{25}^B. The radial extents of these various regions are also given in Table 1.

4.2. NGC 1738/9

Figure 3a shows a *B* image of the Sbc pair NGC 1738/9. The symmetry of this system is not good enough to do an opacity analysis in the same detail as for AM1316-241. Instead, the two regions indicated in Figure 3a are investigated: a foreground arm region at 0.65 R_{25}^B and an interarm region at 0.55 R_{25}^B. Symmetric regions in the foreground and background galaxies are used to infer the apparent optical depths in *B* and *I* in these

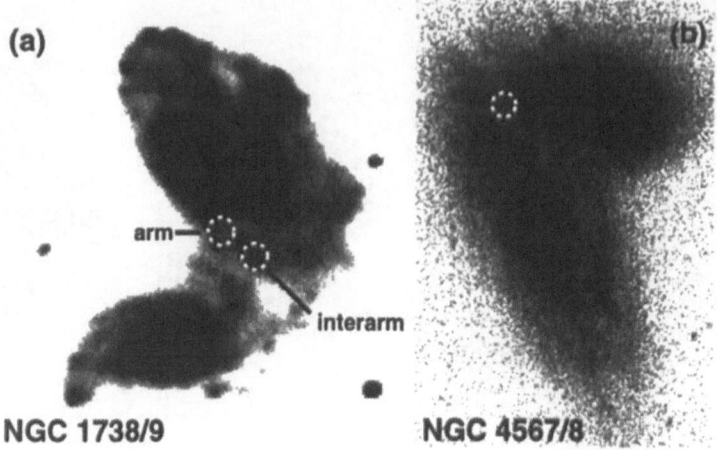

Figure 3. *B*-band images of: a) NGC 1738/9; b) NGC 4567/8.

two regions. Since the axial ratio of the foreground galaxy (NGC 1739) is 1.95 (indicating an inclination of 59°), the apparent values of τ are divided by 1.95 to give the face-on-corrected values listed in Table 1. The face-on optical depths are again quite low: in the arm region, $\tau_B = 0.3$ and $\tau_I = 0.2$, while in the interarm region, $\tau_B = 0.2$ and $\tau_I = 0.15$.

TABLE 1. Face-on Optical Depths τ

Galaxy Pair	τ R/R_{25}^B	arm	interarm	disk average
AM 1316-241	τ_B	0.35	0.075	0.175
	τ_I	0.15	0.05	0.125
	R/R_{25}^B	0.75	0.4-0.7	0.4-0.75
NGC 1738/9	τ_B	0.3	0.2	
	τ_I	0.2	0.15	
	R/R_{25}^B	0.65	0.55	
NGC 4567/8	τ_B	1.1		
	τ_I	0.6		
	R/R_{25}^B	0.5-0.85		

4.3. NGC 4567/8 (UGC 7777/6)

The Sbc pair NGC 4567/8 (UGC 7777/6), shown in Figure 3b, is another case where the analysis is limited by the general lack of symmetry. Here we concentrate on the dark lane (indicated by a dotted circle in the figure) cutting across a brighter background galaxy arm. The comparison region for the foreground arm is taken from a region along the arm but beyond the projected bulk of the background galaxy; the comparison region for the background arm is along the background arm, just away from where it is blocked by the foreground galaxy. The foreground galaxy (NGC 4568/UGC 7776) has an axial ratio of 2.29, implying an inclination of 64°. The assessed region in the foreground galaxy samples, in projection, a range of radii spanning 0.5-0.85 R_{25}^B. We calculate face-on optical depths of $\tau_B = 1.1$ and $\tau_I = 0.6$ for this region (see Table 1), which are substantially larger than in the previous two systems.

Four additional sample pairs are shown in Fig. 4. NGC 1232B is a small background system (at $z = 0.1$) seen through interarm regions in the disk of the bright Sc I galaxy NGC 1232. AM 0500-620 shows a face-on spiral with outer dust structure traced by a background elliptical, in nearly ideal circumstances for extinction measurements. NGC 3314 is a central superposition of a face-on spiral in front of a more inclined spiral, allowing us to trace the dust structures to radii inward of those available in the other systems. UGC 2942/3 illustrates the use of a background edge-on spiral; the dust lane in the background galaxy provides a sharp marker for the presence of transmitted light in the overlap region.

5. Summary

At this point, we have absolute extinction measures for eight galaxy pairs. For each pair, there is some range in radius for which we can measure the residual intensity due to background light transmitted by the foreground disk. We translate these measures into arm and interarm extinctions (where such a distinction is possible) in both B and I bands. In all cases, there is a large difference between arm and interarm values. The systems described here have face-on optical depths ranging a factor of $\sim 3 - 4$. In arm regions, $\tau_B \approx 0.3 - 1.1$ and $\tau_I \approx 0.15 - 0.6$, while in interarm regions, $\tau_B \approx 0.08 - 0.2$ and $\tau_I \approx 0.05 - 0.15$. Table 2 summarizes the pairs and regions for which extinction measurements have been made. The radius range refers to the region over which useful extinction measurements can be made, in units of the isophotal radius (to the mean level $B = 25$ mag arcsec^{-2}) as measured from our CCD data.

The interarm ("disk") extinction shows a systematic trend with radius (Fig. 5), reaching 1 magnitude at B only within about 0.3 R_{25}. In contrast,

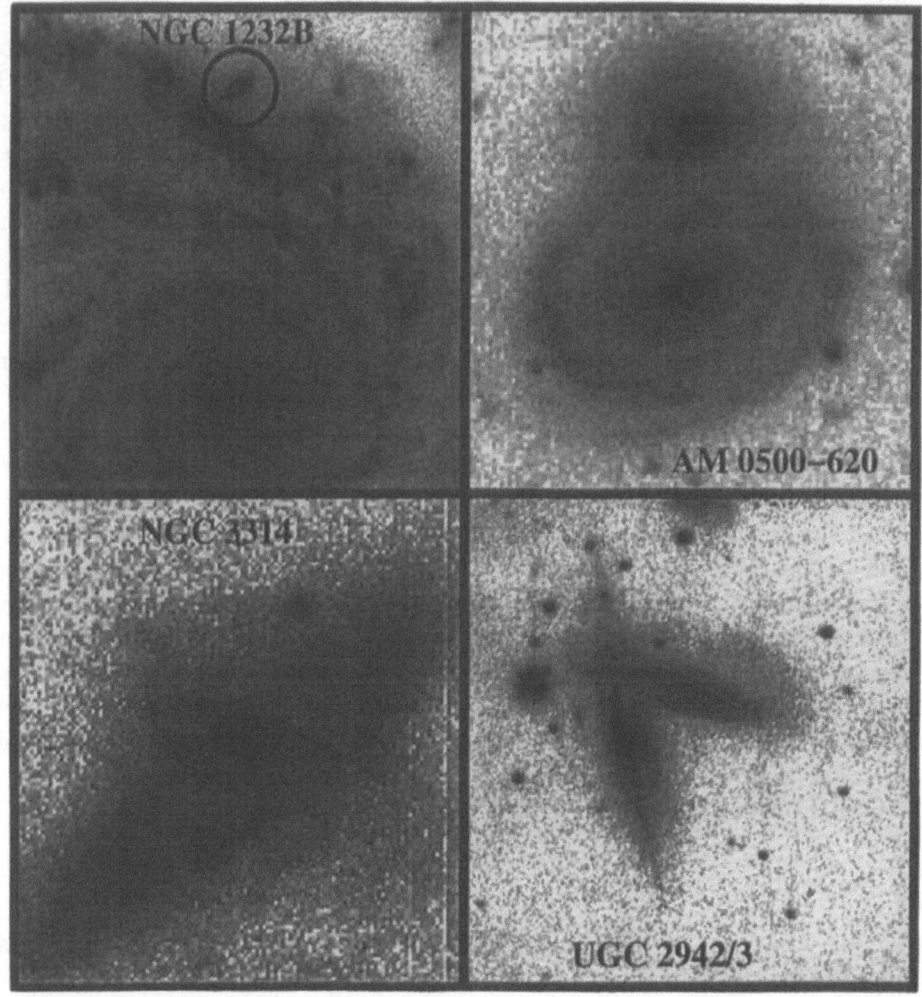

Figure 4.   Sample galaxy-overlap regimes, in images from the CTIO 1.5-m telescope.

spiral arms can be optically thick at almost any radius. We do not see evidence for substantial extinction in the outer parts of disks, even though such dust should have been very prominent in the initial screening of pairs for our study (as well as in our results themselves).

It is also common to see an extinction curve in backlit parts of galaxies which is flatter ("greyer") than the Galactic curve, with $A_B/A_I \sim 1.5$ in some systems. This may be a signature of clumped absorbing material, and in fact we already resolve some structure at the level of hundreds of parsecs. The distribution of dust in nearby galaxies leads us to expect structure on considerably finer scales as well, so that the effective extinction averaged over large regions may bear little relation to the intrinsic properties of the

TABLE 2. Overlapping Galaxy Pairs with Extinction Measures

Foreground	cz, km s^{-1}	Background	R/R_{25}
AM1316-241S	10365	AM1316-241E	0.45–0.6
NGC 4567	2199	NGC 4568	0.5–0.8
NGC 4647	1386	NGC 4649	1.0–1.1
UGC 2942	6270	UGC 2943	0.5–0.6
NGC 1739	3892	NGC 1738	0.5–0.6
NGC 450	1851	UGC 807	0.95–1
NGC 3314A	2763	NGC 3314B	0.3–0.4
AM0500-620S	–	AM0500-620E	0.5–0.6

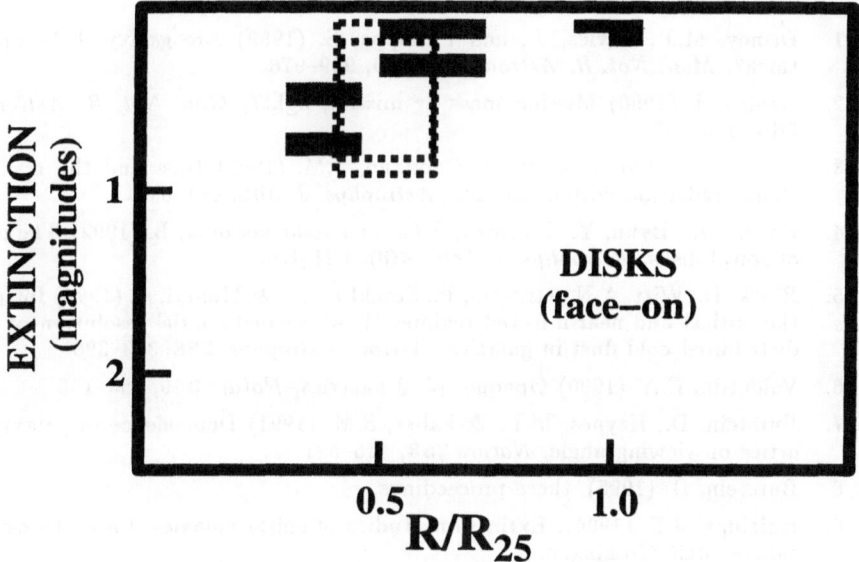

Figure 5. Interarm extinction values from backlit spirals. *B* band values are in black, with *I*-band values in gray.

grains.

Our initial results on AM1316-241 led us to conclude that opacity is concentrated in spiral arms, and that interarm regions are fairly transparent. Our newer work is generally consistent with this picture. Therefore, the distribution of absorption tends to be spatially correlated with particularly emissive regions, since spiral arms are brighter than interarm regions. We suggested in [10] that this spatial correlation between internal extinc-

178

tion and emission may account for the statistical results reported in earlier studies — that surface brightness is roughly independent of inclination.

We will report elsewhere on our studies of "peeking bulge" systems, since their analysis is more subtle. To avoid underestimating the optical depths in the intervening disks, one must be sure to scrape off the emission due to the intervening disk, which is difficult to estimate from the far side; furthermore, such systems are likely to have forward-scattered bulge light "fill in" much of the true absorption.

We acknowledge support from EPSCoR grant EHR-9108761, and allocations of telescope time at Kitt Peak and Cerro Tololo (operated by AURA, Inc., under cooperative agreement with the National Science Foundation); at Lowell Observatory, and at the European Southern Observatory.

References

1. Disney, M.J., Davies, J., and Phillipps, S. (1989) Are galaxy disks optically thick?, *Mon. Not. R. Astron. Soc.* **239**, 939–976

2. Davies, J. (1990) Missing mass or missing light?, *Mon. Not. R. Astron. Soc.* **245**, 350–357

3. Witt, A.N., Thronson, H., & Capuano, J.M. (1992) Dust and the transfer of stellar radiation within galaxies, *Astrophys. J.* **393**, 611–630

4. Bosma, A., Byun, Y., Freeman, F.C., and Athanassoula, E. (1992) The opacity of spiral disks, *Astrophys. J. Lett.*, **400**, L21–L24

5. Block, D., Witt, A.N., Grøsbol, P., Stockton, A., & Moneti, A. (1994) Imaging in the optical and near-infrared regimes II. Arcsecond spatial resolution of widely distributed cold dust in galaxies, *Astron. Astrophys.* **288**, 383–395

6. Valentijn, E.A. (1990) Opaque spiral galaxies, *Nature* **346**, 153–155

7. Burstein, D., Haynes, M.P., & Faber, S.M. (1991) Dependence of galaxy properties on viewing angle, *Nature* **353**, 515–521

8. Burstein, D. (1995), these proceedings.

9. Huizinga, J.E. (1994), Extinction studies of spiral galaxies, Ph.D. thesis, Rijksuniversiteit Groningen.

10. White, R.E., III, & Keel, W.C. (1992) Direct measurement of the optical depth in a spiral galaxy, *Nature* **359**, 129–131

11. James, P.A. & Puxley P.J. (1993), A measurement of the optical depth through a galaxy disk, *Nature*, **363**, 240–242.

12. Keel, W.C. (1983) Dust in backlit galaxies: properties of the foreground systems in NGC 3314 and NGC 1275, *Astron J.* **88**, 1579–1586

13. van Houten, C.J. (1961) Surface photometry of extragalactic nebulae, *Bull. Astron. Inst. Neth.* **16**, 1–69

14. Simien, F., Morenas, V., & Valentijn, E.A. (1993) On the transparency of the inner regions of early-type spiral galaxies, *Astron Astrophys.*, **269**, 111–118

15. Bruzual A., G., Magris, G., & Calvet, N 1988,. *Astrophys J.* **333**, 673–688

16. Karachentsev, I.D. (1972) Catalog of isolated pairs of galaxies in northern galactic hemisphere, *Soobsch. Spets. Astrof. Obs.* **7**, 3–91

17. Zhenlong, Z., Jiansheng C., Xiaoying T., Yulin, B. (1989) Catalogue of Double Galaxies in the South Galactic Cap, *Publ. Beijing Astronomical Observatory*, **12**, 8-37.

18. Arp, H.C. & Madore, B.F. (1987) *A Catalogue of Southern Peculiar Galaxies and Associations* (Cambridge)

19. Arp, H.C. (1966) Atlas of peculiar galaxies, *Astrophys. J. Suppl.* **14**, 1-20

20. Schweizer, F. & Thonnard, N. (1985) The two superposed galaxies of NGC 3314, *Publ. Astron. Soc. Pacific*, **97**, 104–109

21. McMahon, P.M., Richter, O.-G., van Gorkom, J.H.& Ferguson, H.C. (1992) H I imaging of NGC 3312 and NGC 3314a: a foreground group to the Hydra cluster?, *Astron. J.* **103**, 399–404

Question
Valentijn

As you find that much structure in the 'opacity maps' it is probably more reasonable to summarise your results without applying a 'screen model' face-on correction. Some of your studies apply to spiral arms which might have a cylindrical structure and the correction does not hold.

Answer
Keel

At least for the more diffuse interarm material, some kind of slab geometry seems more plausible than an alternative. I agree that the spiral-arm dust features have poorly-known geometry, and for the arm, it is not clear how we should relate observed and "face-on" properties. We used the simple recipe for purposes of comparison.

Question
D Zaritsky

Have you used images of spiral galxies to measure the uncertainty introduced by your reflection or rotation techniques of determining the flux in the obscured region?

Answer
Ray White

We compared regions on either side of the symmetry axis, but well away from the overlap region, to estimate the systematic error induced by departures from perfect symmetry. We estimated this systematic error in $e^{-\tau}$ to be ≤ 0.1 in B and ≤ 0.03 in I.

THE OPTICAL DEPTH THROUGH NGC 3314A

PHIL JAMES
Liverpool John Moores University

AND

PHIL PUXLEY
Royal Observatory, Edinburgh

1. Introduction

Most observational studies of galaxy optical depths have been statistical
in nature, based on large samples of galaxies, but generally indirect and
model-dependent. By contrast, we have applied a much more direct test of
optical depth to a single object, exploiting an unusual aligned pair of spiral
galaxies, NGC 3314a and b, to measure the optical depth through the disc
of the nearer member. This galaxy, NGC 3314a, is a face-on Sc, with a
redshift of 2835 kms^{-1}, through which can be seen NGC 3314b, an inclined
galaxy of somewhat earlier type, possibly Sb, and which has a redshift of
4641 kms^{-1}. Both galaxies appear undisturbed, indicating that they are
well-separated in space, as indicated by the difference in their redshifts.
They are remarkably well aligned: Keel (1983) resolves the nucleus of NGC
3314b 2-3″ NW of the nucleus of NGC 3314a.

We used the Balmer decrement, Hα/Hβ, of HII regions in the background galaxy to determine the optical depth through NGC 3314a. The
redshift difference of the two galaxies is large enough to separate completely
their emission in these lines. Thus we are measuring extinction using light
that must have travelled right through a galaxy disc, removing any uncertainties due to star and dust distributions.

2. Data reduction and analysis

The spectra presented here were obtained at the University of Hawaii 2.24
metre telescope on Mauna Kea with the Faint Object Spectrograph. The
slit was aligned with the major axis of NGC 3314b, 140° east of north, and

181

J. I. Davies and D. Burstein (eds.), The Opacity of Spiral Disks, 181–183.
© 1995 *Kluwer Academic Publishers.*

was centred on the optical nucleus. There was no attempt to position the slit on specific HII regions in the background galaxy.

Hα emission was detected from four regions in NGC 3314b, the nucleus and 3 HII regions. Two of these regions were to the NW of the nucleus, by 10.5″ and 26″, and one 11.5″ to the SE. These correspond to radial distances of 1.45, 2.6 and 1.6 h^{-1} kpc at the distance of NGC 3314a, where h is the Hubble constant in units of 100 kms^{-1}Mpc^{-1}. Hβ emission was clearly detected from all 3 off-nuclear regions, but not from the nucleus.

The line ratios were converted to extinction values by assuming that the intrinsic ratios are given by Case B with an electron density of 100 cm^{-3} and temperature of 10^{4}K, giving an emitted ratio of 2.86. Assuming a standard extinction law, the total extinction between the telescope and the HII regions in NGC 3314b can be calculated. This is listed in column 4 of Table 1. The extinction external to the Galaxy, which is listed in the final column of Table 1, was calculated by subtracting a Galactic component, taken from Burstein and Heiles (1984).

TABLE 1. Line fluxes (10^{-19}Wm^{-2}) and A$_V$ values.

Region	Hα	Hβ	A$_{V,TOT}$	A$_{V,EXT}$
1	55.6	10.0	1.72±0.33	1.58±0.33
2	105.7	17.4	1.94±0.33	1.80±0.33
3	62.8	<7.9	>2.64	>2.5
4	31.7	6.0	1.58±0.34	1.44±0.34

We accounted for extinction internal to the source galaxy by comparing the extinction values towards NGC 3314b with those for HII regions in well-studied galaxies from the literature. Belley and Roy (1992) present CCD narrow-band photometry at Hα and Hβ for 130 HII regions in NGC 628 and 160 HII regions in NGC 6946. We calculated A$_V$ values for each of the HII regions, using the same method as for NGC 3314.

The corrected A$_V$ distributions for NGC 628 and NGC 6946 both have the bulk of their points between 0 and 0.5 magnitudes, with a high-A$_V$ tail stretching to 1.5 magnitudes. However, all 3 of the off-nuclear HII regions in NGC 3314b have A$_V$ values of >1.4, and thus seem highly unlikely to come from the same distribution (probabilities of 2×10^{-4} and 8×10^{-4}). In both cases, it is clear that the extinction values toward NGC 3314b are significantly higher than for the comparison galaxies (Figure 1).

We did a further test to check whether the inclination of NGC 3314b could be contributing to the extinction. Garnett & Shields (1987) give line

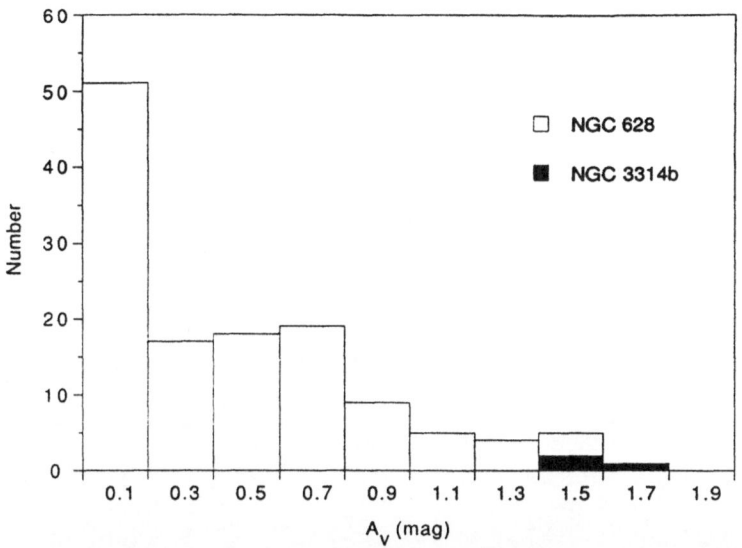

Figure 1. The distribution of A_V values for 130 HII regions in NGC628 (unshaded) compared with the 3 HII regions in NGC3314b (solid shading).

fluxes for 18 HII regions in M81, which has an inclination of 60°. The derived A_V values were found to be entirely consistent with the distributions for NGC 628 and NGC 6946, showing that inclination effects are small.

3. Conclusions

The excess extinction towards NGC 3314b appears to be at least 0.8 magnitudes for marginal statistical consistency, with the best-fit value being much larger, 1.3 magnitudes. The obvious interpretation is that this excess represents the extinction through the disc of NGC 3314a. This value, corresponding to an A_B of about 1.7 magnitudes, is in agreement with the large extinction values preferred by Disney et al. (1989) and Valentijn (1990), and is certainly discrepant with optically thin models. The lines-of-sight observed are possibly biased towards low A_V values by selection effects, and there may be regions in the disc with significantly higher extinction. Adapted, with permission, from *Nature* **363**, 240. Copyright 1993 Macmillan Magazines Limited.

References

Belley, J. & Roy, J.-R., 1992. *Astrophys. J. Suppl.*, **78**, 61.
Burstein, D. & Heiles, C., 1984. *Astrophys. J. Suppl.*, **54**, 33.
Disney, M., Davies, J. & Philipps, S., 1989. *Mon. Not. R. astr. Soc.*, **239**, 939.
Garnett, D.R. & Shields, G.A., 1987. *Astrophys. J.*, **317**, 82.
Keel, W.C., 1983. *Astron. J.*, **88**, 1579.
Valentijn, E.A., 1990. *Nature*, **346**, 153.

DUST EXTINCTION IN HIGHLY INCLINED SPIRALS

J.H. KNAPEN[1] AND J.E. BECKMAN
Instituto de Astrofísica de Canarias, E-38200 La Laguna,
Tenerife, Spain
[1] *Present Address: Université de Montréal, Dépt. de Physique,*
C.P. 6128, Succ. Centre Ville, Montréal, Québec, H3C-3J7
Canada

AND

R.A. JANSEN, R.F. PELETIER AND R. HES
Kapteyn Astronomical Institute, Postbus 800,
NL-9700 AV Groningen, The Netherlands

Abstract.
We recently studied a small number of highly inclined spiral galaxies, and used their special geometry to determine absolute extinction values of the dust in their dust lanes. We apply a simple model of a uniform mixture of stars and dust to the data, and find that the relative extinction values in these galaxies are the same as those making up the Galactic extinction law. This Galactic law seems to be universally applicable.

An important question relating to the subject of opacity of spiral discs is whether the properties of the dust in external galaxies differ widely, or even at all, from those in our own Galaxy. It is not obvious that one can assume that the Galactic extinction law is in fact universal when discussing dust in external spirals.

In a previous study of the Sombrero galaxy (Knapen *et al.* 1991) we looked at the dust properties in the dust lane that made the galaxy so well-known. The high degree of symmetry and the special geometry of this galaxy: highly inclined, but not edge-on, so that the dust lane does not obscure the centre of the galaxy, allowed us to obtain absolute dust extinction values in the dust lane by simply subtracting the unobscured half of the minor axis light profile from the half that is obscured by the dust lane, only assuming intrinsic underlying symmetry in the light distribution. By

185

J. I. Davies and D. Burstein (eds.), The Opacity of Spiral Disks, 185–188.
© *1995 Kluwer Academic Publishers.*

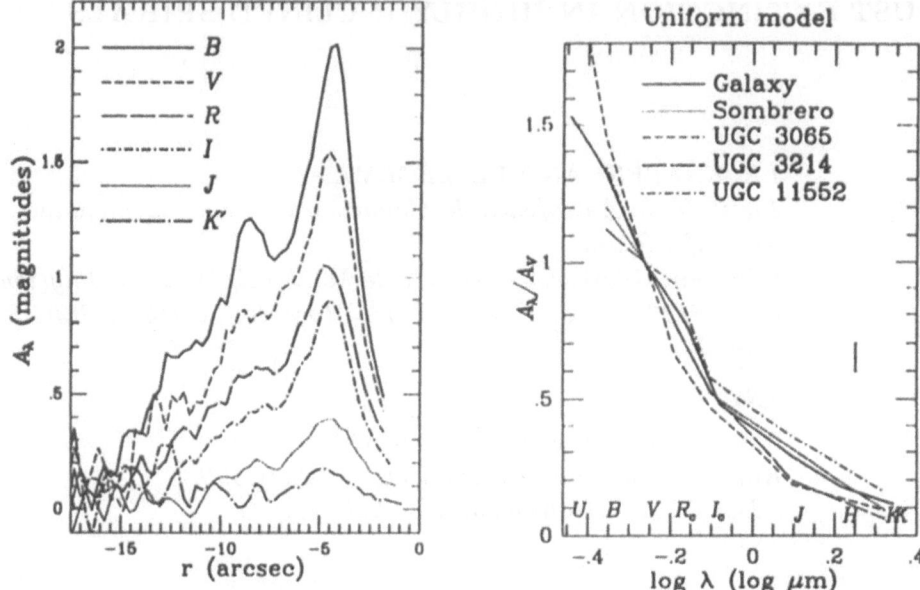

Figure 1. (left): Extinction profiles perpendicular to the major axis, over the dust lane, in UGC 3065, as a function of distance from the centre r. From top to bottom: optical B, V, R and I and NIR J and K' bands. *Figure 2.* (right): Extinction ratios A_λ/A_V as functions of the logarithm of the wavelength for the observed data of UGC 3065, UGC 3214 and UGC 11552, fitted with a model where stars and dust are uniformly mixed in the dust lane. Results are compared to the extinction law for our Galaxy (solid line) and for the Sombrero galaxy.

doing this in optical and near-infrared (NIR) bands, and assuming a simple model where the dust is uniformly mixed with the stars (see below), we could conclude that the dust properties were in fact very similar to those of the dust in the Galaxy.

Recently we studied a few more highly inclined spiral galaxies with a geometry resembling that of the Sombrero galaxy: a dust lane along the major axis, but slightly offset from the centre of the galaxy (Jansen *et al.* 1994). We obtained images in the optical B, V, R and I bands, and in the NIR J and K' bands of the four galaxies, UGC 3065, UGC 3214, UGC 11552 and MCG 02-10-009. In all the galaxies it is clearly seen that although the dust lane cuts out significant emission from one half of the galaxy in the bluer optical bands, the effects of dust extinction gradually become less important toward the red, and almost completely disappear in the K band.

We aligned the frames using positions of foreground stars, and after determining the correct major axis position, subtracted the unobscured part of the minor axis light profile from the dust-obscured part, thus obtaining extinction profiles across the dust lanes, for each photometric band that was studied. In Figure 1 we show these differential profiles for one of the galaxies, UGC 3065. The extinction is given in absolute magnitudes, which are derived as an intrinsic product of the differential method used to produce the extinction profiles. The extinction is strongest, as expected, in the B band, with a peak value of $A_B = 2.3$ mag, and gradually weakens while moving towards longer wavelengths. From these dust extinction profiles we then derived, for each galaxy, ratios A_λ/A_V at different distances from the centre, and compared these values to those for the Galactic extinction law.

In fact, measured values of A_λ against A_V for these three galaxies are along lines that significantly deviate from the "Galactic" line in some cases. To explain these deviations, we have invoked a simple "uniform" model, following Walterbos & Kennicutt (1988) and Disney *et al.* (1989), where we assume that the dust and the stars in the dust lane are uniformly mixed. The results of fitting the data points with such a model are shown graphically in Figure 2, where we plot the fitted values of the extinction ratios A_λ/A_V as a function of the logarithm of the wavelength, indicating along the abscissa the location of the photometric bands used in our imaging. We have also included in the plot the Galactic extinction law, and the results found in our earlier work for the Sombrero galaxy. The main conclusion is that for all galaxies, within the uncertainties, well fitting, simple, models can be made that follow the Galactic extinction law. There are some deviations, notably in the blue, which could well be due to scattering of light within the galaxies, an effect we have not studied in detail in the present work. In the case of UGC 11552, the slight deviations from the Galactic law are possibly a result of a less-than-perfect determination of the position of the centre. In that galaxy, the dust lane shows its effects on the light distribution up to quite small distances from the centre. But in general the results confirm that we can use the Galactic reddening law in external galaxies, so that it is almost justifiable, and certainly tempting, to talk of a "universal dust extinction law", at least in the visible and NIR ranges.

References

1. Disney, M., Davies, J.I. and Phillips, S., 1989, *Mon. Not. R. Astr. Soc.* **239**, 939
2. Jansen, R.A., Knapen, J.H., Beckman, J.E., Peletier, R.F. and Hes, R., 1994, *Mon. Not.R. Astr. Soc.* **270**, 373
3. Knapen, J. H., Hes, R., Beckman, J. E. and Peletier, R. F., 1991, *Astr. Astrophys.*, **241**, 42
4. Walterbos, R.A.M. and Kennicutt, R.C., 1988, *Astr. Astrophys.* **198**, 61

Question
Valentijn

As your method is a differential method, would you detect any hypothetical grey absorbing component or is the method most sensitive to a dust component obeying a reddening.

Answer
J Knapen

A grey absorbing component with the same geometrical distribution as the "normal dust", i.e. concentrated in the dust lane, would give a residual absorption in the K-band, which is not observed. In order not to be seen with this method, any significant grey component would have to be distributed extremely smoothly over the whole galaxy. This is not likely to be the case given the normal colours of our sample galaxies as compared to other galaxies in general.

Question
Witt

1.   By not including scattering in your model analysis, you probably underestimated the optical depths for extinction by about a factor of two.

2.   The result showing the wavelength variation of optical depths in these galaxies to be very similar to the extinction law in our galaxy, is quite consistent with an essentially constant dust albedo ranging from the B-band to the K-band.

3.   Models for uniform mixtures of stars and dust including both absorption and scattering (with Draine-Lee dust albedos) are available in published form in the paper by Witt, Thronson and Capuano (1992).

Answer
J Knapen

In the present study, we limited ourselves to simple geometric dust models without scattering, which in fact work surprisingly well. In future work, especially at short wavelengths, scattering effects should be included.

Comment
Byun

Scattering would have been a lot more important if the light was mixed with stars. However, although Knapen calls his model a "uniform model", the geometry involved is that of "screen" and the bulk of (bulge) light is either in the front or behind the dust layer.

AN OPTICAL SEARCH FOR DUSTY DISKS

Techniques and first results

M. NÄSLUND AND S. JÖRSÄTER

Stockholm Observatory
S-133 36 Saltsjöbaden, Sweden

1. Introduction

We are doing deep surface photometry of some selected nearby spiral galaxies, addressing several questions of large-scale nature. Among those are the study of the outer parts of these galaxies in search for signs of dust influence at optical wavelengths.

We have, in particular, been looking for faint excess light associated with dusty disk components with longer scalelengths than the ordinary stellar disk. The existence of such disks have been proposed by Valentijn (this volume, and references therein). If such disks exists they will cause an apparent increase of the scalelength at the outermost parts of the optical disks. Here we briefly discuss the first results for NGC 4565 in the V band.

2. Observations and reductions

The present observational material was obtained in two runs with the Nordic Optical Telescope (NOT) at La Palma, Spain. During the second run, for which the first results are presented here, we used the IAC Thomson 1024^2 CCD, yielding a $7'$ effective field and $0''.462$/pixel.

We used a focal reducer to widen the field, thereby allowing the sky background to be estimated more accurately. The resolution is, however, high enough to facilitate the removal of stars. We acquire the flatfields in connection with the object frames, not only to have equal light paths, but also to get the same *colour* in object frames and flatfield frames, since the background sky flux is totally dominating the galaxy light at the outer parts of the galaxies.

J. I. Davies and D. Burstein (eds.), The Opacity of Spiral Disks, 189–191.
© *1995 Kluwer Academic Publishers.*

The images were carefully reduced and combined using the program COMB-CCD, which does a careful noise analysis pixel by pixel to form a complete map of the galaxy. The field has in this way been extended to 25′ x 7′. Details about the reduction procedures can be found in [1].

3. First results

We have calibrated our data of NGC 4565 using existing aperture photometry. We have traced the profile down to $\mu_V = 28$, and the disk seems to remain a good exponential at both ends, although with different scalelengths. There is no clear hint in these data of any change in disk scalelength beyond the cutoff radii, i.e. we have found no excess light above $\mu_V = 28$ by tracing the major axis at a constant position angle.

By following the warp, however, we see some faint light beyond the 'optical' disk. NGC 4565 exhibits a warp at the NW side that is clearly seen in HI, but is less prominent in optical bands (see e.g. [2] and [3]). The V band data reveal a faint structure at a level of about $\mu_V = 27$. In figure 1 the smoothed V image is represented by contours, while the HI observations by [4] are in grey scale. The faint feature is seen to align well with the HI warp, and shows a similar extent, which makes us believe that we have found optical emission outside the disk cutoff. This indicates that our photometry down to $\mu_V = 28$ is reasonably accurate. In order to get some information of the nature of this faint extended structure we will measure the B-V and V-R colours, a work that is in progress ([5]).

4. Conclusions

We do not see any light associated with disks with longer scalelengths than that seen in the inner parts. The rather steep gradient of the disk surface brightness which starts at $r = 7′$ from the nucleus continues to $\mu_V = 28$. We have been able to identify a (complete) optical counterpart to the NW HI warp in NGC 4565.

References

1. Näslund M. (1994) Deep Surface Photometry of Nearby Spiral Galaxies, Licentiate thesis, Stockholm University
2. Sancisi, R. (1976) Warped HI Disks in Galaxies, *Astronomy and Astrophysics* **53**, 159-161
3. Kruit P. C. van der (1979) Optical Surface Photometry of Eight Spiral Galaxies Studied in Westerbork, *Astronomy and Astrophysics Supplement* **38**, 15-38
4. Rupen M. P. (1991) Neutral Hydrogen Observations of NGC 4565 and NGC 891, *Astronomical Journal* **102**, 48-106
5. Näslund M., Jörsäter S. (1995), in prep.

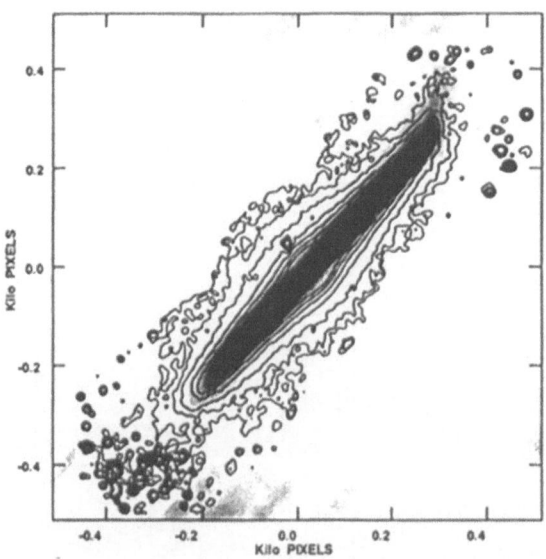

– Fig. 1. The smoothed combined V band image as contours and the HI data as grey scale. HI data courtesy of M. Rupen.

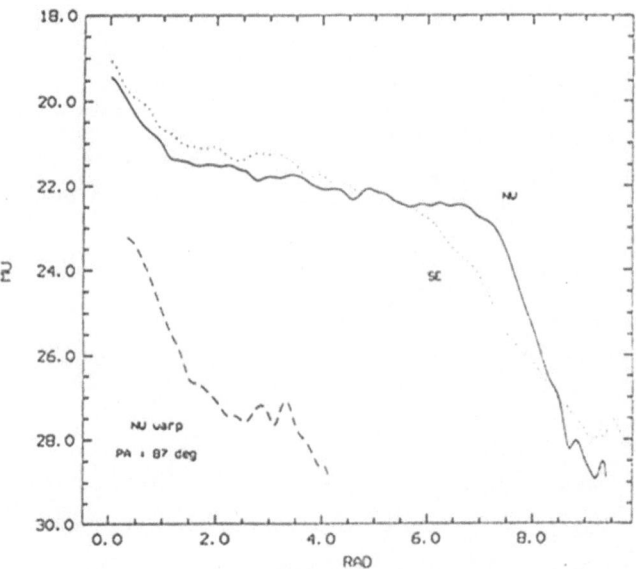

– Fig. 2. The semimajor axis profiles of NGC 4565. 7 pixels perpendicular to the axes are averaged at each radii. PA = 0° corresponds to the positive x axis. At the lower left we have a similar 7 pixel average profile of the NW warp (radial scale not valid). The radii are in arcminutes and the surface brightness in mag/□ ".

PHOTOMETRIC ASYMMETRY AND
DUST OPACITY OF SPIRAL GALAXIES

Y. I. BYUN
Institute for Astronomy, University of Hawaii
2680 Woodlawn Drive, Honolulu, HI 96822, U.S.A.

1. Introduction

The dust extinction of spiral galaxies is best seen in highly inclined spiral galaxies. For the galaxies which are not exactly edge-on, the extinction results in asymmetry in the light distribution; one side of a galaxy divided by its major axis is brighter than the other half. The dust distributed in the main plane attenuates the light behind it, and this is much more effective in the half of galaxy which is closer to us. These galaxies offer a good opportunity to study the dust extinction and its distribution in a very direct manner, because the brighter side can be used as a reference for the heavily obscured side.

2. Asymmetry Analysis of Individual Galaxies

For a given line-of-sight in the near side of a galaxy, the incoming surface brightness can be described as the sum of the following three components : the fractional light in front of dust (ξ), light mixed with dust (ζ), and the light behind dust $(1-\xi-\zeta)$. If we assume that (a) the system is intrinsically symmetric, (b) the scattering effect is not important, and (c) the stars and dust are uniformly mixed within the dust, the total line-of-sight surface brightness at the near and far side can be expressed by $I_{\text{near}}(\lambda) = \xi + \zeta(1 - e^{-\tau_\lambda})/\tau + (1-\xi-\zeta)e^{-\tau_\lambda}$, $I_{\text{far}}(\lambda) = (1-\xi-\zeta)+\zeta(1-e^{-\tau_\lambda})/\tau+\xi e^{-\tau_\lambda}$, and the luminosity asymmetry is defined as $A_{\text{sym}}(\lambda) = -2.5 \log[I_{\text{near}}(\lambda)/I_{\text{far}}(\lambda)]$.

The amplitudes of extinction and asymmetry vary with wavelength. It is in principle possible to solve the above equation for opacity, ξ, and ζ using multiwavelength measurements of the asymmetry. In practice however, an assumption of negligible ζ (i.e. infinitely thin dust layer) simplifies the pro-

J. I. Davies and D. Burstein (eds.), The Opacity of Spiral Disks, 193–195.
© *1995 Kluwer Academic Publishers.*

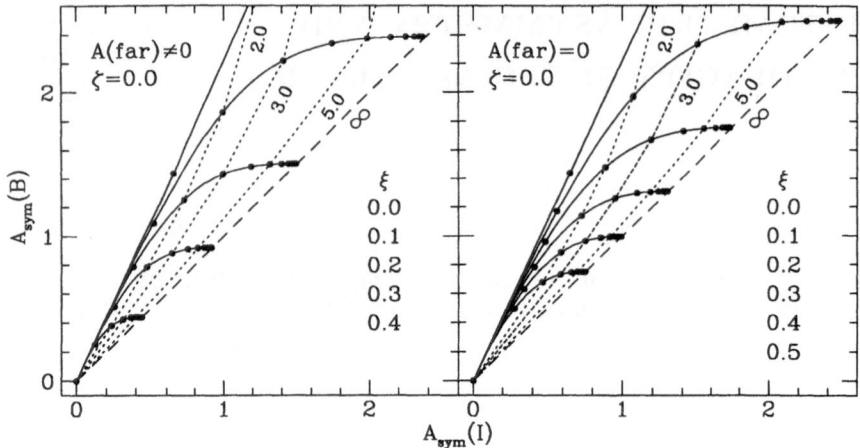

Figure 1. Selective asymmetry as a function of ξ and τ.

cedure and guarantees more stable solutions. Asymmetry in B and I-band is compared for several cases of ξ and τ in Figure 1. The left diagram is when the far side extinction is taken into account and the right one is when the far side extinction is neglected.

For individual galaxies, one can determine the radial distribution of opacity by using the selective nature of asymmetry. Our recent studies show that, for a sample of Sa and Sb galaxies, the face-on optical depth is almost negligible in the outer parts and also rather small ($\tau_V < 1.0$) even at its maximum (Byun 1993).

3. Radiative Transfer Calculations and Its Application

Mingsheng Han (University of Wisconsin) and I am working together on the application of the asymmetry method to larger sample of galaxies. Numerical simulations of radiative transfer can be very useful for this purpose, because they predict general behavior of luminosity distribution for various cases of dust opacity (see Byun, Freeman, and Kylafis 1994).

We define a convenient measure of asymmetry, color asymmetry Γ, as the difference in $V - I$ between far and near sides of a galaxy. Γ is measured for each elliptical annulus and therefore forms a radial profile of color asymmetry. The ellipse parameters are pre-determined from the I-band images and used for both V and I images. Note that Γ is simply a measure of the V luminosity asymmetry on the I-band isophote. True color asymmetry between near and far sides is difficult to measure without knowing the exact location of intrinsic galaxy center. The apparent luminosity peak is often displaced due to the dust obscuration.

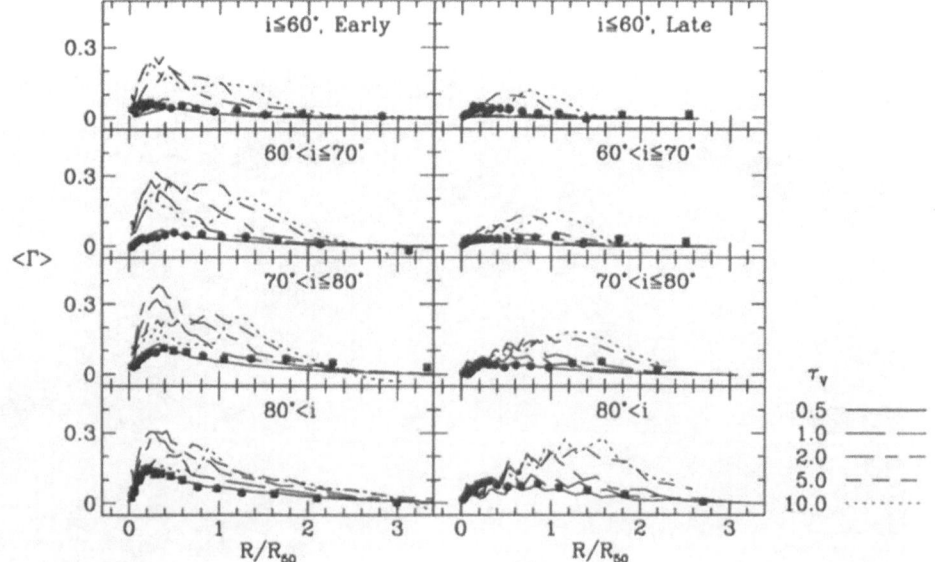

Figure 2. Mean Γ profiles for 284 spiral galaxies compared with model

We have investigated the behavior of Γ profiles using a number of model galaxy images which are constructed by radiative transfer calculations. Γ profiles are also measured for a sample of 284 spiral galaxies from Han (1992). The galaxies are divided into early and late type groups and their Γ profiles are averaged for several inclination bins. The mean Γ profiles from real galaxies are compared with model profiles in Figure 2. We find that there is a general agreement between the observations and models if the central optical depth τ_V is 0.5 ∼ 1.0. This is consistent with the view that the dust extinction within spiral galaxies is very modest.

References

1. Byun, Y. I. (1993), Tests for Dust Opacity of Spiral Galaxies, *Publications of the Astronomical Society of the Pacific*, **105**, 993-995.
2. Byun, Y. I., Freeman, K. C., and Kylafis, N. D. (1994), Diagnostics of Dust Content in Spiral Galaxies : Numerical Simulations of Radiative Transfer, *The Astrophysical Journal*, **432**, 114-127.
3. Han, M. (1992), I-band CCD Surface Photometry of Spiral Galaxies in 16 Nearby Clusters, *The Astrophysical Journal Supplement Series*, **81**, 35-47.

The dust

THE SCALE-LENGTH TEST FOR DUST IN FACE-ON SPIRALS

J.E. BECKMAN, R.F. PELETIER[1], J.H. KNAPEN[2], M.J. MATE ANI
L.J. GENTET

*Instituto de Astrofísica de Canarias, E-38200 La Laguna,
Tenerife, Spain*
[1] *also: Kapteyn Astronomical Institute, P.O. Box 800, 9700
AV Groningen, The Netherlands.*
and
*Isaac Newton Group, Apartado 321, 38700 Santa Cruz de la
Palma, Spain.*
[2] *Present Address: Université de Montréal, Dépt. de Physique,
C.P. 6128, Succ. Centre Ville, Montréal, Québec, H3C-3J7
Canada*

Abstract.
    We have used a scale-length vs. wavelength test to measure the dust optical depth in the (assumed) exponential discs of three nearly face-on Sb/Sbc spirals: NGC 3631, M51 and NGC 4321 (M100). The test was applied to each galaxy after dividing the discs into two separate regimes: the arms and the interarm zones. To a first approximation the scale-lengths of the interarm discs are wavelength independent over the infrared and optical bands. For NGC 3631 and M51 the arm scale lengths increase systematically with decreasing wavelength, while for NGC 4321 they remain constant. Models homogeneous in dust give fair fits to the data with little interarm dust, and considerably more in the arms: projected on-axis visual dust optical depths of $\tau_{V,c} \leq 0.5$ and $\tau_{V,c} \sim 1.5 - 2$ respectively. If the dust in the arms is distributed in narrow lanes the extinction can be locally greater than the above averages would imply. In the blue and ultraviolet bands the observations cannot be fitted by the models used, and scattering, or large stellar population gradients, or a mixture of the two, must be invoked to solve the discrepancy.

J. I. Davies and D. Burstein (eds.), The Opacity of Spiral Disks, 197–208.
© 1995 *Kluwer Academic Publishers.*

1. Introduction

When we look at the image of a face-on spiral in the V or B band it is obvious (see Plate 1) that the distribution of dust is non-uniform. The strongest concentration of dust lies close to the spiral arms. In the context of the general question whether or not most spirals are optically thick in dust we decided that it was worth addressing the particular question whether there are measurable differences in the amount of extinction in the arms and the interarm disc. By looking at one galaxy behind another White & Keel (1992) already demonstrated notable differences between arm and interarm behaviour in this respect. Here, however, we decided to follow up a line of work which we had begun using scale-lengths in discs. The original technique consisted simply in comparing the scale lengths of the disc measured in different photometric bands, in a similar way to that derived in Evans (1994). If the data are of good enough quality out to a sufficiently large radius (say 0.5 R_{25}) it may be possible to distinguish between discs optically thin and optically thick in dust, using appropriate numerical modelling. In the present paper we have set ourselves a more modest aim: to see what are the differences, if any, between scale-lengths in the arms and interarm discs. This enables us to look at several different questions. Firstly the global behaviour - does the whole disc scale length vary with wavelength, and if so, what does this suggest about the optical depth? Secondly are there any important differences from the arm regime to the interarm regime, and if so, do they render meaningless any attempt to use whole disc colours to look at dust? Finally, whatever the answer to the second question, can we model the observations uniquely, or are we left with ambiguities which require external data to resolve?

2. The Data

We decided to work with three disc galaxies for which we could obtain imaging over a reasonably wide wavelength baseline, and which were large enough to make separation of an arm from the rest of the disc relatively straightforward. They are face-on grand-design objects: NGC 3631, NGC 4321 (M100) and NGC 5194 (M51). For M51 we used broad band images in the optical B, R and I and the near-infrared (NIR) K-band, kindly supplied to us by H.-W. Rix. The acquisition and reduction of these images is described in detail by Rix & Rieke (1993). The field of view of the images is some 7' × 7', and they cover the inner disc of the galaxy, out to a level of $\mu_B = 23$ mag arcsec^{-2}. We supplemented these images with our own U-band observations using an image obtained by D. Carter at the 4.2m William Herschel Telescope on La Palma.

B, V, R and I-band images of NGC 4321 were obtained at the INT using the Prime Focus Camera. The basic reduction of the images and the subsequent positioning was described in Knapen *et al.* (1994). The photometric calibration was performed by comparing the data with aperture photometry in the literature. The field of view is almost 10', large enough to cover a large part of the visible disc. We show a real-color image of M100, composed of the separate images in B, V and I-bands, in Plate 1.

Images of NGC 3631 were obtained for us by P. Rudd and M. Asif in service time at the 1m JKT at La Palma in January 1994. They were taken with a 1280 × 1180 EEV CCD, with pixels of 0.31", which implies that the field of view was about 6' × 6'. Since NGC 3631 is smaller than NGC 4321 this is enough to cover the galaxy down to the same surface brightness.

For each galaxy, after removing foreground stars, we produced two separate images: one with arm regions alone, and the other with only interarm regions. The separation was effected using a mask derived from the I-image. First an azimuthally averaged profile was obtained, in annuli, and the arm regions were then defined as those with intensity higher than 7% above the average within the given annulus. Figure 1 shows, in isophote format, the arm-interarm images thus obtained, for M51, to show the principle of the method.

We went on to use these images: the arm, the interarm, and the whole disc, as inputs for the profile fitting routine Galphot (see Jørgensen *et al.* 1992). For the whole disc, isophotes of constant ellipticity and major axis position angle were computed, and Galphot then used separately on the part of the image within the arm-defining mask, and outside the mask respectively, as well as the whole disc, to obtain the 'arm' profile, and the profiles of the interarm region and the whole galaxy. As all three galaxies are Sb or Sbc objects, we did not need to perform a bulge-disc decomposition, but measured only over the regions of the discs well outside the bulges, keeping the same radial range in each photometric band used, in the arm regions, the interarm regions, and the averaged whole disc, from which scale lengths were obtained using least-square fits to the intensity as a function of linear distance.

3. The observed scale lengths

3.1. NGC 4321

The scale lengths in the available bands: B, V, R and I for this galaxy are given in Fig. 2 in three regions: arm, interarm and whole disc. Errors in observed scale lengths consist of random and systematic errors, of which the latter clearly dominate. Since real discs of galaxies generally don't have surface brightness profiles that are exactly exponential the derived scale lengths may vary by 20% depending on the radius range that is used. The

M51

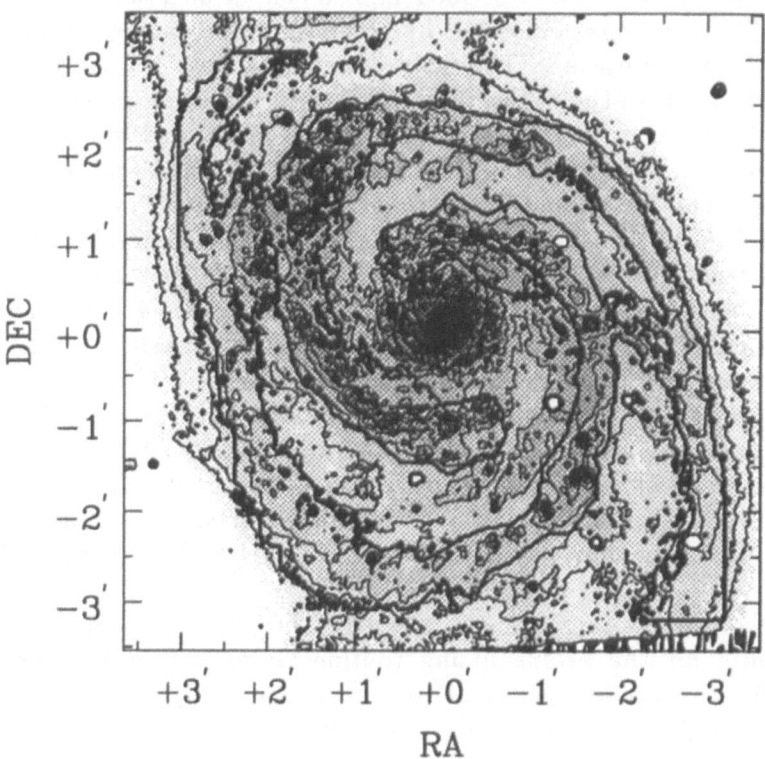

Figure 1. Contour map of M51. The thick lines indicate the boundaries between the regions that we define as arm and interarm region.

errors from least squares fitting are generally ~ 5-10%. For this reason we have used the same radius range for the fits in all bands. For the scale lengths presented here one should assume individual errors of about 10%, It is striking that while the scale-lengths are constant with wavelength in both arms and interarm disc, the arm scale lengths are 25% longer, while those of the whole galaxy behave as an average, weighted slightly towards the arm values. This is not surprising, since in NGC 4321 the arms

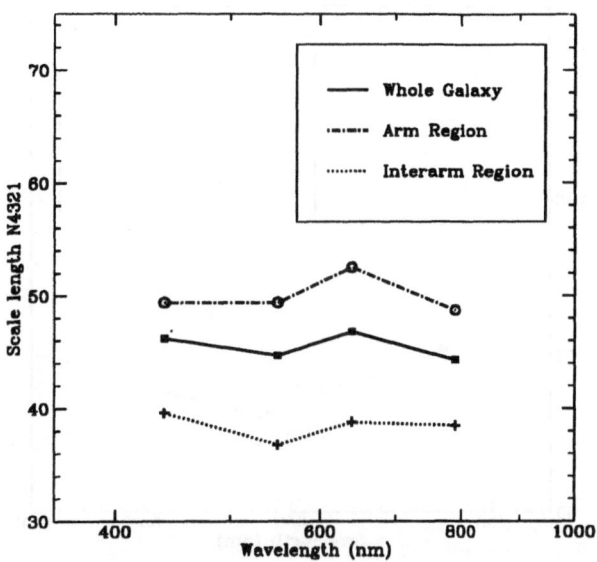

Figure 2.   Radial scale lengths for NGC 4321 in *B,V,R* and *I*

are quite diffuse and occupy at least 50% of the disc area. A qualitative interpretation of these observations is that there can be relatively little dust in NGC 4321, but that there is a systematic difference in the stellar population distribution between the arms and the rest of the disc. A glance at Plate 1 shows, however, that to claim the presence of little dust in this galaxy is to fly in the face of the evidence. In the next section we will show how invariant scale lengths can be reconciled with the presence of dust.

3.2. NGC 3631

In Fig. 3 we have plotted the *B, V, R* and *I* scale-lengths for the arms, the interarm disc, and the averaged disc of NGC 3631. These data tell a different story from those of NGC 4321, in that we see a clear trend of decreasing scale-length with increasing wavelength, which is what would be expected if there were significant dust present. Further, the trend is much steeper in the arm regions than in the interarm regions, which is, again, what one might predict from the general idea that there should be more dust in the arms. Again the fact that the averaged disc follows the interarm disc rather closely is what would be expected from the fact that in NGC 3631 the arms occupy less than 25% of the projected disc area. We can see also that to use the averaged disc only would be to miss the

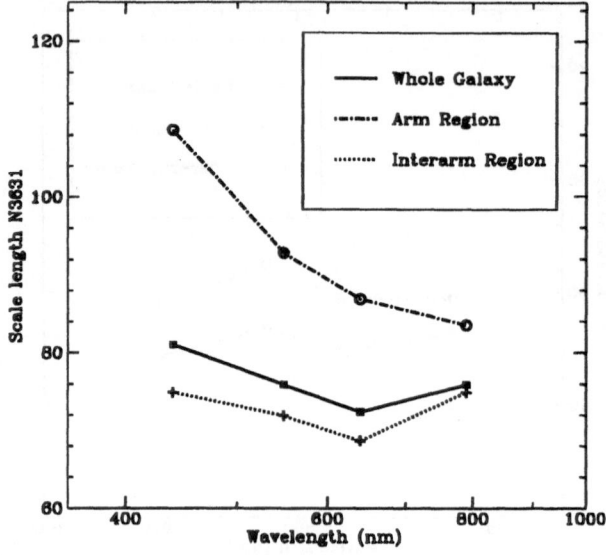

Figure 3. Radial scale lengths for NGC 3631 in B,V,R and I

striking difference in behaviour between the arm and the interarm zone. In the next section we will attempt to quantify these trends in term of dust optical depth.

3.3. M51

Our results for M51 are more significant than those of the other two galaxies because the baseline in wavelength range is much longer, extending from the near-infrared K-band through the visible bands to the near UV U-band. In fact for the wavelength range between B and I the behaviour of M51 is very similar to that of NGC 3631, as can be seen in Fig. 4. The addition of the K-band strengthens the idea that the changes in scale length seen, notably in the arms, are due principally to the presence of some dust rather than significant stellar population changes, since the scale-length changes very little between the visible bands and K. The B and R bands strengthen this picture, since steep scale length increases are seen between B and R for both arms and interarm disc, and this is very unusual for stellar population gradients. We see that for M51, as well as for NGC 3631, the whole disc average follows the interarm disc, as expected, since the arms occupy a low fraction of the disc area.

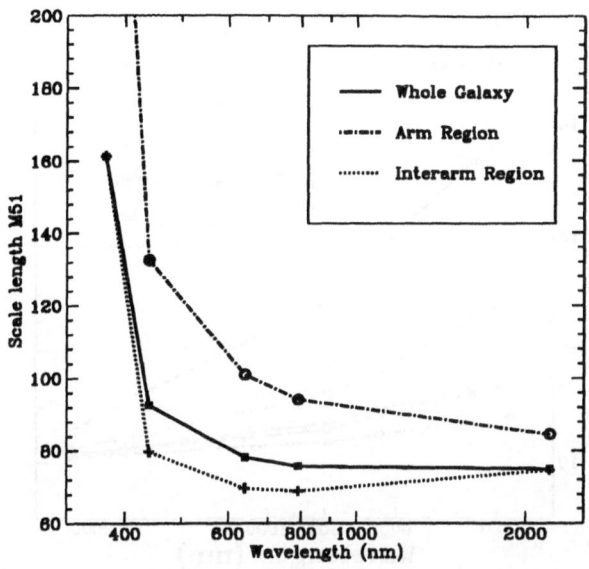

Figure 4. Radial scale lengths for M 51 in U,B,R,I and K

4. Dust models for the scale-lengths

In order to see how the dust can be producing the observed scale-length variations it is important to model the data presented in Figs. 2-4. We developed a simple set of models, analogous to the 'triplex' models of Disney *et al.* (1989). The stars and the dust are distributed with cylindrical symmetry (or along a single radial coordinate, which is equivalent), and each system is characterized by a scale height above the disc plane, and an intrinsic radial scale length parallel to the plane. We have fixed the scale height of the dust at half that of the stars, as suggested by recent observations (Peletier & Willner 1991). The ratio of scale lengths of dust and stars is still a matter of debate (see Valentijn 1994), so that we have varied it between 1 and 5. One sees that this can materially affect the predicted scale length ratios. In Fig. 5 we give a set of model predictions, aimed at explaining the M51 observations. The parameter $\tau_{V,c}$ is the dust optical depth in the V-band integrated along the central axis of the model, i.e. the optical depth the dust distribution **would** have if the exponential disc in fact extended right to the centre of the galaxy. Obviously the optical depths over the observed parts of the disc will always be less than this. The figure shows how the optical scale-length would vary with wavelength for a range of $\tau_{V,c}$ and ratios of dust to stellar scale lengths, between 1 and 5. Note that pure extinction without scattering was assumed, and this is most likely to

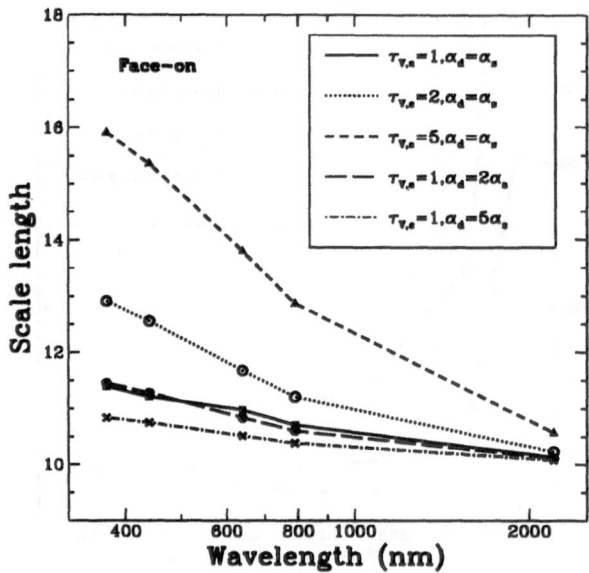

Figure 5. Observed scale lengths as a function of wavelength for a model with a stellar scale length of 10 for various dust geometries.

break down at the shorter wavelengths.

We can see the limitations, but also some of the utility of the modelling technique looking at Figure 6, where the observed arm and interarm scalelengths are compared with model predictions, in Fig. 6a and 6b respectively. The striking lack of agreement at the blue end of the range is somewhat compensated by the trend at the red end. In particular in Fig. 6a a value for $\tau_{V,c}$ of close to 2 fits reasonably well the dependence of scale length on wavelength in the red and infrared for the arms. If we assume dust to be compressed into the plane the best value turns out to be closer to 2.5. A similar exercise for the interarm zones puts $\tau_{V,c}$ well below 1. In fact, a good fit for the interarm zone would have to rely on a slightly bluer stellar population near the centre to get a reasonable fit even in the infrared. All we can see with clarity is that there is less than ~30% of dust optical depth in the interarm zones compared with the arms. This simple model fails to bring out the steep rise in the observed scale lengths to the blue and UV. The rise is so steep that different dust geometries will not help in explaining it, so new, more 'physical' models will have to be made that include the effects of scattering and stellar population changes.

As a final contribution to the scale-length argument for M51 we show in Fig. 7 the effects of compressing the dust into a lane. We have in fact taken two extremes: dust covering the whole arm, and dust covering only

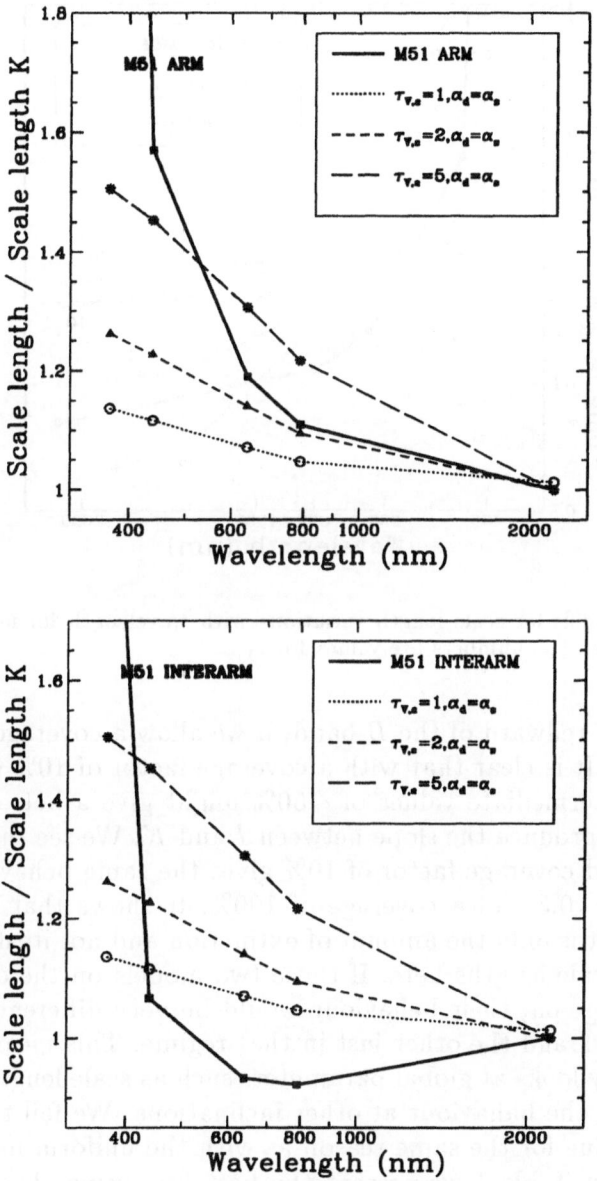

Figure 6. Observed scale lengths for M51 in (a) arms and (b) interarm regions, compared with various models.

10% of the arms as seen face-on. The real situation may well be somewhere between these, as we saw for NGC 4321 in Plate 1. The Figure has been arbitrarily normalized for the *V*-band. Our models here fit the data quali-

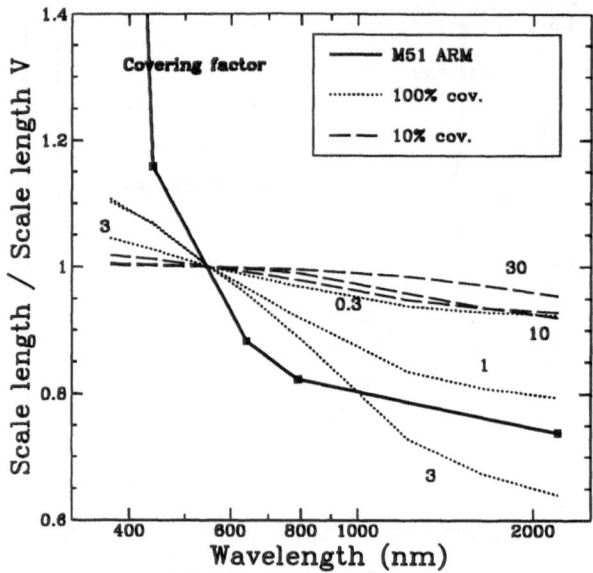

Figure 7. Models for scale length variations with wavelength for models with various covering factors. The numbers are values for $\tau_{V,c}$.

tatively well, redward of the *B*-band, if we allow a coverage factor of 100% in the arms. It is clear that with a coverage factor of 10% tha data cannot be fitted. Intermediate values of $\sim$50% might give a better fit and at the same time reproduce the slope between *I* and *K*. We see that a model with $\tau_{V,c} = 3$ and coverage factor of 10% gives the same behaviour as a model with $\tau_{V,c} = 0.3$ and a coverage of 100%. It shows that in the optically thin regime it is only the amount of extinction and not its distribution that affects the scale lengths here. If these two models on the other hand were turned to edge-on, their behaviour would be very different, with one very optically thick and the other just in that regime. This means that an analysis that only looks at global parameters such as scale length is not capable of predicting the bahaviour at other inclinations. We fail to reproduce the rise to the blue for the same reason as with the uniform models.

Finally, in Table 1 we present the best fits, using the present models, to the on-axis optical depths $\tau_{V,c}$, in the V-band, for each of the systems described. Since these values are representative rather than definitive, we have kept the models simple: the scale height of the dust is half that of the stars. However for models with the dust scale height set at that of the stellar scale height the values increase by no more than a few tenth of dex, and there would be no qualitative changes. We can see from Table 1 that none of these galaxies could be considered optically thick in dust.

TABLE 1. On-axis optical depths

Galaxy	Sub-system	On-axis optical depth
NGC 4321	Arms	≤ 0.7
	· Interarm disc	≤ 0.7
	Averaged disc	≤ 0.7
NGC 3631	Arms	~ 2
	Interarm disc	≤ 0.7
	Averaged disc	~ 1.4
M51	Arms	~ 2.5
	Interarm disc	≤ 0.7
	Averaged disc	≤ 1.0

In the strongest case, the arms of M51, the on-axis dust optical depth is no more than 3, and a characteristic value for an arm would be between 1 and 2: a corresponding characteristic value for the interarm disc would be less than 0.5, and for the averaged disc less than 0.6. For NGC 3631 the averaged disc must also be not optically thick in dust, and for NGC 4321 the averaged disc is clearly optically thin in dust. We could attempt to substitute azimuthally averaged quantities by values adapted to dust-lane geometry, but in the present state of our modelling this is clearly not worthwhile.

5. Conclusions

We have measured scale-lengths in photometric bands between B and I of these Sb/Sbc galaxies, and also in U and K for one of them. These measurements have been made separately for the subsystems comprising the arms and the interarm disc, as well as for the whole disc for all these objects. The results tell us that the scale-lengths vary very little with wavelength for the interarm discs of all three galaxies for the V-band and bands redward of it. The simplest explanation of this is the absence of dust in optically thick quantities in these zones (dust highly compressed in the plane and opaque throughout the disc even in K could also yield this result but we dismiss this as improbable). Within the arms of two of the galaxies: NGC 3631 and M51, we have systematic variations of scale-length with wavelength consistent with moderate amounts of dust: central optical depths in the range 1 to 2. The averaged disc would be optically thin in all three cases, but one result of the present study is to show that the averaged disc result is not valid for either of its sub-systems, and must be used with caution.

Modelling each sub-system as uniform, we find reasonably good fits of simple exponential models to the data for the infrared and red bands, and even a fair fit up to V, but for the B-band and a *fortiori* the U-band (in M51) the fits are bad. It is possible that this can be accounted for by incorporating scattering, but until attempting to do so any opinion here is necessarily speculative.

Finally we have made one further tentative step in de-homogenizing our models by assuming that the dust in the arms does not fill their projected surface area, but occupies a defined fraction of it. We find that by multiplying the characteristic on-axis optical depth by the reciprocal of this fraction, the inhomogeneous model gives, as might well be expected, similar behaviour to the corresponding homogeneous model, as long as the amount of extinction is not too large. A similar exercise for the interarm region gives similar results. If, as our physical assumptions are improved, we can obtain improved agreement between prediction and measurement, we should be able to test hypotheses about the total optical thickness of dust in discs, about its global distribution, and about how this will affect the photometric properties of galaxies which are more nearly edge-on to us.

References

1. Disney, M., Davies, J.I. and Phillips, S., 1989, MNRAS, 239, 939
2. Evans, R., 1994, MNRAS, 266, 511.
3. Jørgensen, I., Franx, M. & Kjaergaard, P., 1992, A&A Suppl. 95, 489.
4. Knapen, J.H., Beckman, J.E., Heller, C.H., Shlosman, I. and de Jong, R.S., 1994, ApJ, submitted.
5. Peletier, R.F. & Willner, S.P., 1991, ApJ, 382, 382.
6. Rix, H.-W. & Rieke, M.J., 1993, ApJ, 418, 123.
7. Valentijn, E.A., 1994, MNRAS, 266, 614.
8. White, R.E. & Keel, W.C., 1992, Nature, 359, 129.

The friends

COLOR GRADIENTS IN SPIRAL GALAXIES

STÉPHANE COURTEAU[1] and JON HOLTZMAN[2]
1. *NOAO/Kitt Peak National Observatory, Tucson, AZ*
2. *Lowell Observatory, Flagstaff, AZ*

ABSTRACT. We present first results from a subset of a multicolor imaging survey of ~350 late-type spiral galaxies. BVRH images are extracted for 50 galaxies. We show that B-H color gradients can be as large as 2 magnitudes. In the outer regions, colors are roughly indepedent of inclination which suggests that disks are optically thin near the Holmberg radius. Finally, dust appears to be a likely explanation for Freeman Type II disks. We also caution that azimuthally averaged profiles should never be used for truly edge-on galaxies.

1. Introduction

The goal of this study is to assemble a large, statistical sample of high-quality, multicolor photometry to study color gradients and dust extinction in spiral galaxies. We have extracted a magnitude- limited sample ($m_B \leq 15.5$) of UGC Sb/Sc galaxies in three inclination bins covering face-on ($i \leq 6°$), edge-on ($i \geq 78°$), and intermediate ($50° < i < 60°$) projections. All galaxies have known redshifts to isolate selection effects. The BVR images are obtained at the 72in telescope of Lowell Observatory with a TI 800×800 chip (at an image scale of $0.''5$/pix); the H images are acquired with the KPNO 84in telescope equipped with either a HgCdTe (IRIM) or an InSb (COB) 256x256 array (scales are $1.''09$/pix and $0.''5$/pix resp.) Typical exposure times are: 300s at R, 400s at V, 1200s at B and on-target integration of 1200s at H. Typical seeing is $1.''8$ at Lowell and $1.''0$ at Kitt Peak. The surface brightness profiles are reliable down to fairly deep levels: $\mu_B = 26.0$, $\mu_V = 25.5$, $\mu_R = 25.0$, and $\mu_H = 22.0$. Sky values are measured on the same frame as the galaxy both in the optical and the IR; typical sky errors are 0.15%.

Color gradients are estimated by comparing major axis cuts of exactly the same region of the galaxy. While surface brightness (SB) profiles are in principle adequate, one must insure that their isophotes allow comparison of the same physical areas in all given bands and inclinations for a proper measurement of color gradients. This condition can only be met if the same isophotal solution (usually at H or K) is used for all given bands and centered properly with respect to reference field stars (the center of the galaxy is to be avoided due to dust assymetric shift). Presentation of the surface photometry and color gradient profiles using this latter technique as well as two-dimensional bulge-to-disk decomposition will be given elsewhere (Courteau and Holtzman, in preparation).

J. I. Davies and D. Burstein (eds.), The Opacity of Spiral Disks, 211–213.
© *1995 Kluwer Academic Publishers.*

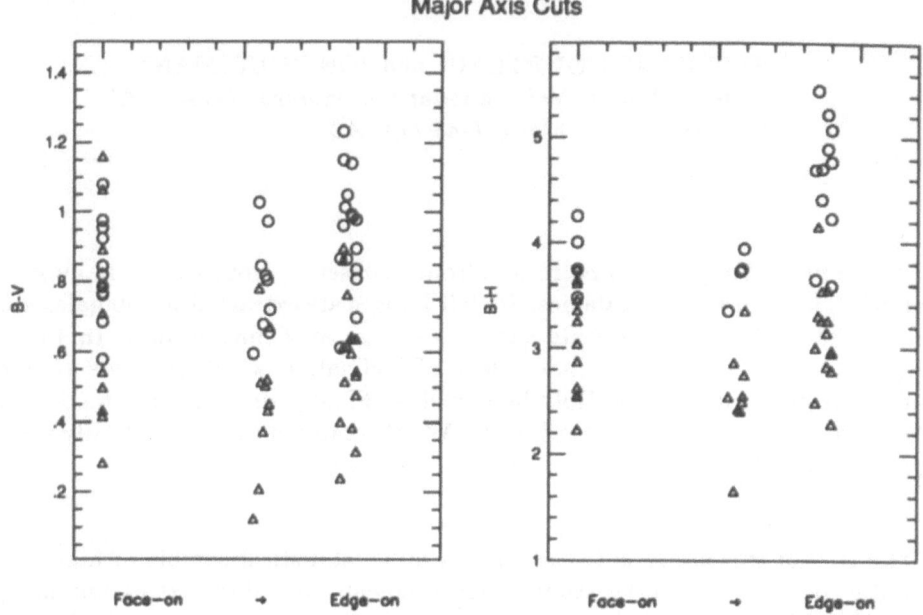

Major Axis Cuts

While this technique works well at essentially all projections, it is worth pointing out that averaging geometry breaks down completely for edge-on galaxies (where one essentially measures SB(z) instead of SB(r)). Therefore, for the purpose of comparing color gradients at ALL inclinations, major axis cuts, in spite of their low S/N levels, are the only viable solutions and the one we adopt here. We note also that direct comparison of B and H-band SB profiles for a few galaxies in common with Bernstein *etal* (1994) and de Jong and van der Kruit (1994) yields excellent agreement.

2. Results And Discussion

The figure above shows mean bulge (circles) and disk (triangles) colors for the sub-sample of 50 Sc galaxies with BVRH photometry. In a few cases, the bulge was sampled so sparsely that a mean color could not be determined. All colors are corrected for Galactic extinction following Burstein-Heiles (1984). One sees that the inner regions ("bulges") of edge-on late-type spirals are significantly redder than face-on "bulges". In the outer regions(triangles), colors are roughly independent of inclination. Disk color gradients are relatively small ($\leq 0^m\!.5$) in B-V and V-R (not shown here) but can be as large as 2 mags in B-H. This range in color gradients is somewhat larger than that reported in Kent (1986) or Terndrup *etal* (1994) but in closer agreement with results presented by the Elmegreen's (1984, 1985) or, more recently, Peletier (this conference) and Jansen *etal* (1994). A significant consequence of large color gradients in spirals, if they are not solely the result of extinction, would be the revision of the assumed constant bulge and disk mass-to-light ratios in mass modeling (Broeils and Courteau, in preparation).

A scenario which can explain most of our observational results is that spiral disks are everywhere transparent at H but significantly opaque in the inner and perhaps intermediate regions at bluer colors. If the inner regions are optically thick at B and thinner at H, the B surface brightness is roughly independent of inclination, but at H, more inclined galaxies get brighter (redder). This would also explain constant colors versus inclination far out into the disk, with disks becoming essentially transparent at both B and H. We note finally that many of our SB profiles exhibit a transition of Freeman Type II at B to Freeman Type I at H which would lend support to the idea that Type II's are dust-dominated near their bulge (Davies 1990, Barnaby *etal.* 1995). Stellar population effects can still not be excluded.

The nature and explanation of color gradients requires a clear knowledge of stellar population effects as well as the combined effect of dust absorption (reddening) and multiple scattering (brightening) to the overall spectra of spiral galaxies. Current studies suggest that the two effects are nearly degenerate in ellipticals (Witt *etal.* 1992, Wise and Silva 1994) but modeling parameters may be more easily disentangled in spirals (Block *etal.* 1994).

We plan our own investigation with a radiation transfer code including N-phase scattering and clumpy distribution of dust. Spectral line data may also be needed to differentiate age from metallicity effects. Finally, we are preparing a joint program of sub-mm observations in order to constrain dust models.

Minor-axis (assymetric) cuts of highly inclined galaxies while be used in a subsequent paper to assess the vertical stellar distribution and the influence of dust versus thick disks.

REFERENCES

Barnaby, B., *etal.*, 1995, preprint.

Bernstein, G.M., *etal.*, 1994, AJ, 107, 1962.

Block, D.L., *etal.*, 1994, AA, 288, 383.

Davies, J.I., 1990, MNRAS, 245, 350.

de Jong, R.S., and van der Kruit, P.C. 1994, AA Supp, 106, 451.

Elmegreen, B.G., and Elmegreen, D.M. 1985, ApJ, 288, 438.

Elmegreen, D.M., and Elmegreen, B.G. 1984, ApJ Supp, 54, 127.

Freeman, K. 1970, ApJ, 160, 811.

Jansen, R.A., *etal.*, 1994, preprint.

Kent, S. 1986, AJ, 91, 1301.

Terndrup, D., *etal.*, 1994, ApJ, 432, 518.

Witt, A., Thronson, H., and Capuano, J. 1992, ApJ, 393, 611.

Wise, M. and Silva, D. 1994, preprint.

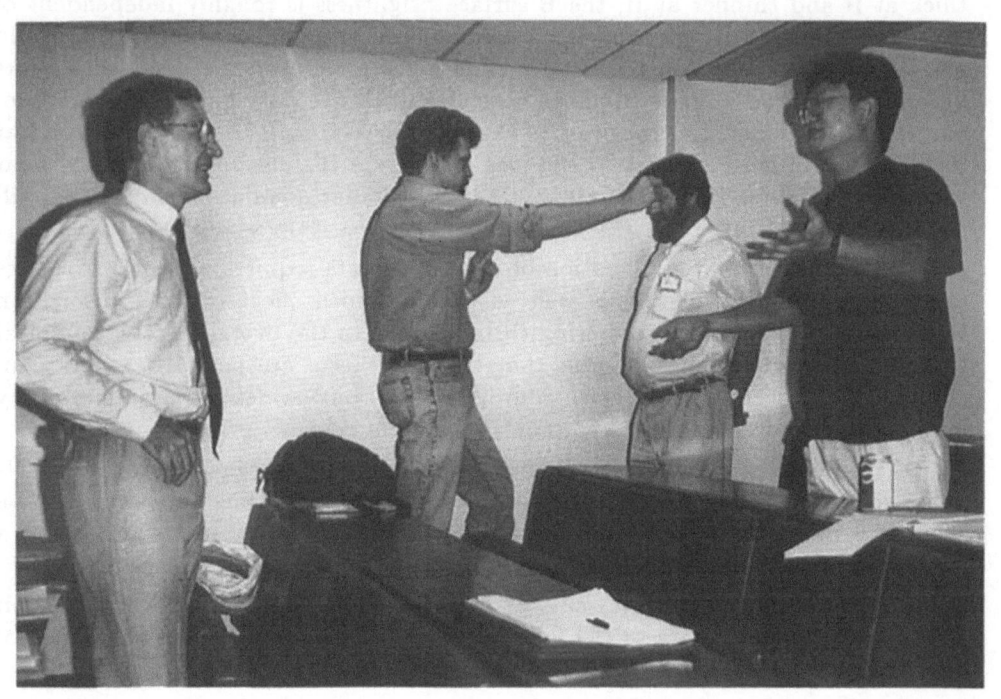

The fight

CONSTRAINTS ON THE OPACITY OF SPIRAL DISKS FROM NEAR-INFRARED OBSERVATIONS

HANS-WALTER RIX

Institute for Advanced Study, Princeton, NJ 08540, USA
and
Max-Planck-Institut für Astrophysik, 85740 Garching, Germany

1. Introduction

In this paper I review how near-infrared (NIR) observations can constrain the opacity of spiral disks. Basic considerations show that NIR photometry provides a powerful probe of the optical depths in spiral galaxy disks in the regime of interest $\tau_V \sim 1$. I review the existing opacity constraints from the analysis of dust lanes in edge-on and face-on galaxies. The "internal extinction correction" in the NIR-Tully-Fisher relation deserves particular attention as the most powerful constraint on the impact of dust on the total luminosity of spiral galaxies. All observations for luminous spirals point towards an effective, face-on optical depth of $\langle \tau_V \rangle = 0.5 - 1$.

2. Which Near-Infrared: 0.8μm or 2.2μm?

The "near-infrared" wavelength region spans a factor of three in wavelength, $0.8\mu m < \lambda < 2.4\mu m$. This region encompasses a wide range of difficulties in the data taking and a comparably wide range in potential for the data interpretation. Basically, data beyond 1μm are much harder to acquire, but the greatly reduced dust opacities at larger wavelength often provide a crucial advantage in the data interpretation.

2.1. GOING TO $2.2\mu M$: THE PAIN

An overview of the differences in the data acquisition between I(0.8μm) and K(2.2μm)is given in Table 1. The first striking difference between CCDs and the HgTeCd arrays (such as the NICMOS3) is their size: currently

J. I. Davies and D. Burstein (eds.), The Opacity of Spiral Disks, 215–226.
© *1995 Kluwer Academic Publishers.*

available CCD have almost 100 times more pixels! This becomes a decisive advantage when imaging nearby galaxies, where the angular size of the target is larger than the field-of-view of the detector and mosaicing may be necessary. Both types of detectors have comparable quantum efficiency, and for both detector types does the photon noise dominate all sources of detector noise. The second large difference between the observational set-ups is independent of the detectors: at one exponential disk scale length, the galaxy surface brightness at $0.8\mu m$ is still 1/2 that of the sky. For the K-band, this ratio is only 0.005, the sky signal dominates everywhere. Fortunately, this large background signal can be used very effectively to "flat-field" the image. Sensitivity variations among the pixels are routinely corrected to a few parts in 10^4.

The bottom line of all these considerations is given at the end of Table 1: for a given amount of observing time, the S/N achievable at $0.8\mu m$ is ten times higher than at $2.2\mu m$. Put differently, to achieve a comparable signal-to-noise one would need to observe 100 times longer in the K-band.

TABLE 1. Detector Comparison for Near-IR Imaging

Comparison	CCD($0.8\mu m$)	NICMOS3($2.2\mu m$)	Ratio
Nr. of Pixels	2048×2048	256×256	65
Quantum Efficiency	$\gtrsim 40\%$	$\gtrsim 50\%$	~ 1
Detector Noise	negligible	negligible	~ 1
Flat-Fielding	$\gtrsim 10^{-3}$	$\gtrsim 10^{-4}$	0.1
Gal. / Sky Contrast[1]	0.5	0.005	100
S/N $(asec^{-2}min^{-1})$[1]	57	5	11

[1] Measured at $R \sim R_{exp}$

2.2. GOING TO $2.2\mu M$: THE GAIN

The above comparison raises necessarily the question why one should bother taking data beyond $1\mu m$. The answer is illustrated in Figure 1. While the dust opacities and albedos in the I band are still very similar to the other "optical" wavelength bands, they are dramatically different at K($2.2\mu m$). The opacity data in Figure 1 are taken from Laor and Draine (1993 and *pers. comm.*); note that the albedo only refers to the isotropic part of the scattering cross section. Most theoretical calculations, and many empirical determinations, show that in spiral disks the dust optical depth in I($0.8\mu m$) is six times higher than in K($2.2\mu m$). For the interpretation of data this difference has several consequences: (1) A lower albedo makes the interpretation of data by radiative transfer models easier. (2) Combining optical and NIR data provides a long wavelength baseline for color and reddening analyses. (3) In many cases (see contributions throughout these proceedings) the dust extinction is negligible only at K(2.2μm), but not yet at I(0.8μm). Therefore, data at 2μm can often provide an empirical template for a galaxy's appearance without dust. (4) In dust lanes and in edge-on galaxies the optically thin regime ($\tau_\lambda < 1$) is only probed at wavelengths as long as $\lambda \sim 2\mu$m.

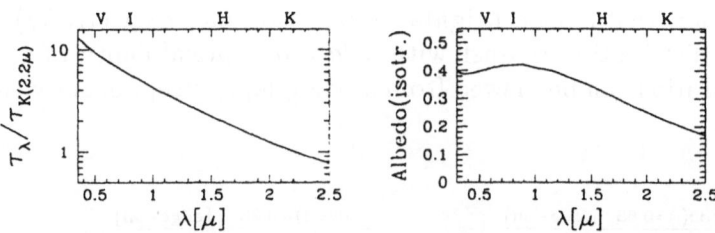

Fig.1: The left panel shows the dramatic drop in the dust absorption cross-section from 0.8μm to 2.2μm, taken from Laor and Draine (1993). The right panel shows the albedo, to isotropic scattering, for the same wavelength range; it peaks near the I-band.

3. For What Dust Signatures Should One Look?

3.1. OBSERVATIONAL SIGNATURES

Most observational test for the presence of dust, using optical-IR data, fall into three categories:

• Study the surface brightness in the presence of dust, $\mu(\vec{R}, \tau_\lambda)$ in highly inclined galaxies, if the dust-free surface brightness, $\mu(\vec{R}, \tau_\lambda = 0)$, is known independently. Most commonly this independent information is obtained through the (assumed) point-symmetry of the dust-free galaxy.

• Study the surface brightness $\mu(\vec{R}, \tau_\lambda)$, but obtain $\mu(\vec{R}, \tau_\lambda = 0)$ directly from NIR observations, if dust extinction can be neglected there. This test, applicable e.g. to dust lanes in face-on galaxies, requires the assumption that the colors of the dust-free galaxy can be determined empirically, e.g. from patches next to the dust lanes.

• Study the surface brightness, total luminosity or color of intrinsically identical galaxies as a function of their inclination $\cos i$. These tests exploit the disk geometry of the dust distribution, which results in increasing extinction towards edge-on orientations. A suitable, dust-independent, property to find identical galaxies is, e.g., their HI rotation speed (see Section 4.3).

3.2. A RADIATIVE TRANSFER MODEL

As discussed throughout this meeting, the conversion of any observations into a statement about the optical depth, τ_λ, is impossible without a radiative transfer model. To relate the observational signatures quantitatively to optical depth estimates, we use here a model, described by Rix and Rieke (1993). [For other radiative transfer models, see Elmegreen (1980), Kylafis and Bahcall (1987) and Witt *et al.* (1992).] Briefly, this model assumes plane-parallel geometry, a Gaussian distribution of dust and stars in the vertical direction (with scale heights of H_D and H_S, respectively) and a homogeneous distribution of dust with a *face-on* optical depth of τ_V. This model configuration can be viewed from any angle, i, except exactly edge-on ($i = 90°$).

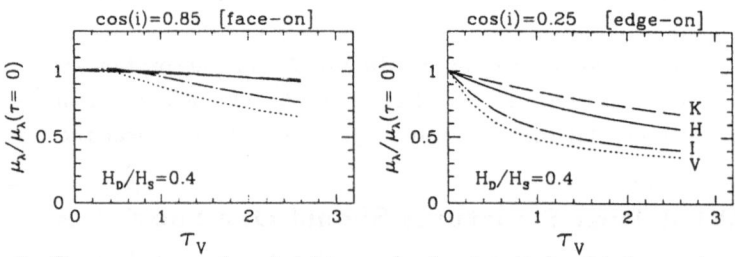

Fig 2.: Decrease in surface brightness (or luminosity) with increasing optical depth, τ_V, predicted by the radiative transfer models for the V, I, H and K band. The left panel shows the case of a nearly face-on galaxy, the right panel a nearly highly inclined galaxy.

Dust scattering is treated in the single scattering approximation and the absorption and scattering properties of the dust are taken from Laor and Draine (1993, and *pers. comm.*). Each model is completely specified by the three parameters: H_D/H_S, τ_V and $\cos i$. The model predictions are applicable in two regimes: (a) when making a local comparison of surface brightnesses from a nearly face-on direction (e.g. when analyzing a dust lane in a face-on galaxy), and (b) when estimating changes in the global luminosity or color due to dust. The model provides a good estimate of the "effective" optical depth (i.e. the optical depth of a homogeneous dust distribution causing the same observational effects), as long as the dust is spatially more extended than the luminosity sources (stars).

Figures 2 and 3 illustrate the model expectation for the observational tests in Section 3.1. Figure 2 shows how the surface brightness (or total luminosity) changes in various pass-bands as the optical depth increases. Figure 3 shows how the observable surface brightness deviates from the dust-free case as the inclination increases. Note that in both figures the I band (dashed-dotted line) is much closer in its behaviour to the V band (dotted line) than the K band (dashed line).

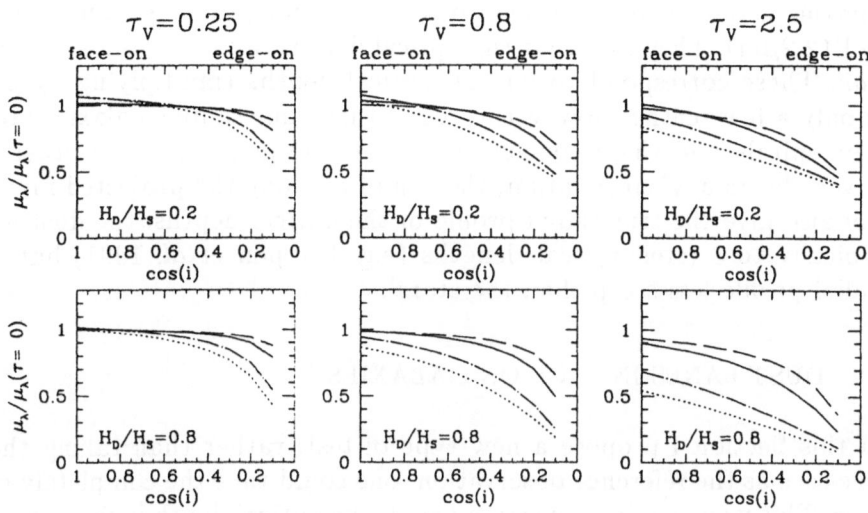

Fig 3.: Decrease in surface brightness (or luminosity) with increasing inclination for three different values of the (face-on) optical depth, τ_V, and for two different scale heights of the dust. The four different lines refer to the V, I, H and K band, as in Figure 2.

4. Observational Signatures

4.1. DUST LANES IN EDGE-ON GALAXIES

Optical and NIR photometry of dust lanes in nearly edge-on galaxies provides one of the most straightforward probes of the dust opacity in disks (Knapen *et al.* 1991, Barnaby and Thronson, 1992, Byun, 1993, Jansen *et al.* , 1994). Such opacity tests rely only on the reflection-symmetric distribution of the stellar luminosity sources, and on the assumption that the dust is distributed in a disk co-planar to the stars. In its simplest form the test consists of comparing the brightness at a point on the near side of the minor axis, $\mu_{near} = \mu_{inside}e^{-\tau} + \mu_{outside}$ to the corresponding point on the far side of the minor axis $\mu_{far} = \mu_{outside}e^{-\tau} + \mu_{inside}$. Here μ_{near} (or μ_{far}) is the surface brightness arising from radii smaller (or larger) than the dust annulus under consideration. If scattering is neglected and if $\mu_{inside} \gg \mu_{outside}$ holds (e.g. because all the bulge luminosity is inside the dust seen in projection), then the optical depth can be obtained immediately by ratioing μ_{near} and μ_{far}. NIR observations are crucial because for a simple, model-independent μ_{near}/μ_{far} test one has to assume not only that $\mu_{inside} \gg \mu_{outside}$ but also that $\mu_{inside}e^{-\tau_\lambda} \gg \mu_{outside}$, which is more likely to holt for small τ_λ. Most recently, Jansen *et al.* (1994) observed four galaxies with edge-on dust lanes in various bands, ranging from $B(0.44\mu m)$ to $K(2.2\mu m)$. They find that the optical depths, at V, are $0.2 - 2$ *in projection*. These correspond to face-on optical depths (multiplying by $\cos(i)$), of only a few tenths. However, three of their four sample galaxies are early type spirals and these conclusions cannot be generalized to other types. Given the galaxy's inclination, the run of τ_λ along the projected minor axis can also give the true radial profile of the optical depths: the dust extends typically to 3 (stellar) scale-lengths (e.g. Knapen *et al.* 1991) but with a radial profile less steep than the stars'.

4.2. DUST LANES IN FACE-ON GALAXIES

In this Section I propose a new type of test: rather than taking the dust free case as the reference observation, one could take the completely opaque case. This way, one can constrain *locally* the optical depth in face-on disks at any point next to a dark dust lane. The constraint comes from a quantitative answer to the question "How much dust can there be in the rest of the disk before the contrast (in intensity or color) to the dust lane is spoiled?".

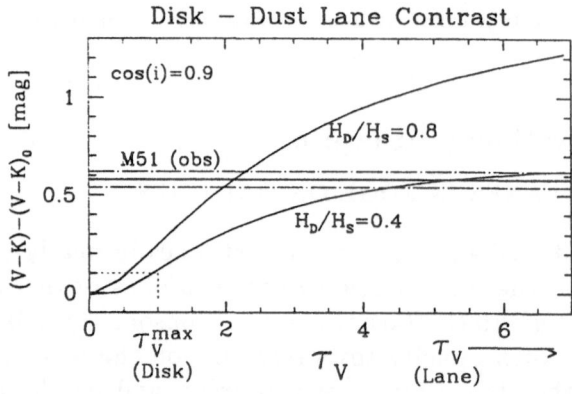

Fig. 4: V-K color contrast between a point in the disk [with τ_V(Disk)] and a dark dust lane [with τ_V(Lane)], predicted by models with $H_D/H_S = 0.4$ and 0.8. The dust free color of the galaxy is denoted by $(V - K)_0$. The observed dust lane - disk color contrast in M51 (RR93) of 0.58 ± 0.04 is well matched by τ_V(Lane) ≥ 5 and τ_V(Disk) $\ll 1$. The observed width of the dust lane < 100pc provides an independent constraint on its scale height $H_D/H_S \lesssim 0.4$. If the disk had an optical depth τ_V^{max}(Disk) $\gtrsim 1$ throughout (i.e. $(V - K) - (V - K)_0 > 0.1$ for $H_D^{Disk}/H_S = 0.4$), no amount of dust in the lane could lead to the observed disk-lane color difference. If the "diffuse" dust throughout the disk has a scale height higher than the lane's, then the limit on τ_V^{max}(Disk) tightens.

It is further necessary to assume that the color of the stellar population in and next to the dust lane changes only little. If suitable color bands (e.g. R and K) are chosen, and if regions of intense star formation are avoided, this should be an acceptable assumption (see e.g. RR93). The test is strengthened if the "diffuse" dust throughout the disk is assumed to be no flatter than the dust lane. As a sample application, we select one of the dust lanes ("Lane 45") in M51 from RR93. The projected width of this dust lane (in the disk plane) is only 80pc, and we assume that its vertical scale height is not much larger than that. Given the independent information (from other galaxies, e.g. Elmegreen, 1980) on the scale heights of the V and K band stellar light, this leads to $H_D^{lane}/H_S \lesssim 0.4$ and $H_D^{lane} < H_D^{diff}$. Figure 3 now shows that (using the models from Section 3) a disk − dust lane color difference of $V - K = 0.58$ is only possible as long as the optical depth outside the lane is less than $\tau_V \lesssim 1$. Even though this technique deserves more detailed modelling than presented here, it clearly is a way to rule out uniformly large optical depths in the inner portions of spiral disks whenever prominent dust lanes are visible.

TABLE 2. Comparison of "Internal Extinction" Corrections

Authors	Wavelength	Relation	Fitted Param.	Remarks
RC3	$B(0.45\mu)$	$A_i = \alpha \log(\frac{a}{b})$	$\alpha = 1.45$	type dependence
PW 92	$H(1.6\mu)$	$\mu = \mu_0 + 2.5\,c\,\log(\frac{a}{b})$	$c = 0.8$	with apert. corr.
Han 92	$I(0.8\mu)$	$\mu = \mu_0 + \tilde{c}\,\log(\frac{a}{b})$	$c = 1.9$	$R \geq 3R_{exp}$
B 94	$I(0.8\mu)$	$m = m_I(0) + a_I\left(1 - \frac{b}{a}\right)$	$a_I = 1.4$	23 galaxies
	$H(1.6\mu)$		$a_H = 0.3 \pm 0.2$	
G 94	$I(0.8\mu)$	$m = m_I(0) + \gamma \log(\frac{a}{b})$	$\gamma = 1.15 \pm 0.2$	

RC3: de Vaucouleurs et al. 1991; PW 92: Peletier and Willner, 1992,

B 94: Bernstein et al. 1994; G 94: Giovanelli et al. 1994

4.3. THE "INTRINSIC EXTINCTION" CORRECTION FOR THE NEAR-IR TULLY FISHER RELATION

The tight relation between the stellar luminosity of a galaxy and its rotation velocity (known as the "Tully-Fisher-Relation", hereafter, TF relation), provides an excellent test for the impact of dust on a galaxy's apparent *total* magnitude. If spiral galaxies were dust-free, there should be no correlation between the apparent magnitudes, $m(i)|_{W_{20}}$, and the inclinations, i, for galaxies with identical rotation speed $W_{20}/\sin(i)$ (as measured through their HI linewidth W_{20}). However, such a correlation, in the sense that more edge-on galaxies appear fainter, has been known from the first studies using B-band photometry. If this B-band "internal extinction correction" (e.g. from RC3) were to scale to the near-IR as the standard extinction curve, it should be negligible there. However, all recent TF work in the NIR has found this correction to be significant (see Table 2).

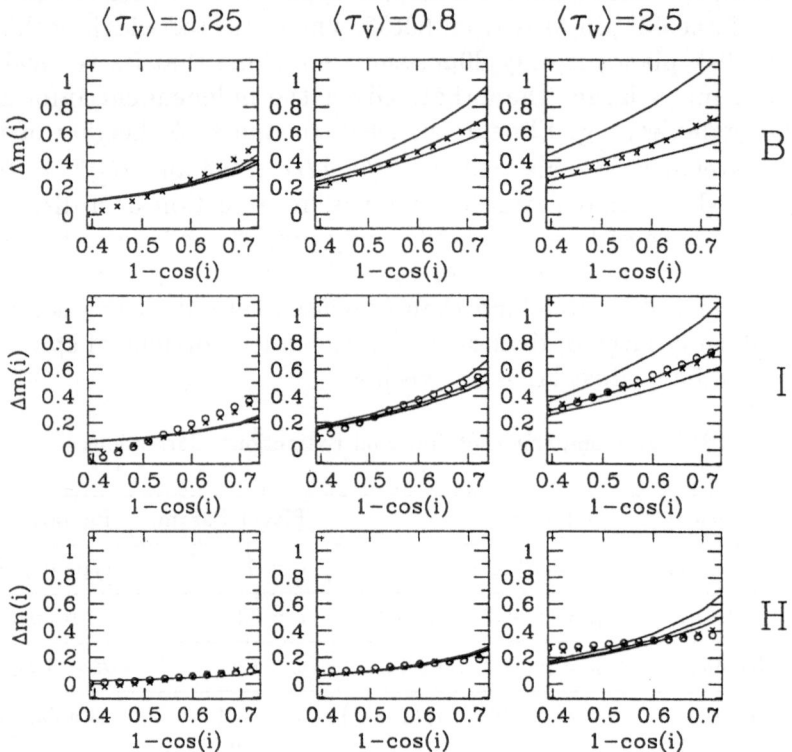

Fig. 5: Comparison of the observed "internal extinction corrections" (points) with radiative transfer models (solid lines). Each panel shows the increasing magnitude difference, $\Delta m(i)$, between a dust free galaxy and a dusty galaxy, as the galaxies become more edge-on (increasing $1 - \cos i$). The panels show $\Delta m(i)$ for three different optical depths and for three different wavelengths. The three lines in each panel represent (top to bottom) models with $H_D/H_S = 0.8, 0.4, 0.2$. The circles represent the I and H band extinction corrections from Bernstein et al. (1994); the crosses are the fits to the data from RC3 in B, Giovanelli et al. (1994) in I and Peletier and Willner (1992) in H. Note that the any vertical offset between the models and the data is arbitrary, because the dust-free luminosity is not an observable. Models with an effective optical depth of $\langle \tau_V \rangle \sim 0.8$ match the data well in all three wavelength bands.

In practice, this correction is determined as follows. Choose a parametriz functional form for the change of apparent brightness with inclination for intrinsically identical galaxies; all authors have used *ad hoc* forms, which are listed in Table 2. Then determine the parameters that minimize the scatter in the TF relation [now $m \equiv m(i, W_{20}/\sin i)$] from the data, usually over the range $0.2 \lesssim b/a \lesssim 0.6$. The results are listed in Table 2. Even though various authors use different parametrizations and differing samples, the resulting "internal extinction corrections", $m(i)|_{W_{20}}$, are mutually consistent, both in the I-band (Bernstein *et al.* , 1994; Giovanelli *et al.* , 1994) and the H-band (Peletier and Willner, 1992; Bernstein *et al.* , 1994). This is illustrated by the circles and crosses in Figure 5.

Using the models from Section 3.2, we can estimate the luminosity-weighted, effective, face-on optical depth, $\langle \tau_V \rangle$, of the galaxies. We consider models with three different optical depths ($\langle \tau_V \rangle = 0.25, 0.8, 2.5$) and with three different scale height ratios ($H_D/H_S = 0.2, 0.4, 0.8$). The predictions for all these models are compared to the observed relations in Figure 5. While the data cannot distinguish between models with different scale height ratios, they can discriminate between models of different optical depths. The model with $\langle \tau_V \rangle = 0.25$ clearly under-predicts the slope of the $m(i)|_{W_{20}}$ *vs* $\cos i$ relation in the I (and B) band and the model with $\langle \tau_V \rangle = 2.5$ clearly over-predicts the slope in the H-band. However, the model with $\langle \tau_V \rangle = 0.8$ matches all bands well.

These results lead to the conclusion that the Sb-Sc galaxies used in the TF studies have an internal dust extinction *equivalent to a uniform, homogeneous dust layer of $\tau_V \sim 0.8$.*

4.4. THE INCLINATION DEPENDENCE OF COLORS AND COLOR GRADIENTS

Rather than considering only the total luminosity as a function of inclination, one could design stronger tests by considering simultaneously the change in luminosity, optical-NIR color and color gradient as a function of inclination. The color gradients can probe the radial profile of effective optical depth, and thus provide a new piece of information. Even though optical-NIR photometry for samples of highly inclined (Terndrup *et al.* 1994) and face-on (de Jong and van der Kruit, 1994) galaxies have been published, no consistent, large data set, sufficient to test $\partial(\text{color})/\partial(\cos i)$ and $\partial(\text{color grad.})/\partial(\cos i)$ exists to date. However, such data sets are currently being assembled.

5. Conclusions

The purpose of this talk has been two-fold. First it tried to give some practical guidelines for designing (and interpreting) observational opacity tests using NIR observations. Second, it tried to present some of the opacity

224

constraints obtained to date.

Several points are worth stressing again in conclusion:

• NIR data beyond $1\mu m$ [especially at $K(2.2\mu m)$] are much more difficult to obtain, but may provide clues that cannot be obtained from data at shorter wavelengths. Even within the NIR the choice of the wavelength region merits thorough consideration.

• The study of a small number of nearly edge-on galaxies with dust lanes shows that in these objects the optical depths (along the line of sight) range from $0.2 \lesssim \tau_V \lesssim 2$.

• The existence of a large color and intensity contrast between the prominent dust lanes (e.g. in M51) and the rest of the disk shows that even the inner portions of disks must have $\tau_V \lesssim 1$.

• The "internal extinction corrections" in the Tully-Fisher relations measured by various authors and in various bands (B, I, H) are *mutually consistent*. All these corrections are matched well by a simple radiative transfer model assuming a mean, face-on optical depths of $\langle \tau_V \rangle \sim 0.8$.

Each of these tests is in need of further modelling. A proper radiative transfer model should produce a robust estimate of the radial profile of τ_V from the nearly edge-on galaxies. Modelling the disk - dust lane color contrast more carfully will provide *localized* measurements of τ_V in the inner disk portions. Finally, it should be checked whether these local constraints are consistent with the global ones derived from the Tully-Fisher relation.

References

Bernstein, G. M., *et al.*, 1994, A.J., **107**, 000.
Barnaby, D. and Thronson, H. A. J., 1992, A.J., **103**, 41.
Byun,Y.-I., 1993, PASP, **105**, 993
de Jong, R. S. and van der Kruit, P. C., 1994, A.A., *in press.*
de Vaucouleurs, G. et al. 1991, "Third Reference Catalogue of Bright Galaxies," (Springer-Verlag: New York), [RC3].
Giovanelli, R., Haynes, M. P., Salzer, J. J., Wegner, G., Da Costa,L. N. and Freundling, W., 1994, A.J., **107**, 2036.
Elmegreen, D., 1980, Ap.J.S., **43**, 37.
Han, M., 1992, Ap.J., **391**, 697.
Jansen, R. A., Knapen, J. H., Beckman, J. E., Peletier, R. F. and Hes, R., 1994, MNRAS, *in press.*
Knapen, J. H., Hes, R., Beckman, J. E., Peletier, R. F., 1991, A.A., **241**, 42.
Kylafis, N. and Bahcall, J., 1987, Ap.J., **317**, 637.
Laor, A. and Draine, B. T., 1993, Ap.J., **402**, 441.
Peletier, R. F. and Willner, S. P., 1992, A.J., **103**, 1761.
Rix, H.-W. and Rieke, M. J., 1993, Ap.J., **418**, 123.
Terndrup, D. *et al.*, 1994, Ap.J., *in press.*
Zaritsky, D., Rix, H.-W. and Rieke, M. J., 1993, Nature, **364**, 313.
Wainscoat, R. J., Hyland, A. R. and Freeman, K. C., 1990, Ap.J., **348**, 85.
Witt, A., Thronson, H. A. J. and Capuano, J. M. J., 1992, Ap.J., **393**, 611.

Question
M Greenberg

I would expect the scale height of the dust to be the same as the scale height of the young stars. Is this the same as the scale height of the background stars? Would this affect your analysis?

Answer
Walter Rix

I did my analysis at a point along the dust lane where there is no OB association, in order to reduce the impact of young stars. Second, I did not use the U and B bands, which will be more affected by young stars. This not-with-standing, I agree that this warrants better modelling. By the way, Elmegreen (1980), using only optical data, has done such modelling, accounting for the different scale heights of different stellar pops.

Question
B Madore

Why do you feel that the physical sizes of individual dust lanes necessarily tell you about the scale height of the ensemble? After all the physical size of a molecular cloud does not tell you the scale height of the distribution of molecular clouds.

Answer
Rix

I don't think that such an inference can be made. However, for my argument I only require that the dust lane at the point I am considering is not much higher than wide. Further, observations of the molecular layer in e.g. the Milky Way show that the molecular clouds have a much lower scale height than the stars.

Question
Zaritsky

How do your estimates of τ change as the scalelength of the dust within the dust lanes of M51 changes?

Answer
Rix

If the dust's scale height is closer to the star's, then the upper limits on the optical depth outside the dust are weaker.

Question
Burstein

Hans, in your graph of "colour W.R.T.K." vs "position offset", I am puzzled by the apparent lack of colour change outside the spiral arm. Please explain.

Answer
Rix

First, in my graph all colours w.r.t. k are set to a mean of zero. So, more accurately the plot shows the colour difference between the lane and the area surrounding it.

Second, there are colour variations outside the lane, typically 0.05-0.1 mag. However, the colour change across the lane is much larger than any of the colour changes seen.

ARCSECOND RESOLUTION OF COLD DUST IN SPIRAL GALAXIES USING OPTICAL AND NIR IMAGING – DUST MASSES INCREASE BY NINE HUNDRED PERCENT

DAVID L. BLOCK
Department of Computational and Applied Mathematics,
Witwatersrand University
Private Bag 3, Wits 2050, South Africa

ADOLF N. WITT
Ritter Astrophysical Research Center,
The University of Toledo
Toledo, OH 43606, USA

AND

PREBEN GROSBøL
European Southern Observatory,
Karl-Schwarzschild Strasse 2,
D-85748 Garching, Germany

Abstract. The extinction cross-section of a dust grain is independent of its temperature. We propose that radiative transfer models (invoking the full effects of multiple scattering) combined with ground-based, digital imaging in the optical and near-infrared (NIR) bands is sensitive to dust grains of *all* temperatures and can be used to map the distribution of both warm *and cold* dust with unprecedented spatial resolution: one order of magnitude better than that possible with the largest sub-mm/mm telescopes and two orders of magnitude better than with IRAS. The method provides a powerful tool for unambiguously identifying and exploring the *widespread* distribution of cold interarm dust illuminated by the general interstellar radiation field of old disk stars. The power of the new method's quantitative accuracy is that, by relying on radiative transfer models invoking continuous media, the cold dust masses we infer are *lower* limits. The transmission coefficient in clumpy, heterogeneous, N-phase scattering ISMs will be even higher. Furthermore, our method does not require a specification of the

J. I. Davies and D. Burstein (eds.), The Opacity of Spiral Disks, 227–242.
© 1995 *Kluwer Academic Publishers.*

wavelength dependence of the far-infrared absorption efficiency. As a lower limit, NGC 4736 and NGC 4826 require a dust mass increase by a factor of $\sim 900\%$ in the form of cold (and very cold) dust. An important conclusion from our work is that the Galactic dust- to-gas ratio, rather than being exceptionally high, may again be rather representative, at least for the spiral galaxies studied here.

1. Introduction

Apart from the detection of cold cirrus dust by IRAS within our own Milky Way Galaxy (eg. Sodroski *et al.* [50]), the presence of cold dust in extragalactic systems has been inferred from far-IR and sub-mm/mm observations (from the early work of Smith [47] to more recent papers eg. Devereux and Young [17]; Andreani and Franceschini [2]; Engargiola and Harper [19]; Fich and Hodge [21]; Clements, Andreani and Chase [15]; Chini and Krügel [14]), but not without hotly debated and controversial conclusions.

Firstly, sub-mm/mm observations themselves *cannot* unambigously identify the presence of cold (15K - 25K) and very cold ($<$ 15K) dust. The inversion of a spectral energy distribution (SED) into a dust temperature distribution is a badly conditioned problem: Hobson and Padman [28] show that it is possible to find markedly *different* dust temperature distributions from an *identical set* of 30 - 1300μm flux measurements. The dilemma concerning the detection of (cold) dust in extragalactic systems and the determination of its mass may be summarised as follows:

1. Cold (15K - 25K) dust is not seen by IRAS (see Figure 8 in Sauvage and Thuan [46]) and further exemplified by Devereux elsewhere in this Volume. Since the scaling of the total infrared emissive power is $\sim T^6$ where T is the grain temperature (Andriesse [3]), cold grains do not emit much energy and add only a small contribution to the low-frequency end of the emission spectrum, even though these cold grains may constitute by far the bulk of the proportion of the dust mass. IRAS dust masses are usually derived from the measured fluxes at 60 and 100 μm and are not sensitive to the flux at longer wavelengths, where the cold dust would radiate.

2. *Very* cold dust ($<$ 15K) will not be seen *even* at sub-mm to mm wavelengths, as beautifully demonstrated by Pajot *et al.* [38]. In some instances has led to claims that very cold dust is simply not present (Clements *et al.* [15]; Carico *et al.* [12]; Eales, Wynn-Williams, and Duncan [18]; Stark *et al.* [52]).

3. Apart from the SED inversion problem being badly conditioned, there are other important reasons why sub-mm observations are far from ideal for determining even the amount of dust which *is detectable* at those wavelengths. Chini and Krügel [14] enumerate these as follows:

   (a) The wavelength dependence of the dust absorption efficiency in the far-IR and sub-mm region is poorly known (Hildebrand [25], Clements *et al.* [15]). Kwan and Xie [34] conclude that the biggest uncertainty in the determination of dust masses is associated with the uncertain dust emissivity law.

   (b) Observations at different wavelengths with different beam sizes are difficult to compare.

   (c) The 'error beam' (extended side lobes) of sub-mm/mm telescopes operating near their minimum working wavelength can be very important, resulting in large calibration uncertainties and difficulties in comparing point and extended sources (J. Lequeux and C. Purton: private communications).

4. The largest mass fraction of interstellar dust resides in the biggest grains, often found in the densest and coldest environments where accretion and coagulation may be important and where radiative heating is weak.

5. The inference of dust masses from CO observations relies on a *very* uncertain derivation of H_2 column density from the CO line intensities (see especially section 4.3 in Allen and Lequeux [1] and section 3 in Lequeux, Allen and Guilloteau [35]).

6. Furthermore, dust-to-gas ratios themselves still need to be confirmed. A full discussion of this problem for the SMC is presented by Lequeux [36]. It is well known that dust-to-gas mass ratios in spiral galaxies observed by IRAS are exceptionally low [46], by $\sim$ an order of magnitude, compared to the standard value in our Galaxy. The mean [warm dust]-to-gas mass ratio determined by Devereux and Young [17] for 58 spirals is $\sim 1/1080$ (with a similar conclusion reached by Sanders, Scoville and Soifer [45]), compared to the canonical Galactic value of $\sim 1/150$.

That the ISM should contain large, cold grains follows from detailed theoretical studies (for example, Mayo Greenberg [this Volume]) and from observational studies both within our Galaxy (star-forming regions, reflection nebulae) and in external systems (for example, Witt *et al.* [61]). The high effective albedo of 0.8 at $2.1\mu m$ measured by us result from scattering by dust grains at least $0.5\mu m$ in radii – *twice as large* as assumed by standard models. Another recent effort by Kim, Martin and Hendry [33] to derive the size distribution of interstellar grains has led to a similar conclusion.

At this NATO Workshop, we present an independent tool which can be used to map the distribution of both warm *and* cold dust with unprecedented spatial resolution: one order of magnitude better than that possible with the largest sub-mm/mm telescopes, and two orders of magnitude better than with IRAS observations. Full details may be found in Block *et al.* [8].

Our method provides information pertaining to dust structures which cannot be secured from IRAS or sub-mm observations. Furthermore, the method does *not* require a specification of β, where $\lambda^{-\beta}$ describes the wavelength dependence of the FIR absorption efficiency: β is, as noted earlier, a crucial but uncertain exponent, when FIR energy distributions are decomposed.

Apart from our optical/near-infrared method being independent of such assumptions, the availability of near-IR array detectors such as NICMOS (HgCdTe) at major observatories will also give a vastly greater portion of the astronomical community the opportunity to study the spatial distribution of warm and cold dust – compared with, for example, only limited access to, and a restricted number of pointings with, the Kuiper Airborne Observatory.

2. A tale of identical tails

It has on many occasions been argued that the presence of large amounts of dust can only be inferred on account of observed colours being redder than, for example, the colours of the reddest ellipticals: that antiquated inference totally neglects the effects of scattering in the radiative transfer process: very large amounts of dust can, on account of multiple scattering, produce almost neutral broad-band colours. Scattering effects are exceedingly important in the optical because of the combination of high dust albedo and high dust extinction.

The distribution of dust has (traditionally) been inferred from colour excesses *shortward* of the 1 micron window, where stellar Population II disks are not yet transparent eg. B(blue; 0.44 μm) minus I (0.83 μm), or R (0.70 μm) minus I. At a wavelength such as I, the extinction optical depth is still large, a full 50% that in V ($\tau_I = 0.48; \tau_V = 1.00$ – see Martin and Whittet [37]). One does not actually know how much of the extinction at I is due to scattering (Witt, Oliveri and Schild [59]) and how much is due to absorption. As a result, quantitative estimates of the dust amount will be uncertain. Not so at K$'$ (the K$'$ filter is similar to the 2.2 μm K filter, but shifted 0.1 μm towards the blue, to reduce the thermal background [54]).

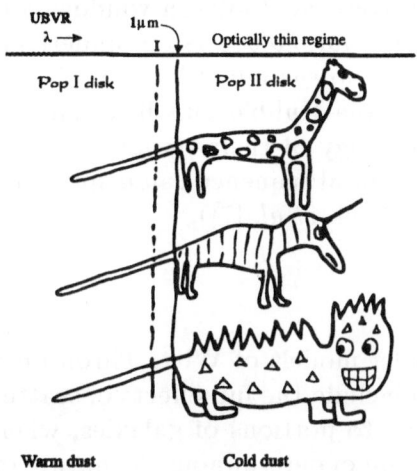

Figure 1. Vast amounts of cold dust, illuminated by the interstellar radiation field of old disk stars, could be entirely missed in galaxies still optically thick at I.

All previous observational probes of the internal extinction at optical wavelengths (from the early work of Holmberg [30], [31] to more recent papers eg. Valentijn [53]; Peletier and Willner [39]; Huizinga and van Albada [32]; Boselli and Gavazzi [10]; Walterbos, Braun and Kennicutt [56]) do not incorporate the effects of multiple scattering by dust. Radiative transfer models involving scattering by dust show that it is imperative – *not optional* – to consider B-K (or B-K′) color maps [not B-I]. Earlier observational probes have invariably accounted for *absorption* (not extinction) when quantifying the optical depths which are consistent with the observed colours.

The accompanying cartoon above (adapted from Conti [16]), nicely illustrates the difficulty of using optical SEDs to characterize the amount of dust in galaxies presenting identical 'tails' in the BVRI regime. While optical colour indices quickly saturate in the presence of optically thick dust embedded thoroughly within a stellar system, the B-K colour index retains its sensitivity to differing amounts of dust because of the relative transparency of such systems at the K wavelength – at K, the extinction optical depth is only $\tau_K = 0.095$.

The *scattering optical depth at K* is therefore very small. For any possible value of the albedo at infrared wavelengths – and this may be high (Witt *et al.* [61]) – dust scattering is far less important than at wavelengths BVRI. The great power of imaging galaxies with NICMOS and other near-infrared camera arrays at K, is that the opacity (be it due to scattering or absorption) is always low.

Indeed, if one crosses the 2 micron window, a spiral galaxy can present a completely different *morphology* as the underlying stellar population disk becomes almost *fully transparent* (Block and Wainscoat [5]). Near-infrared images confirm that the Hubble classification of spiral galaxies does *not* constrain the morphology of their stellar Population II disks: galaxies on opposite ends of the spiral sequence can display remarkably similar evolved disk morphologies (Block *et al.* [7]).

3. The Method

The radiative transfer models of Witt, Thronson and Capuano [60] (hereafter WTC), which include the full effects of scattering as well as of absorption, may be applied to portions of galaxies, where the effects of dust are apparent, with the aim of determining the nature of the star/dust geometry and the amount of dust likely to be present. Colour probes such as V-K are almost totally dominated by the extinction at V, allowing V optical depths to be properly and correctly evaluated. Our methodology may be delineated as follows:

1. Optical and near-infrared colour excesses in dusty galaxies are determined.
2. Using the multiple-scattering radiative transfer models of WTC, these colour excesses are used to estimate the V extinction optical depth.
3. These optical depths are converted into dust column densities and hence dust masses.

The quantitative power of the new technique is that the ratio of our computed dust masses to the IRAS dust masses can never be decreased by *a priori* invoking clumpy interstellar media, even though the dust in the WTC models is assumed to be continuously distributed (although the 'starburst' and 'cloudy' models have dust filling only 33% of the volume occupied by stars).

It is well known that the interstellar medium has a complex, *clumpy* structure; detailed observations of giant molecular clouds (hereafter GMCs) have, without exception, revealed a high degree of internal structure, even in the absence of star formation (Williams and Blitz [58]; Falgarone, Puget and Pérault [20]). Using the methods of unsharp masking and image amplification (see [4]), Block, Dyson and Madsen [6] showed that 'primordial' GMC clump-interclump molecular structure can be *optically* probed. Now even a two-phase (clump-interclump) heterogeneous medium is only a first approximation to a realistic ISM where GMCs contain a rich hierachy of structure (Plate L2 of Block *et al.* [6]). Furthermore, the GMCs themselves are not homogeneously distributed within interstellar space.

But the effects of a multiphase, hierachical density structure on the penetration of continuum radiation into clumpy molecular clouds has now been carefully studied: from the pioneering paper of Boissé [9] to the more recent investigation of Hobson and Scheuer [26]. As also further elaborated by Witt [this Volume], *clumpy media with the same amount of dust will always have a higher transmission coefficient than continuous media do.* Consequently, by relying on continuous media models, our computed dust masses are always *lower* limits to the actual amount of dust present: if our radiative transfer models yield an additional dust mass increase (by a factor of 900%) in the form of cold dust, at sufficiently low temperatures to have been undetected by IRAS, this percentage cannot be decreased.

In the WTC models, differentiations between screen and embedded dusty geometries are prescribed by the ratio of surface brightness reduction in the V-band, ΔV, to the corresponding reddening E(V-K), at unit E(V-K). This ratio has the value 1.18 (or less, if the K' albedo exceeds 0.2) for screen geometries, while embedded geometries have values of 1.30 and greater [60]. Similar contrasts are evident if V is replaced by B surface brightness data.

4. Analysis

We demonstrate the power of subtracting K' images from optical images by presenting such colour maps for two early type spirals, NGC 4736 and NGC 4826.

4.1. EMBEDDED COLD INTERARM DUST IN NGC 4736

NGC 4736, with its preferentially face-on orientation (i $\approx$ 35 degrees) and post-starburst mode, is well suited in attempts to probe cold, interarm dust. Extensive amounts of observations, at different wavelengths, of this early (species: ab) spiral, abound in the literature. The galaxy is famous for its inner, 'knotty ring' of HII regions at a radius of $\sim$ 50 arcseconds (see Figure 1(a) in [8]). Several airborne observations of NGC 4736 have been secured with the Kuiper Airborne Observatory (KAO), the most recent being important new, high spatial resolution 50 and 100 μm measurements reported by Smith *et al.* [49]. Because of the absence of spectral signatures pointing to young massive star formation, Smith *et al.* [49] conclude that the FIR emission from within the inner HII ring, radius $\sim$ 50″, *is not powered by young massive stars.*

NGC 4736 = IRAS 12485+4123 has been detected in the sub-mm by Chini, Kreysa and Mezger [13] (see their Table 1; note that the listing of NGC 4736 there appears under its UGC designation, which is UGC 7996). Chini *et al.* [13] fitted a 2-component model to the FIR-submm spectrum,

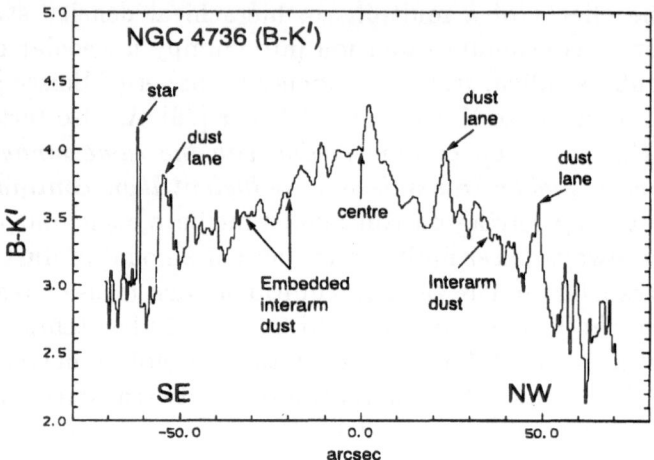

Figure 2. A SE-NW colour cut in B-K′, through the galaxy centre of NGC4736.

yielding two temperatures: 15K and 52K. This is certainly more accurate than the 41.5K of Sage and Isbell [42], which, being based on IRAS fluxes, is intermediate, as expected. A temperature of 41.5K supports our contention of a skewed perception of dust emission as seen by IRAS, which misses the evidence for cold dust in this galaxy entirely.

Our B minus K′ colour cut is presented in Figure 2 while the complete B-K′ colour map is presented in Figure 3. Prominent dust *lanes* demarcate the spiral arms itself – compare the dust lanes in our optical photograph ([8]) to the black dust lanes seen in Figure 3 here. The positive B-K′ enhancements in Figure 3 clearly betray the presence of dust rather than viewing a population of red, metal rich stars (see Section 5 for a full discussion); there is no reason why such a population should coincide exactly with the dust lanes demarcating the spiral structure in optical images.

The internal extinction rises sharply in the warm spiral arm component associated with the OB stars and HII regions: values for B-K′ in the dust lane ∼ 20 arcsec NW from centre attain 4 magnitudes. To determine intrinsic colours, we empirically decomposed our K′ image (azimuthally averaged and inclination corrected) into bulge and exponential disk components, and assigned empirically determined B-K′ colours for galactic disks (NGC 4449: B-K′ = 2.9) and for bulges (B-K′ = 4.1) following Bruzual and Charlot [11]. We then added the colours, weighted by the relative contributions from bulge and disk, as a function of radius. This suggests an intrinsic B-K′ colour for the underlying stellar population in NGC 4736 of 3.2 to 3.4. Hence, reddenings of the order E(B-K′) = 0.6 to 0.8 are inferred here. For a 'dusty galaxy' environment (wherein stars and dust are

Figure 3. Uniform, homogeneous, plane-parallel ISMs in dusty disks? – caution! The widespread distribution of embedded cold *interarm* dust, seen at arcsecond resolution, in this B-K′ colour map of NGC 4736. Positive B-K′ colour excesses are coded black. The IRAS dust mass (sensitive to warm dust grains delineating the spiral arms) underestimates the total dust mass by a factor of ∼ 900%.

uniformly mixed [60]), these reddening values imply optical depths in V in the range 2 to 4.

For NGC 4736, the Gérin, Casoli and Combes [22] CO (1-0) observations imply extinctions in the bulge area of $A_V \sim 3$ magnitudes. The extinctions have also been estimated from the KAO observations of Smith *et al.* [49], who again find 'best' V extinction estimates, for the central bulge area, of ~ 3 magnitudes. From their near-infrared spectroscopy, Walker, Lebofsky and Rieke [55] give a central extinction of ~ 3 magnitudes (see their Table 1). The extinctions in V estimated using our radiative transfer models are between 2×1.086 to 4×1.086, in excellent agreement with NIR, FIR and CO estimates for A_V.

Secondly, in Figure 3 we affirm the existence of widespread, cold *inter-arm* dust within the inner HII knotty ring. That one can now study the actual distribution (at arcsecond resolution) of cold dust grains is because the extinction cross section of a dust grain is independent of its temperature: therefore, the presence of such grains in the interarm regions is betrayed through illumination by the interstellar radiation field of old disk stars. The pattern in the distribution of the interarm dust in NGC 4736 is reminiscent of the distribution of discrete dust clouds in M31 (which do not appear to show any affinity to the spiral structure of M31) mapped by Hodge [29]. On the basis of observed B and K' surface brightness profiles (the reader is referred to Section 5 for a discussion of how optical and NIR surface brightness profiles are used to uniquely identify dust-caused colour excesses from red star-produced colour excesses) and the B-K' colour map – which includes the entire HII ring from -50 to +50 arcsec – we conclude that the dust is *embedded* rather than contained in an overlying, foreground screen. This conclusion is based on the widespread presence of filamentary structures in the B-K' colour cut Figure 2, the ratio of visual extinctions to colour excesses, and the relatively large ratio of L(IR)/L(Opt) (see below). Typical reddening values are E(B-K') = 0.2 or more in the interarm regions. That the cold interarm dust is illuminated by the old stellar Population is in excellent accord with Figure 5 of Smith *et al.* [49], who finds that the distribution of the 100 μm emission does *not* trace the spatial distribution of young stars delineated from Hα+[NII] observations but *does* trace the distribution of old stars seen through their broadband red (F) image.

For a 'dusty galaxy' model, this colour excess corresponds to V optical depth values of order 0.75. Adopting a distance to NGC 4736 of 7 Mpc (Sandage and Tammann [44] with H=50 km s^{-1} Mpc^{-1}) and integrating over the inner disk out to a radius of 60 arcseconds, yields a dust mass of $3 \times 10^6 M_\odot$. If the 'cloudy' dust model of WTC is used – where there is a filling factor of 0.33 – the dust mass approaches $6 \times 10^6 M_\odot$. For the radiative transfer in a multiphase, clumpy medium in NGC 4736, all dust estimates will be even *larger* (Hobson and Scheuer [26]; Hobson and Padman [27]). By contrast, the dust mass of NGC 4736 derived from IRAS data is is only $7 \times 10^5 M_\odot$ (See Figure 7b of Sage [41]). This clearly illustrates that the bulk of the dust mass is apparently at low enough temperatures to have been undetected by IRAS and that the IRAS dust mass underestimates the total dust mass by $\sim$ 900%.

While optical colour indices quickly saturate in the presence of optically thick dust embedded thoroughly within a stellar system (Witt *et al.* [60]), the B-K$'$ colour index retains its sensitivity to differing amounts of dust because of the relative transparency of such systems at the K$'$ wavelength.

The key point is that embedded cirrus dust (not associated with any star forming regions) *can* be deconvolved, its existence affirmed shortward of 2.5 μm and studied using appropriate optical and 2.1 μm colour enhancements. Grosbøl [23] determined an optical (R) scale length for the exponential disk in NGC 4736 of 35.1$''$; *cold dust in Figure 3 is distributed over more than one (optical) disk scalelength.*

Furthermore, the disks of spiral galaxies can be both optically thick or optically thin, depending on whether the warm spiral arm dust, or cold interarm dusty areas, are being probed. This inhomogeneity in extinction with galactocentric radius is attested to by earlier studies (for example, White and Keel [57]), where the observed extinction in the galaxy pair AM1316-241 is found to be largely confined to the spiral arms of the foreground galaxy.

4.2. COLD DUST IN NGC 4826

We now turn our attention to NGC 4826, described by Sandage as the earliest type Sb galaxy in the *Hubble Atlas* (Sandage [43]). The designation of this galaxy in the IRAS catalogue is 12542+2157. NGC 4826 is detected in the sub-mm – see Table 1 of Chini *et al.* [13], where the galaxy appears under its UGC 8062 designation. Cold and warm dust temperatures are given by Chini *et al.* [13] as 16K and 48K respectively. Sage and Isbell [42] derive an intermediate dust temperature of 32.7K for NGC 4826, based on IRAS fluxes and a dust emissivity proportional to ν^{-1}. 3mm HCN and CS emission, tracing high density gas (10^5 cm^{-3}) in the bulge of NGC 4826, has been recently reported by Helfer and Blitz [24].

Our V-K$'$ colour map is presented in [8], while Figure 4 shows a NE-SW cut through the galaxy centre. Earlier (optical) investigations into the NGC 4826 dust/stellar disk geometry have extended up to I; in their BVI study, Walterbos, Braun and Kennicutt [56] do not incorporate the effects of scattering by dust. Our radiative transfer models, with multiple scattering, firstly allow us to distinguish a *foreground dusty screen* in the NE

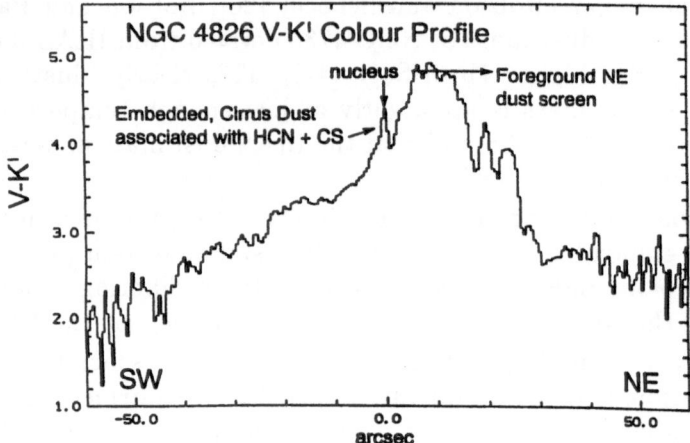

Figure 4. A NE-SW colour cut in V-K', through the centre of NGC 4826.

quadrant. We systematically surveyed the screen in NGC 4826 by producing surface brightness profiles oriented through the centre of the galaxy but sequentially spaced 10 degrees each in position angle. By assuming symmetry with respect to the galaxy's center for both the V-surface brightness profile and the V-K' colour profile of the underlying bulge and disk, a least squares fit (for 108 ΔV, E(V-K') measurement pairs) yields ΔV / E(V-K') =1.04 ± 0.05 for the NE dark lane.

The WTC models then allow us to identify this value with the geometry of a foreground screen with an average V-optical depth, $\tau(V) \approx 2.1$. The most optically opaque regions in the screen have $\tau(V) \sim 3$.

An important remark is that our determination of colour excesses here does *not* depend on the intrinsic stellar population of the galaxy but simply on symmetrically opposite on-screen (NE) and off-screen (SW) positions (see Witt *et al.* [61] for further details). Because of the screen geometry, E(V-K') is simply a differential colour excess produced by the dust screen alone. As seen in Figure 4, observed values of V-K' are $\sim$ 5 magnitudes in the dense NE screen. The optically thick, overlying screen is juxtaposed right up to the nucleus of this spiral. Indeed, the nucleus is clearly revealed for the first time (see Figure 3b in [8]).

Secondly, we identify an envelope of dust which contains the inner nuclear region of NGC 4826 inside a galactocentric radius of 20 arcsec, corresponding to $\sim$ 700 pc. The dust density increases gradually toward the nucleus and corresponds to an optical depth at V of order 5 if stars and dust are well mixed, or of order 2, if a dust screen simply envelopes the nucleus. The dust masses associated with the nuclear region of NGC 4826, corresponding to these two possible scenarios, range from 1.4×10^6 M$_\odot$ to

6×10^6 M$_\odot$, respectively, and refer to the region with r < 700 pc.

We remarked earlier that the arcsecond spatial resolution attainable by our method, in probing dust grains of all temperatures, is one order of magnitude better than that attainable with the largest sub-mm telescopes, such as JCMT and CSO. We consider it very likely that the central cirrus dust which we can see in our colour maps within a radius $\sim$ 700 pc of the nucleus is associated with the source for the HCN and CS emission reported by Helfer and Blitz [24].

The ratio of L(IR)/L(Opt) is at least twice as large for NGC 4736 as it is for NGC 4826 (see Rice *et al.* [40], Soifer *et al.* [51]) and attain values of 0.33 and 0.16 respectively. These numbers, coupled with far less visible dust presence in NGC 4736 compared to NGC 4826, speak strongly for a widely embedded geometry there; in the case of NGC 4826, the 50% lower L(IR)/L(Opt) ratio despite a very prominent dust feature supports the screen model for this particular feature. The dust embedded in NGC 4826 may only be 25% of that found in NGC 4736.

Assuming a distance for NGC 4826 of 7 Mpc (Sandage and Tammann [44]), we compute a dust mass for the screen of 10^6 M$_\odot$. The total mass of dust is then $(1.4 - 6 \times 10^6$ M$_\odot$) [nuclear region] + 1×10^6 M$_\odot$ [for the screen]. In contrast, the total amount of dust derived from the IRAS data of NGC 4826 is a full order of magnitude less $- 3 \times 10^5$ M$_\odot$ (see Sage [41]). Again, we stress that the geometry we have adopted in our radiative transfer models leads to a lower limit for the deduced dust mass from the given (observed) amount of reddening. Dust distributed in clump-interclump (N=2) or yet larger (multiphase) scattering media leads to higher dust mass estimates. The crucial point is that even this lower limit is almost an order of magnitude larger than the IRAS dust mass. *As a lower limit, our analysis does require substantial quantities of cold dust in the interstellar medium of this galaxy.*

5. Stellar Population Gradients and Luminous Concentrations of Red Stars

It is well known that stellar population gradients occur in disk galaxies, because with increasing galactocentric radius, the projected densities of the bulge population and of the disk population diminish according to their own *specific* laws (usually generalized as $r^{1/4}$ for the former and an exponential law for the latter, although morphological signatures such as prototypical oval galaxies may have non-exponential disks). This leads to a gradual bluing of the galactic light with increasing radius. Our method of detecting widely distributed cold interarm dust in disk galaxies is cognisant of this gradient and our estimates of V-K' or B-K' colour excesses are solely based upon *localized* deviations from the overall colour gradient.

Furthermore, the effect in our method of a blueward radial stellar population gradient with increasing galactocentric radius, would be to increase localised V-K' or B-K' colour excesses, which in turn would increase derived optical depths and therefore yield yet larger dust masses. We reiterate that our method strictly produces lower dust mass limits.

A final caveat: a localized excess in V-K' or B-K' could also be due to a concentration of luminous red stars. However, this would then lead to an excess in the surface brightness of the galactic disk by a *greater* amount at K', compared to a lesser amount at an optical wavelength. In contrast, for the colour excesses we attribute to dust, the corresponding optical (V or B) surface brightness profiles are locally depressed with respect to neighbouring positions, while the K'-band surface brightness profile shows a *much smaller or non-detectable* surface brightness depression. Surface brightness profiles, used with colour maps, offer a powerful handle to uniquely identify star-produced colour excesses from colour excesses caused by dust.

6. Conclusions

1. The identification, quantification and distribution of widespread cold (15K - 25K) dust in galaxies has remained a widely debated and controversial issue. Definitive answers cannot come from inversion of 30-3000μm SEDs into dust temperatures.

2. The extinction cross-section of a dust grain is independent of its temperature. Therefore, we propose that optical and near-infrared (K or K') surface brightness profiles and colour maps provide a unique opportunity to actually identify and 'see' the distribution of large, cold dust grains – at arcsecond resolution.

3. We have illustrated the technique for two early type spiral galaxies, NGC 4736 and NGC 4826, for which cold dust has been inferred by Chini *et al.* [13] from sub-mm observations. Invoking radiative transfer models with multiple scattering, dust masses are computed which are higher than IRAS dust masses by a factor of $\sim$ 900%.

4. Our method of combining high resolution digital imaging with radiative transfer models (wherein the effects of multiple scattering are fully incorporated) does *not* require a specification of the wavelength dependence of the far-infrared absorption efficiency.

5. The disks of spiral galaxies such as NGC 4736 can be *both* optically thick *or* optically thin, depending on whether the warm spiral arm dust, or cold interarm dusty areas, are being probed.

6. The Galactic dust-gas ratio may not be exceptionally high, but may again be representative, at least for the spiral galaxies studied here. (A

similar conclusion has recently been reached by Fich and Hodge [21] –
but for their sample of elliptical and SO galaxies).

7. The power of our new method's quantitative accuracy is that, by relying on the transfer of radiation in continuous media, the dust masses
   we infer are always *lower limits*. For clumpy, heterogeneous multiphase
   scattering ISMs, computed dust masses are even higher. *Neither NGC
   4736 or NGC 4826 can be succesfully modelled without invoking cold
   dust in the interstellar media of these galaxies, necessitating an increase in their computed dust masses by at least 900%.*

Acknowledgements. We dedicate this paper to Mike Disney, for his un-
ending *determination* to shoot crow during the Cardiff Workshop!

The hospitality extended to DLB as a Visiting Scientist at ESO (Garching and Chile) and at the IFA (Hawaii) is deeply appreciated, as is the
generous allocation of telescope time. Our warm appreciation is expressed
to our collaborators A. Stockton and A. Moneti. It is a great pleasure to
thank J. Lequeux for his incisive comments on our optical/NIR method.
Financial support to DLB for this entire program has been received from a
Witwatersrand University Council Fellowship, from Professors J.P.F. Sell-
schop and P.D. Tyson, from the Mellon Foundation (New York) and the
Office of the Director General (ESO). We warmly thank M. Greenberg, N.
Devereux, M. Sears, G.C. Buric, B. Stobie and H. Thronson for their en-
couragement and the Workshop Organizers for a truly memorable week in
Cardiff.

References

1. Allen, R.J., Lequeux, J.: 1993, *ApJ.* **410**, L15
2. Andreani, P., Franceschini,A.: 1992, *A & A.* **260**, 89
3. Andriesse, C.D.: 1977, 'Radiating cosmic dust' in *Vistas in Astronomy* **21**, 107
4. Block, D.L.: 1990, *Nat.* **347**, 452
5. Block, D.L., Wainscoat, R.J.: 1991, *Nat.* **353**, 48
6. Block, D.L., Dyson, J.E., Madsen,C.: 1992, *ApJ.* **390**, L13
7. Block, D.L., Bertin, G., Stockton, A., Grosbøl, P., Moorwood, A.F.M., Peletier,
   R.F.: 1994, *A & A* **288**, 365 (Paper I)
8. Block, D.L., Witt, A.N., Grosbøl, P., Stockton, A., Moneti, A.: 1994, *A & A* **288**,
   383 (Paper II)
9. Boissé, P.: 1990, *A & A.* **228**, 483
10. Boselli, A., Gavazzi, G.: 1993, *A & A.* **283**, 12
11. Bruzual, G., Charlot, S.: 1993, *ApJ.*, **405**, 538
12. Carico, D.P., Keene, J., Soifer, B.T., Neugebauer, G.: 1992, *PASP.* **104**, 1086 1193
13. Chini, R., Kreysa, E., Mezger, P.G.: 1986, *A & A.* **166**, L8
14. Chini, R., Krügel, E.: 1993, *A & A.* **279**, 385
15. Clements, D.L., Andreani, P., Chase, S.T.: 1993, *MNRAS.* **261**, 299
16. Conti, P.S.: 1986, in *Luminous Stars and Associations in Galaxies* IAU Symp. No
    116 (ed. C.W.H. de Loore, A.J. Willis and P. Laskarides, Reidel, Dordrecht), p.
    199.

242

17. Devereux, N.A., Young, J.S.: 1990, *ApJ*. **359**, 42
18. Eales, S.A., Wynn-Williams, C.G., Duncan, W.D.: 1989, *ApJ*. **339**, 859
19. Engargiola, G., Harper, D.A.: 1992, *ApJ*. **394**, 104
20. Falgarone, E., Puget, J.L., Pérault, M.: 1992, *A & A*. **257**, 715
21. Fich, M., Hodge, P.:1993, *ApJ*. **415**, 75
22. Gérin, M., Casoli, F., Combes, F.: 1991, *A & A*. **251**, 32
23. Grosbøl, P.: 1985, *A & A Supp*. **60**, 261
24. Helfer, T.T., Blitz, L.: 1993, *ApJ*. **419**, 86
25. Hildebrand, R.H.: 1983, *QJRAS*. **24**, 267
26. Hobson, M.P., Scheuer, P.A.G.: 1993, *MNRAS*. **264**, 145
27. Hobson, M.P., Padman, R.: 1993, *MNRAS*. **264**, 161
28. Hobson, M.P., Padman, R.: 1994, *MNRAS*. **266**, 752
29. Hodge, P.W.: 1980, *AJ*. **85**, 376
30. Holmberg, E.: 1958, *Medn. Lunds. Astr. Obs.* **2**, 136.
31. Holmberg, E.: 1975, in Sandage, A., Sandage, M., Kristian, J. (eds.) *Stars and Stellar Systems*. Univ. Chicago Press, Chicago, p. 167.
32. Huizinga, E., van Albada, T.: 1992, *MNRAS*. **254**, 677
33. Kim, S.-H., Martin, P.G., Hendry, P.D.: 1994, *ApJ*. **422**, 164
34. Kwan, J., Xie, S.: 1992, *ApJ*. **398**, 105
35. Lequeux, J., Allen, R.J., Guilloteau, S.: 1993, *A & A*. **280**, L23
36. Lequeux, J.: 1994, *A & A*. **287**, 368
37. Martin, P.G., Whittet, D.G.B.: 1990, *ApJ*. **357**, 113
38. Pajot, F., Boissé, P., Gisbert, A., Lamarre, J.M., Puget, J.L., Serra, G.: 1986, *A & A*. **157**, 393
39. Peletier, R.F., Willner, S.: 1992, *AJ*. **103**, 1761
40. Rice, W., Lonsdale, C.J., Soifer, B.T., Neugebauer, G., Kopan, E.L., Lloyd, L.A., de Jong, T., Habing. H.J.: 1988, *ApJS*. **68**, 91
41. Sage, L.J.: 1993, *A & A*. **272**, 123
42. Sage, L.J., Isbell, D.W.: 1991, *A & A*. **247**, 320
43. Sandage, A.: 1961, *Hubble Atlas of Galaxies*. Washington DC: Carnegie Institution
44. Sandage, A., Tammann, G.A.: 1981, *A Revised Shapley-Ames Catalogue of Bright Galaxies* Washington DC: Carnegie Institution
45. Sanders, D.B., Scoville, N.Z., Soifer, B.T.: 1991, *ApJ*. **370**, 158
46. Sauvage, M., Thuan, T.X.: 1994, *ApJ*. **429**, 153
47. Smith, J.: 1982, *ApJ*. **261**, 463
48. Smith, B.J., Lester, D.F., Harvey, P.M., Pogge, R.W.: 1991, *ApJ*. **373**, 66
49. Smith, B.J., Harvey, P.M., Colome, C., Zhang, C.Y., DiFrancesco, J., Pogge, R.W.: 1994, *ApJ*. **425**, 91
50. Sodroski, T.J., Dwek, E., Hauser, M.G., Kerr, F.J.: 1989, *ApJ*. **336**, 762
51. Soifer, B.T., Sanders, D.B., Madore, B.F., Neugebauer, G., Danielson, G.E., Elias, J.H., Lonsdale, C.J., Rice, W.L.: 1987, *ApJ*. **320**, 238
52. Stark, A.A., Davidson, J.A., Harper, D.A., Pernic, R., Loewenstein, R., Platt, S., Engargiola, G., Casey, S.: 1989, *ApJ*. **337**, 650
53. Valentijn, E.: 1990, *Nat.* **346**, 153
54. Wainscoat, R.J., Cowie, L.: 1992, *AJ*. **103**, 332
55. Walker, C.E., Lebofsky, M.J., Rieke, G.H.: 1988, *ApJ*. **325**, 687
56. Walterbos, R.A.M., Braun, R., Kennicutt, R.C.: 1994, *AJ*. **107**, 184
57. White, R.E., Keel, W.C.: 1992, *Nat.* **359**, 129
58. Williams, J.P., Blitz, L.: 1993, *ApJ*. **405**, L75
59. Witt, A.N., Oliveri, M.V., Schild, R.E.: 1990, *AJ*. **99**, 888
60. Witt, A.N., Thronson, H.A., Capuano, J.M.: 1992, *ApJ*. **393**, 611
61. Witt, A.N., Lindell, R.S., Block, D.L., Evans, R.: 1994, *ApJ*. **427**, 227

UNVEILING STARS AND DUST IN SPIRAL GALAXIES

R. F. PELETIER[1,2]
Isaac Newton Group, Apartado 321, 38700 Santa Cruz de la Palma, Spain
[1] *Kapteyn Astronomical Institute, Postbus 800, 9700 AV Groninger The Netherlands.*
[2] *Instituto de Astrofísica de Canarias, 38200 La Laguna, Tenerife, Spain.*

E.A. VALENTIJN
SRON, Postbus 800, 9700 AV Groningen, The Netherlands.

A.F.M. MOORWOOD
ESO, K. Schwarzschildstr. 2, D-85748 Garching bei München, Germany

AND

W. FREUDLING
ST-ECF, K. Schwarzschildstr. 2, D-85748 Garching bei München, Germany

Abstract.
We have obtained deep K-band images for a sample of 37 Sb and Sc galaxies, and determined radial surface brightness profiles. Since for each of these galaxies B and R images (scanned for the ESO-LV galaxy catalog) were available, we have determined radial profiles in B and R as well, and compared them with those in K. Our main conclusion is that these galaxies exhibit color gradients that are much larger than those found in elliptical galaxies: B and K scale length ratios vary between 1.2 and 2 and increase with axis ratio. The large size of these gradients suggests that they are not due to stellar population differences, but primarily due to extinction by dust in the B-band. We find that although $B - K$ colors at the centers and at the effective radii are redder for galaxies of higher inclinations (more edge-on) this is no longer evident in the outer parts. Modeling shows that an average face-on galaxy has an extinction of about $A_B = 1$ - 2 mag in the center and less than 0.4 mag at 3 scale lengths. We do not manage to find

J. I. Davies and D. Burstein (eds.), The Opacity of Spiral Disks, 243–258.
© 1995 *Kluwer Academic Publishers.*

244

a model with $\alpha_d/\alpha_*=1$ that fits the observed scale length ratios, implying that we either need more complicated models, or that these galaxies contain a dust component that is more extended than the stars.

1. Introduction

Knowledge of the optical depth of spiral galaxies is important e.g. for our understanding of galaxy evolution and the application of global distance indicators such as the Tully-Fisher relation. So far, most workers have accepted the classical evidence that spiral galaxies are optically thin, which is based on the analyses of the surface brightness of galaxies at different inclination angles (e.g. Heidmann *et al.* 1972 and references therein). Recently, variations of that test have been applied to new data in several bands (e.g. Valentijn (1990, hereinafter V90), Burstein *et al.* 1991, Huizinga & van Albada 1992, Valentijn 1994 (hereinafter V94), Byun 1992, Giovanelli *et al.* 1994.), most of which have been presented at this conference. However, these papers often differ strongly in their conclusions, even those analyzing the same set of data (for a detailed discussion of this see V94). One reason that these tests are intrinsically difficult is that they are extremely sensitive to selection effects in the galaxy samples used. This motivated us to use a completely independent approach to investigate the same issue. We use K(2.2μm)-band surface photometry of a subsample of the sample analyzed by V90 and Huizinga & van Albada (1992) to determine the underlying stellar distribution and compare it to B and R-band photometry of the same galaxies to directly estimate the extinction in the B-band. Such a comparison is intrinsically less sensitive to selection effects in the catalogue from which the samples are drawn than the previously mentioned tests. This is because both the primary quantity investigated here, color, is not closely related to parameters, such as surface brightness, used to select galaxies for the catalogues and the fact that the restricted size of the sample permits us to study the galaxies in more detail.

Photometry in the optical bands does not directly reflect the stellar mass distribution in spiral galaxies. The presence of dust, if not accounted for, leads to an underestimate of the mass in stars while, on the other hand, the presence of bright young stars tends to cause an overestimate. A virtually extinction-free measure of the intrinsic spatial distribution of the stars can be obtained from surface photometry in the K-band at 2.2 μm, where the extinction is about a factor 10 smaller than in V. For example, Rix & Rieke (1993) showed that the K-band image of one representative spiral, M51, is not significantly affected by dust extinction and, after testing for the possible supergiant contribution (by means of the gravity-sensitive 2.3 μm CO index), that it does reliably trace the mass in the disk.

Here we present surface photometry in the K-band obtained for 37 galaxies with the IRAC2 camera at the ESO/MPIA 2.2m telescope (Moorwood et al. 1992) to study the mass distribution of Sb and Sc galaxies in general. We have selected a sample of randomly oriented, normal Sb and Sc galaxies from the ESO-LV catalog (Lauberts & Valentijn 1989), which is a subsample of the sample investigated by V90. As far as we know, this is the first systematic infrared array detector survey of normal spirals which includes galaxies of all inclinations. The availability of both B and K for the same galaxies provides us with the means for determining the amount of extinction in B, since K is essentially unaffected by it. Since the intrinsic color of the stellar population is unknown, this cannot be done accurately for every galaxy individually. However, one can separate the effects of inclination from those of stellar populations by investigating the behavior of color as a function of inclination, since only extinction is inclination-dependent.

The expected radial behavior of colors for various amounts and geometries of dust has been modeled extensively in Evans (1994), (see also V90, Burstein et al. (1991) and Peletier & Willner (1992)). In order to facilitate the inclination tests in B, R and K our sample has been selected to have a uniform distribution in cos(i), i.e. is representative of the distribution of galaxies on the sky. With this sample we can measure in a statistical way the extinction as a function of radius in Sb and Sc galaxies. Although the sky background in K is considerably larger than in the optical and the K-surface photometry is slightly less deep than ESO-LV, we obtain reliable K-band surface photometry and colors up to radii corresponding to the blue D_{25} diameter.

Here in this paper we will first give a short description of the observations and the data reduction. After that some parameters are presented that follow directly from the observations, and we will discuss what they imply for the dust distribution in these spirals. At the end we present a slightly more complicated model to explain the scale length differences between B and K.

2. Observations and Data Reduction

2.1. THE PROCEDURE

The sample, the observations, the reduction, and the way galaxy parameters such as scale lengths were obtained, are described in much more detail in Peletier et al. (1994). Here we only briefly indicate the various steps.

The sample is a representative sample, uniform in orientation on the sky and of morphological types between 3 and 6.5. The sample was selected on the basis of the ESO-LV catalog, and is diameter limited in B. All galaxies were observed in Nov. 1992 with the IRAC2 camera on the 2.2m telescope at ESO, La Silla. The instrument is equipped with a 256×256 HgCdTe NICMOS3 array. All observations were taken with a pixel size of 0.49", and under photometric conditions, with a seeing smaller than or comparable to 1 arcsec. On-source integration times per galaxy range from 8 to 20 minutes. We observed each galaxy at various position on the chip, interleaved with sky exposures of the same duration. The method of obtaining final mosaiced images is described in detail in Peletier (1993). The photometric calibration was performed using standard stars and is accurate to 0.07 mag including errors due to sky subtraction and flat fielding.

After having obtained the final, calibrated, reduced images in K, we aligned them with the optical images, by rotating and magnifying the latter, and fitted ellipses to them, to obtain radially averaged surface brightness profiles.

Next, the sky background was fitted in each band, by determining the median in various boxes near the edges of the frame and taking the average. The spread in these values, rather than the lack of photons, limits the depth of our K-photometry.

Then, a bulge-disk decomposition was performed using the method developed by Kent (1986). An exponential profile was fitted to the disks after the decomposition, using a range in radius set by eye, but the same for each passband. Since the bulges are small, the decomposition barely affects the scale lengths.

2.2. TWO ILLUSTRATIVE EXAMPLES

Fig. 1 shows the images in B, R and K of two typical galaxies in the sample on the same scale. Taking into account that the effective seeing in the optical images (~ 2" against ~ 1") is higher than in those in the infrared mosaics, we find that

- there are dramatic changes in morphology between B and K.
- For the face-on galaxy the central bar is very pronounced in K, while the spiral arms are prominent in B and R.
- For the edge-on galaxy the inner regions are very asymmetric in B and R, indicating large amounts of dust. In K we suddenly observe that the galaxy as a whole is lopsided.

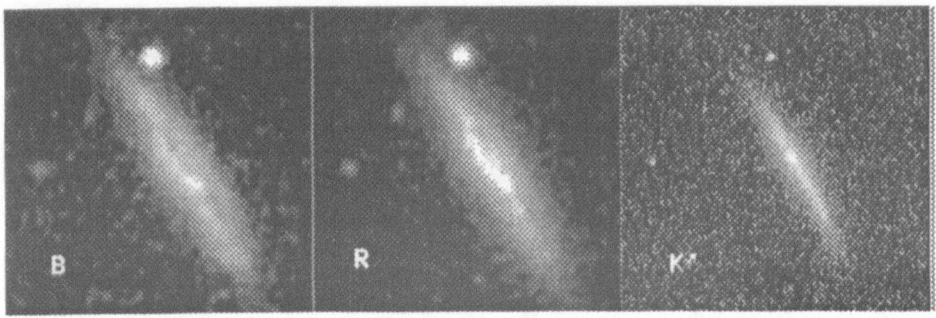

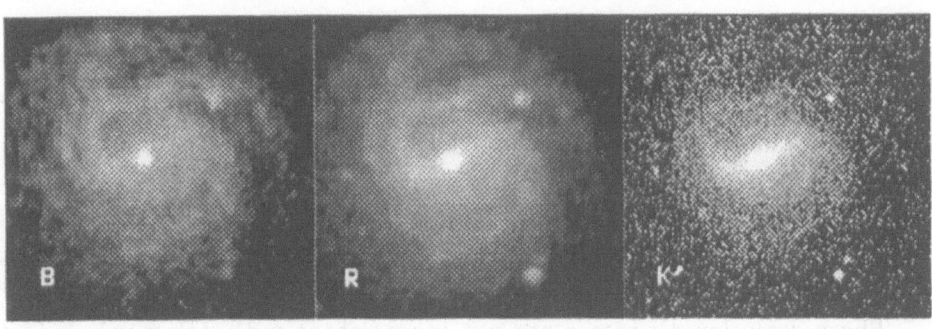

Figure 1. Grayscale images of ESO 157-G-049 (a) and ESO 353-G-048 (b) in B, R and K'. The images have the same scale and orientation. For a discussion see the text.

3. Observational Results

3.1. SCALE LENGTHS AND SCALE LENGTH RATIOS.

In Fig. 2 we show the ratio of the scale lengths of the disks in B and K as a function of axis ratio. The figure shows that the scale length in B is

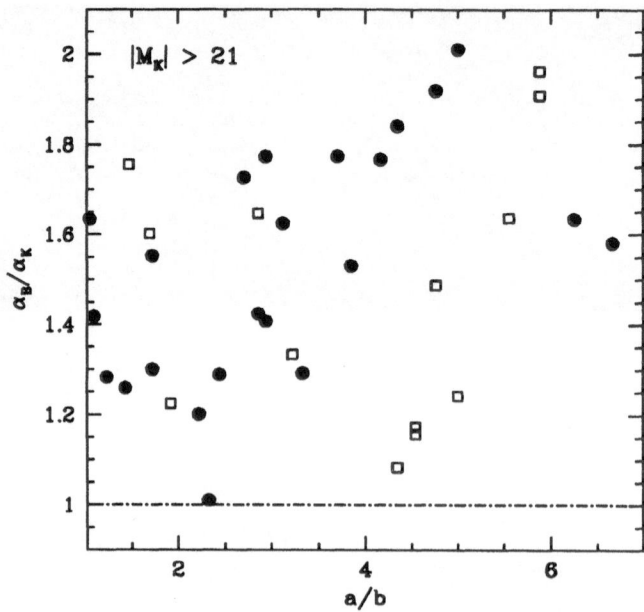

Figure 2. Exponential scale length ratios between B and K, as a function of axis ratio a/b. The galaxies indicated with filled circles are brighter than $M_K = -21$. The others are fainter, or no redshift is known. One galaxy, ESO 157-IG-050 falls below the plot.

between a factor of 1.2 and 2 larger than its counterpart in K, indicating large color gradients in the disks.

These scale length ratios are not inconsistent with other values in the literature. Elmegreen & Elmegreen (1984) find that the scale length ratio between B and I is 1.16 with a scatter of 0.47 for a large sample of face-on galaxies. According to the models of Evans (1994) the scale length ratio (minus unity) between B and H is usually about a factor 2 larger than the ratio (minus unity) between B and I, implying an average α_B/α_K ratio of 1.32, in broad agreement with this paper. It should be noted however that Evans' models are for face-on galaxies. If ζ, the scale height ratio between dust and stars, is more or less 0.5 (see Evans 1994, Peletier & Willner 1992) it is impossible to find a face-on galaxy with $\alpha_B/\alpha_H = 2$ according to his models. However, if a galaxy is tilted with respect to the line of sight the layering properties are different and larger ratios are possible. Moreover, the effects of light scattering, which tend to neutralize somewhat the effects of absorption, are worst for face-on galaxies (Rix & Rieke 1993). As far as the observations of Evans (1994) are concerned: his scale length ratios are generally closer to unity than those of Elmegreen & Elmegreen (1984). The sample that he presents is not well defined, however, and $< \alpha_B/\alpha_I >$ is lower than $< \alpha_V/\alpha_I >$, showing that his scale lengths are probably rather badly determined, either because they vary with the range in radius over which the exponential fit is made, or because they are affected by star formation regions etc. Kent (1986) finds large color differences between B and I of sometimes more than a magnitude between the inner and outer regions for his sample, which includes galaxies of all inclinations, confirming the large scale length ratios found here. A quantitative comparison is difficult, however, because his color profiles are very irregular. Out of the galaxies of type 3-6.5 of Terndrup et $al.$ (1994) the number of galaxies with $positive$ $r - K$ or $J - K$ disk color gradients is almost as large as the number of galaxies with negative gradients. They however use a very small 58×62 detector array, which, most likely, introduces large uncertainties in their disk color gradients.

We have tried to independently estimate the effects of stellar populations on the scale length ratios. For the galaxies of Balcells & Peletier (1994) with type < 1, i.e. without much visible dust, α_B/α_I is found to be 1.04 ± 0.05. This might correspond at most to scale length ratios between B and K of 1.08 ± 0.10 (using e.g. models of Arimoto & Yoshii 1986). For galaxies of the same type one can look at metallicity gradients from HII regions. The average gradient for the sample of normal nearby spirals of Vila-Costas and Edmunds (1992) is $\Delta(O/H) = -0.28$ dex/scale length. Since this is a scale length in B this corresponds to -0.19 dex/(K scale length) for the galaxies measured here. For a simple, single-age old stellar population this corresponds to $\Delta(B - K) = -0.20$ mag/(K scale length) (Peletier 1989) or $\alpha_B/\alpha_K = 1.19$, corresponding more or less to the lower limit of the observations.

The filled circles in Fig. 2 indicate galaxies with $|M_K| > 21$ ($H_0 = 75$ km s^{-1} Mpc^{-1}) while the squares are fainter galaxies, or galaxies for which no redshifts are available. Many of the galaxies indicated with open squares are at high inclinations and have small scale length ratios, and as such cannot contain a lot of dust. To these also belongs ESO 157 IG 500, which does not lie on the plot, since its $\alpha_B/\alpha_K < 1$. The bright galaxies however show a correlation with inclination, in the sense that the larger the axis ratio, the larger is the scale length ratio.

The majority of the galaxies indicated with open squares are galaxies with large axis ratios. We argue that these galaxies are of later types, and as such have been misclassified. Since the classification has been done based on the optical images only, and since for small edge-on galaxies there is not much difference between e.g. an edge-on Sc and Sd, this is quite likely. For those galaxies one should find small scale length ratios, since there is not much dispute that Sds are generally transparent in B (see e.g. V90).

We conclude that the galaxies that are indicated with filled dots in Fig. 2, probably the genuine Sb and Sc galaxies, all (except one) show a scale length ratio between B and K that is larger than 1.2, showing that their profiles in B are notably affected by extinction.

3.2. CENTRAL COLOR AND SURFACE BRIGHTNESS.

Freeman (1970) found that the inclination-corrected central disk surface brightness of spiral galaxies is constant around $\mu_{0,B}^c = 21.65 \pm 0.30$ mag $(arcsec)^{-2}$. The current study offers a good opportunity to understand Freeman's law by investigating the relation in K. In Fig. 3 we plot $\mu_{0,K}$ (the central surface brightness of the disk) against $(B-K)_0$, the extrapolated central disk color. The figure shows that the distribution of central surface brightness in K is different from that in B. There is more that 4 mag difference between the galaxy with the lowest and the highest central surface brightness, against 2.5 in B. The fact that there is a good correlation reflects that the range in $\mu_{0,K}$ is considerably larger than in $\mu_{0,B}$, since otherwise the scatter would have been larger or the slope different. The sample has been selected partly on the basis of the B-surface brightness, so cannot be used to measure Freeman's law in K, but the qualitative behavior of the galaxies can be investigated. The symbols in Fig. 3 are the same as in Fig. 2. The open symbols represent the faintest galaxies in K, and those without redshift. These are the galaxies with the bluest $B-K$ colors, and, since $\mu_{0,B}$ is more or less constant, also the ones with the lowest central K surface brightness.

It is unlikely that this difference in range is due completely to a large range in color in the stars, since more than half the sample has central $B-K$ colors that are redder than those of bright giant ellipticals ($B-K \sim 4.3$). This indicates that a large part of the qualitatively different behavior has to be ascribed to dust extinction.

If one looks at the numbers, the situation is more complicated. find that the average central surface brightness in B is 21.23 mag $(arcsec)^{-2}$, with a scatter of 0.52 mag. Corresponding values for R and K are: $\mu_{0,R} = 19.71 \pm 0.59$, $\mu_{0,K} = 16.62 \pm 0.93$. If we correct for inclination, assuming that the galaxies are transparent, we find that $\mu_{0,B}^c = 22.43 \pm 1.30$, $\mu_{0,R}^c = 20.91 \pm 1.33$, and $\mu_{0,K}^c = 17.82 \pm 1.51$. We find that the scatter in B is smaller than in K, but that the scatter in K does not decrease after correcting for inclination. However, if the stellar surface density has a considerable intrinsic range (i.e. if small, late-type galaxies

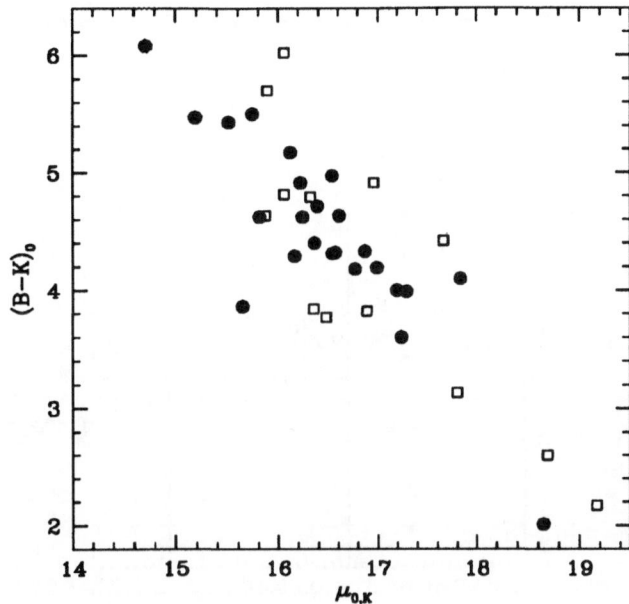

Figure 3. Central disk color against the central disk surface brightness in K (both extrapolated from the exponential fits). The meaning of the symbols is the same as in Fig. 2.

have lower central surface brightness than large, earlier-type galaxies) the scatter depends on the sample, and does not necessarily decrease after correcting for inclination.

We conclude that the central surface brightness in K shows a larger range than in B, and that Freeman's law for Sb's and Sc's is very likely to be due to extinction by dust.

3.3. OPTICAL-INFRARED COLOR DISTRIBUTIONS

To say something about the distribution of extinction in the galaxies we have looked at $B-K$ colors at various places, and their relation as a function of inclination. As opposed to most other colors, $B - K$ data are easier to interpret, since we can safely assume that the extinction in K is zero or almost zero (e.g. Knapen *et al.* 1991). So in the absence of color gradients in the stellar populations a change in B-K only indicates a change in the amount of extinction in B. This is not the case for other colors, which suffer

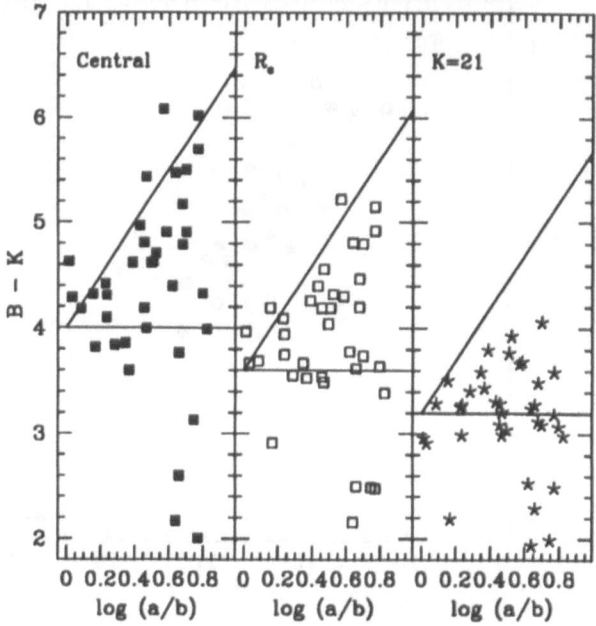

Figure 4. Radially averaged $B - K$ color at various positions in the galaxy. On the left is shown the extrapolated central disk color, in the middle the color at $r_{eff,K}$, and on the right at K=21 mag arcsec^{-2}

from the fact that not only is the color of the underlying stellar population unknown, but also the extinction law, since dust and stars are mixed, so that the extinction law depends on the quantity and on the distribution of the dust (see e.g. Evans 1994, Jansen *et al.* 1994).

In Fig. 4 we present disk colors at various galactocentric radii as a function of inclination. On the left we plot the extrapolated disk color $\mu_{0,B}$ - $\mu_{0,K}$, in the middle the color as measured at the effective radius in K, and at $\mu_K = 21$ in the right panel. Also drawn are two lines indicating the behavior of a very simple models of completely opaque and a completely transparent (horizontal line) galaxies.

Several effects are seen here:

- We find that in the center galaxies are much redder than in the outer parts, (see also the large color gradients in the individual galaxies (Peletier *et al.* 1994)).
- In the center there is a large spread in central color for edge-on galaxies.

The edge-on galaxies with very blue central colors are almost all small galaxies, so if only those galaxies are selected with $M_K < -21$ this large spread disappears. We see that in the center $B - K$ gets redder as a function of inclination, while this is not evident at $D_{K,21}$. This indicates that in the center, and to a lesser extent at r_{eff} the galaxies are optically thick in B, while not at $D_{K,21}$.

– Even after removing the galaxies with $M_K > -21$ the range in color for face-on galaxies is smaller than for edge-on galaxies, demonstrating that the amount of extinction varies from galaxy to galaxy.

Applying the inclination-test, which assumes that face-on and edge-on galaxies are drawn from the same population, seems incorrect for this sample: all the galaxies that are faint in M_K or do not have measured redshifts are edge-on and generally are also much bluer than the rest of the galaxies. The only conclusion to draw from Fig. 4 is that galaxies in the center and at r_{eff} seem to behave in an opaque way, while the behavior at $D_{K,21}$ looks more like what one expects for a transparent galaxy. At $D_{K,21}$ the extinction for an edge-on galaxy is at most 1 mag, implying a maximum optical depth there of $\tau=2$. A galaxy like that, seen face-on, would then have an optical depth at K=21 of at most 0.4, or an extinction $A_{B,21}$ of 0.4 mag.

4. The amount of dust from scale length modeling

The observations show (Fig. 2) that for a typical face-on galaxy the scale length ratio between B and K is about 1.3, and 1.7 for an edge-on galaxy. We now ask ourselves the question how much extinction we would need to obtain these ratios. We will consider two cases: A. the case where we assume that there are no intrinsic stellar population gradients, and B., the case where stellar population gradients are responsible for a ratio of scale lengths of 1.1 (see before). The dust then has to cause scale length ratios of 1.2 for face-on galaxies and 1.6 for edge-on galaxies.

To do this we have constructed radiative transfer models of disks, that contain a uniform distribution of stars, exponential in radial and vertical direction, and a uniform distribution of dust, also exponential in both directions, but with different scale lengths and heights. We project the models, and fit exponentials to the major axis profiles between one and three stellar scale lengths, in the same way as is done for the observations. One set of models has been made for face-on galaxies, and one for galaxies at 80 degrees, close to edge-on.

The scale length ratios are plotted in Fig. 5. We have allowed three different possibilities for the ratio of scale lengths of dust and stars, and three different values for the ratios of the scale heights.

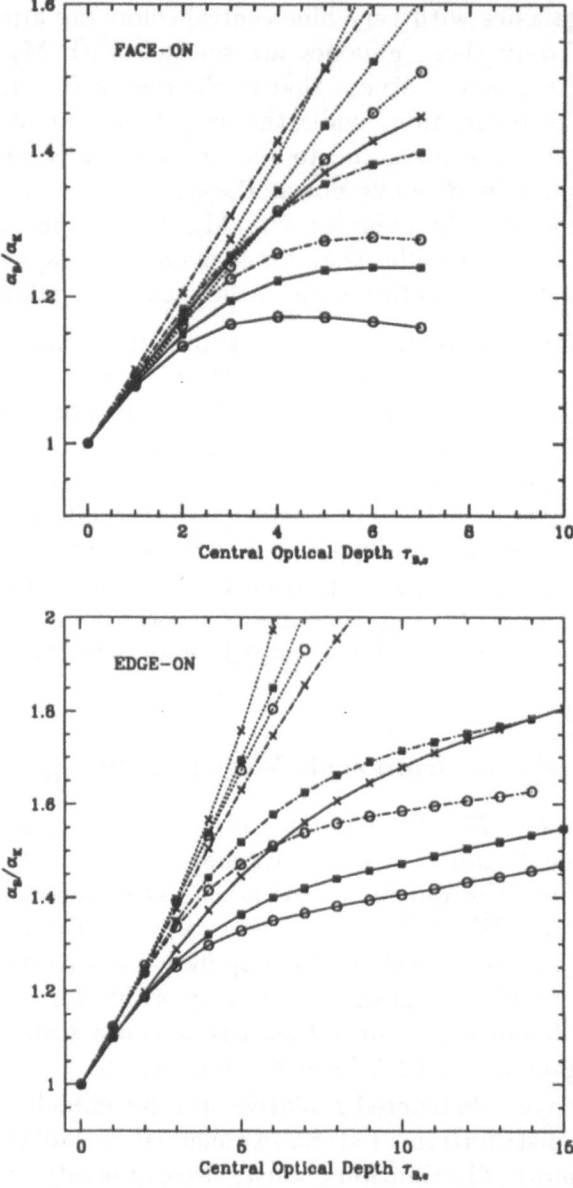

Figure 5. Scale length ratios between B and K as a function of central optical depth for disk models with various ratios of scale lengths and heights of dust and stars. Dotted lines indicate models where the scale lengths of dust and stars are equal. For dashed lines the scale length of the dust is a factor 2 larger, and for drawn lines a factor 3. Filled squares indicate galaxies for which the scale height of the stars is twice as large as that of the dust. Open circles indicate thick models, with equal scale heights of dust and stars, while the crosses represent models where the scale height of the dust is 0.3 times as large as that of the stars.

In order for such a model to look like a real galaxy, in case A, we impose that the scale length ratio between B and K in the face-on case has to be between 1.2 and 1.4. When this model is turned to edge-on, its $\tau_{B,c}$ should be a factor 5 higher, and now the scale length ratios should lie between 1.6 and 1.8. Table 1 indicates for which $\tau_{B,c}$ each of these criteria is fulfilled, and which models fulfill the criteria at the same time for face-ons and edge-ons, i.e. give a τ_{edge} that is a factor 5 larger than τ_{face}. The first two columns of Table 1 list the particular geometry for the assumed dust distribution, the third column gives the range of acceptable central optical depth values, given the observed range of B/K scalelength ratios in face-on galaxies, the fourth column gives a range of acceptable solutions for edge-on galaxies, while the fifth column marks those solutions in which the edge-on solutions of column 3 could be obtained from a projection of the face-on solutions in column 2.

For several geometries no acceptable solutions could be identified in colomn 5. For case A, we see that there are only very few possible solutions. The only ones that are formally acceptable are models with $\alpha_d/\alpha_* = 2$, or 3, but they all need so much extinction in the edge-on case that the resulting surface brightness would be uncomfortably low.

Since it is known from e.g. emission line studies that spiral galaxies show abundance gradients, we argue that our models in case B are much more attractive. Here we find solutions for any thickness, if $\alpha_d/\alpha_* \geq 2$. The central face-on optical depth is around 1 or 2. It is remarkable that the models with $\alpha_d/\alpha_* = 1$ do almost not provide any solutions. In this case the scale length ratios for edge-on galaxies are expected to be larger than observed.

Most acceptable solutions indicate $\tau_{face}^B = 0.2$ to 0.4 at 3 K-band scale lengths, although two other solutions are found (line 2 and line 6 of case B) that correspond to higher optical depth. This exercise demonstrates, in the first place, the difficulty of accepting models with an equal scale length for the dust and the stars. More refined models are required. One possibility is the introduction of two-component models such as presented by Valentijn (this Volume).

5. So how much extinction is there?

We have now obtained solutions using two different methods of analyzing the data - by looking at $B - K$ at various radial distances, and by analyzing the scale lengths. For the center both methods agree reasonably well, with on the average about 1 - 2 mag of extinction in B for a face-on galaxy. In the outer parts, at 3 K-band scale lengths, or at K = 21 mag arcsec^{-2} we find from the $B - K$ colors that the extinction for a face-on galaxy is

TABLE 1. Allowed central optical depths

Models: Case A h_d/h_*	α_d/α_*	$\tau_{B,c}$ (face) $\alpha_B/\alpha_K = 1.2\text{-}1.4$	$\tau_{B,c}$ (edge) $\alpha_B/\alpha_K = 1.6\text{-}1.8$	$\tau_{B,c}$ (face) face + edge
1	1	2.5 - 5.2	4.4 - 6.0	-
1	2	2.4 - inf	11.8 - inf	2.4 - inf
1	3	-	-	-
0.5	1	2.1 - 4.5	4.4 - 5.7	-
0.5	2	2.1 - 7	6.3 - 12	2.1 - 2.4
0.5	3	3.0 - inf	18 - inf	3.0 - inf
0.3	1	2.1 - 4.2	4.1 - 5.2	-
0.3	2	1.9 - 3.8	4.7 - 6.4	-
0.3	3	2.3 - 5.6	7.7 - 14.7	2.3 - 2.9

Models: Case B h_d/h_*	α_d/α_*	$\tau_{B,c}$ (face) $\alpha_B/\alpha_K = 1.1\text{-}1.3$	$\tau_{B,c}$ (edge) $\alpha_B/\alpha_K\ 1.5\text{-}1.7$	$\tau_{B,c}$ face + edge
1	1	1.1 - 3.5	3.8 - 5.2	-
1	2	1.2 - inf	5.5 - inf	1.1 - inf
1	3	1.4 - inf	-	-
0.5	1	1.0 - 3.4	3.8 - 5.1	1.0 - 1.1
0.5	2	1.0 - 3.7	4.7 - 9.0	1.0 - 1.8
0.5	3	1.2 - inf	12 - inf	2.4 - inf
0.3	1	1.1 - 3.2	3.6 - 4.7	-
0.3	2	1.0 - 2.9	4.0 - 5.5	1.0 - 1.1
0.3	3	1.1 - 3.6	5.7 - 10.3	1.1 - 2.1

smaller than 0.4 mag, while the scale length method gives between 0.2 and 0.4 mag of extinction, with possibly higher values when dust-to-stellar scale length or scale height ratios are large.

One can refine these values by making the models a bit more realistic, by making the dust distribution clumpy, or by introducing scattering. It is possible that clumpy models will allow solutions with $\alpha_d/\alpha_*=1$. In this case, less extinction is required in the outer parts, probably only $A_B=0.1$ mag at 3 K-band scale lengths. One could also introduce two-component dust-models (see Valentijn 1994, this conference). They have the advantage that diameters do not change much as a function of inclination, and could still match the data presented here.

Another correction to the amount of extinction we derive has to be made to account for the effects of scattering. This effect is not very large (see Rix & Rieke 1993), given the fact that we are dealing here with very large scale length ratios and radial color differences, and for this reason we have not modeled it. These numbers are zeroth-order estimates, and can be refined by doing more detailed calculations. For example, the fact that dust is present in some areas (the arms) and not so much in others (see Beckman *et al.* 1994) will alter somewhat the results, as well as the fact that stellar population gradients in the outer parts have not been studied very well, and might be larger than assumed here.

References

1. Arimoto, N. & Yoshii, Y., 1986, A&A, 107, 135.
2. Balcells, M. & Peletier, R.F., 1994, AJ, 107, 135.
3. Burstein, D., Haynes, M. & Faber, S.M., 1991, Nature, 353, 515.
4. Byun, Y.I., 1992, Ph.D. Thesis, Australian National University.
5. Elmegreen, D. M. & Elmegreen, B. G., 1984, ApJSupp., 54, 127.
6. Evans, R., 1994, MNRAS, 266, 511.
7. Freeman, K.C., 1970, ApJ, 160, 811.
8. Giovanelli, R., Haynes, M.P., Salzer, J.J., Wegner, G., da Costa, L.N. & Freudling, W., 1994, preprint.
9. Heidmann, J., Heidmann, N. & de Vaucouleurs, G., 1972, Mem. R. Astr. Soc., 75, 121.
10. Huizinga, J.E. & van Albada, T.S., 1992, MNRAS, 254, 677.
11. Jansen, R.A., Knapen, J.H., Beckman, J.E., Peletier, R.F. & Hes, R., 1994, MNRAS, 270, 373.
12. Kent, S.M., 1986, AJ, 91, 1301.
13. Knapen, J.H., Hes, R., Beckman, J.E. and Peletier, R.F., 1991, AA, 241, 42.
14. Lauberts, A. & Valentijn, E.A., 1989, *The Surface Photometry Catalogue of the ESO Sky Survey*, ESO, Garching.
15. Moorwood, A.F.M., Finger, G., Biereichel, P., Delabre, B., van Dijsseldonk, A., Huster, G., Lizon, J.-L., Meyer, M., Gemperlein, H. & Moneti, A., 1992, The Messenger, 69, 61.
16. Peletier, R.F., 1989, Ph.D. Thesis, Univ. of Groningen
17. Peletier, R.F., 1993, A&A, 271, 51.
18. Peletier, R.F., Davies, R.L., Illingworth, G., Davis, L.E. & Cawson, M., 1990, AJ, 100, 1091.
19. Peletier, R.F. & Willner, S.P., 1992, AJ, 103, 1761.
20. Peletier, R.F., Valentijn, E.A., Freudling, W. & Moorwood, A.F.M., 1994, A&ASupp, in press.
21. Rix, H.W. & Rieke, M.J., 1993, ApJ, 418, 123.
22. Sandage, A.R., 1973, ApJ, 183, 711.
23. Terndrup, D.M., Davies, R.L., Frogel, J.A., dePoy, D., and Wells, L., 1994, ApJ, 432, 518.
24. Valentijn, E.A., 1990, Nature, 346, 153 (V90).
25. Valentijn, E.A., 1994, MNRAS, 266, 614.
26. Vila-Costas, M.B., and Edmunds, M., 1992, MNRAS, 259, 121.

Question
Freeman

How about the effects of age gradients rather than chemical gradients.

Answer
Peletier

Early-type spirals (SO - Sab, see Balcells & Peletier 1994) can only have small, insignificant age gradients. For later types the situation is more complicated, and probably unknown. The large U-V gradients in the spiral arms of M51 (see paper by Beckman) shows that age gradients can be important.

Question
James

Are your galaxy light profiles well-fitted by exponentials at both B and K?

Answer
Peletier

Yes, over 3½-4 magnitudes in surface brightness.

AZIMUTHAL DISTRIBUTION OF DUST IN NGC 2997[1]

PREBEN J. GROSBØL
European Southern Observatory
Karl-Schwarzschild-Str. 2, D-85748 Garching, Germany

DAVID L. BLOCK
Dep. of Comp. and Appl. Math., Witwatersrand University,
Private Bag 3, Wits 2050, South Africa

AND

PANOS A. PATSIS
European Southern Observatory
Karl-Schwarzschild-Str. 2, D-85748 Garching, Germany

1. Introduction

The galaxy NGC 2997 is a grand design spiral classified as Sc(s)I in the Revised Shapley-Ames Catalog. Its inclination angle $(30°- 40°)$ and linear scale $(1''\approx 50pc)$ make it well suited for both morphological and dynamical studies. It contains a significant amount of cold dust [2] and is therefore also a good candidate for analyzing the spatial distribution of dust in grand design spirals.

2. Observations

The observations of NGC 2997 in K' $(2.1\mu m)$ were done with IRAC2 and the NICMOS3 array mounted on the 2.2m ESO/MPI telescope at La Silla [1]. The galaxy which has $D_{25} = 8.1'$ was covered with a 3×2 mosaic of $2'$ fields with a pixel size of $0.5''$. Each field consisted of a stack of five 90 sec. exposures with interleaved sky frames yielding a total exposure of 15 min. A signal-to-noise of 10 for a single pixel was achieved at

[1]Based on observations collected at the European Southern Observatory, La Silla, Chile

259

J. I. Davies and D. Burstein (eds.), The Opacity of Spiral Disks, 259–261.

$K'=17.4$ mag/arcsec^{-2}. The frames were calibrated using aperture photometry for the K-band [4].

Visual maps in B and R were obtained with EMMI on the NTT (courtesy of S. D'Odorico) and calibrated using the photometry of Longo & de Vaucouleurs [5]. No color term were included in the transformations. The frames were aligned using common field stars. The ESO-MIDAS system was used for all reductions.

The position angle (PA) and inclination angle (i) based on the outer isophotes in the R-band suggests PA$\approx$92° and $i \approx$46° [7]. These values are however biased by the very prominent spiral arms. A dynamical major axis with PA=102° was derived from radial velocities of HII regions [8]. This corresponds well with the 101° obtained for of the inner disk region 20″-35″ which is outside the bulge but still not effected by the spiral pattern. It suggests that i can be determined from this inner disk since a chance alignment of a central oval distortion is unlikely. Thus, the projection parameters PA=101° and i=32° were adopted.

3. Azimuthal variation of (B-K′)

The surface photometry in K′ represents mainly the old Population II disk [9] and can therefore be used as an estimate of the mass distribution in the disk although some emission from Population I objects (*e.g.* HII regions) also contribute to the K′ intensity. The two armed spiral pattern in the K′-map is detectable at a radius of 50″ and increases in amplitude outward. The phase $\Phi(r)$ of its surface density perturbation as function of radius was defined by m=2 Fourier component of the azimuthal intensity variation of the face-on K′-map from which all foreground stars were removed. It fits well a logarithmic spiral from around 60″ to the edge of the K′ image around 120″. Some systematic errors in the phase are present due to emission from HII-regions (*e.g.* Brγ) which was not removed.

The intensities in B and K′ were sampled with a 2″ Gaussian beam to estimate the (B-K′) color index. Its azimuthal variation in a set of 4″ wide radial strips is shown in Fig. 1. The angular offset from the spiral is given as $\delta\theta = \theta - \Phi(r)$ with increasing angles from North through West.

(B-K′) is almost constant at 4.0 in the inner parts (40″<r<60″) where the spiral is very weak. This corresponds to E(B-K′)$\approx$0.8 assuming an underlying Population II disk with (B-K′)$\approx$3.2 [3]. At larger radii, a clear modulation develops as the spiral amplitude increases. The maximum (B-K′) values still reaches values of 3.5-4.0 while the minima are around 2.5, however, with a large spread due to low K′ surface brightness between arms. A number of sharp dust lanes with (B-K′)>4 can also be seem. A systematic phase shift of the (B-K′) maxima relative to the spiral as function of

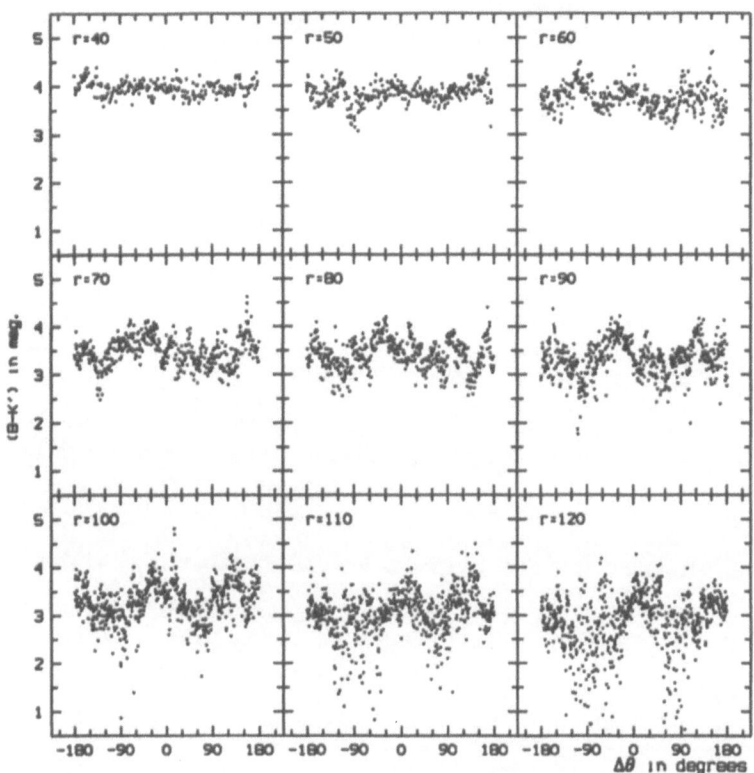

Figure 1. (B-K') as function of azimuthal angle $\delta\theta$ relative to the phase of maximum amplitude of the m=2 Fourier component of the spiral pattern for different radial strips.

radius is suggestive and may be cause by the change in relative speed of the material and a density wave.

4. Conclusions

The azimuthal modulation of (B-K') suggests that dust properties vary significantly between arm/interarm regions and correlate well with the spiral perturbation in agreement with Greenberg [6].

References

1. Block,D.L., Grosbøl,P., Moneti,A., Patsis,P. (1993) The Messenger **71**, 41
2. Block,D.L., Witt,A.N., Grosbøl,P., Stockton,A., Moneti,A. (1994) A&A **288**, 383
3. Bruzual,G. Charlot,S. (1993) ApJ **405**, 538
4. de Vaucouleurs,A., Longo,G. (1988) Monographs in Astron. **5**, Univ. Texas, Austin
5. Longo,G., de Vaucouleurs,A. (1983) Monographs in Astron. **3**, Univ. Texas, Austin,
6. Greenberg,J.M. (1970) IAU Symp. **39**, Ed. H.J.Habing, Reidel, Dordrecht, 306
7. Grosbøl,P. (1985) A&AS **60**, 261
8. Peterson,C.J. (1978) ApJ **226**, 75
9. Rix,H.-W., Rieke,M.J. (1993) ApJ **418**, 123

INTERNAL EXTINCTION IN SPIRAL GALAXIES AT OPTICAL AND NEAR INFRARED WAVELENGTHS

A. BOSELLI
DEMIRM - Observatoire de Paris
61, Av. de l'Observatoire, 75014 Paris, France

AND

G. GAVAZZI
Osservatorio Astronomico di Brera
Via Brera 28, 20121 Milano, Italy

1. Introduction

This work is an update of the analysis presented in Boselli and Gavazzi 1994. We analyse the extinction in spiral galaxies at optical and near infrared wavelengths using a statistical analysis based on a sample of 862 spirals of morphological type Sa-Sc with homogeneous photometrical data available in the literature. The sample is taken from the magnitude-limited (mpg<15.7) CGCG catalogue (Zwicky et al. 1961-1968) in 9 nearby (z<0.03) clusters, and is complete to 90% and 72% in the B and H band respectively at 15.0 mag, and to 68% and 62% at 15.7 mag.

2. The internal extinction in spiral galaxies

The magnitude correction for internal extinction can be written as:

$$\Delta mi = -2.5 D log(b/a) \tag{1}$$

where D is a wavelength dependent coefficient and a and b are the galaxy major and minor galaxy axes. If mo is the observed magnitude, the mean surface brightness μ uncorrected for internal extinction is defined as:

$$\mu = mo + 2.5 log(ab) \tag{2}$$

Since we use a and b measured in the photographic system, equation (2) applies exactly only in the B band. In the other bands we can only derive an hybrid surface brightness. In the hypothesis that the emitting stars and

263

J. I. Davies and D. Burstein (eds.), The Opacity of Spiral Disks, 263–266.
© 1995 *Kluwer Academic Publishers.*

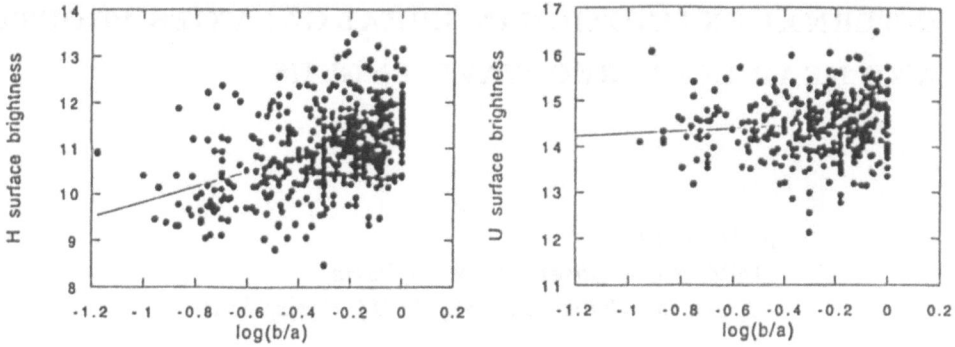

Figure 1. The relation between the surface brightness in the H, and U bands and the axial ratio $log(b/a)$. Lines represent the best fits to the data

the absorbing dust are well mixed all over the disc, μ remains constant with inclination if discs are optically thick, while in transparent discs the longer line of sight through inclined systems approximately brightens μ as:

$$\mu = \mu c - 2.5(1 - D)log(b/a) \qquad (3)$$

where μc is the face-on surface brightness. However if galaxies are optically thin or semi-transparent their observed isophotal diameters change with inclination because the longer line of sight brightens μ at a given radius, or increases the isophotal diameters with inclination. The relationship between surface brightness and inclination can then include the dependence on inclination of both diameters and magnitudes.

In the assumption that galaxies are optically thick in the B, blue diameters do not change with inclination and the the analysis of the $\mu - log(b/a)$ relation is an appropriate tool to derive the extinction coefficients. Since the B optically thick case implies that the observed relation between μ and $log(b/a)$ is all due to a magnitude change with inclination, this hypothesis leads to the determination of an upper limit for the D coefficients.

Fig. 1 shows the distribution of μ as a function of $log(b/a)$ in the H and U bands. The slopes $(1 - D)$ generally decrease from early to late types and from H to U. The D values for all types mixed together (from Sa to Sc) are $< DV >=0.68$, $< DB >=0.79$ and $< DU >=0.89$.

The corrected colour index between the bands $\lambda 1$ and $\lambda 2$ is:

$$m\lambda 1 - m\lambda 2)c = m\lambda 1 - m\lambda 2)o + 2.5(D\lambda 1 - D\lambda 2)log(b/a) \qquad (4)$$

The dependence of $m\lambda 1 - m\lambda 2)o$ on inclination is given in Fig. 2 for $\lambda 1 = V, U$ and $\lambda 2 = H$. Galaxies in three classes of absolute H luminosity are plotted with different symbols to show the effect of the known colour-luminosity relation (Gavazzi, 1993). The colours become redder with increasing inclination and the slope is wavelength dependent, i.e. the extinction is higher

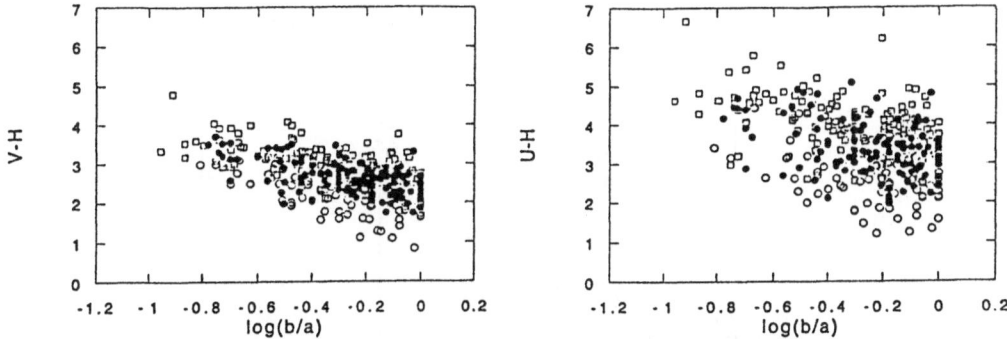

Figure 2. The colour-inclination relation. Points are coded according to the absolute H luminosity as follows: $-17 > H> -22$: empty circles; $-22 > H> -23$: filled circles; H< -23: squares

at shorter wavelengths. The slopes of the relation correspond to the difference between the correction coefficients D in the H band and those in the V, B, U bands respectively.

To check if the above results are biased from the incompleteness of the sample, we have recomputed the surface brightness and colour inclination analysis using different subsamples: magnitude limited; diameter limited; distance limited; absolute H magnitude limited. The colour inclination relation gives consistent results using each different subsample, while we see marginal differences in the $\mu - log(b/a)$ relation only between the various H absolute magnitude subsamples. This suggests that the incompleteness of the sample does not affect our results. The extreme assumption that galaxies are fully transparent in the H band ($DH=0$) leads to a lower limit for the D coefficients: $DV=0.44$, $DB=0.54$ and $DU=0.64$. This assumption is certainly unrealistic even in the H band since H imaging of some edge-on galaxies show prominent dust lanes, suggesting that even at IR wavelengths the extinction is not negligible (Barnaby and Thronson, 1992). Furthermore we do not account for the contribution of the scattered light which should weaken the colour-inclination relation as shown by Bruzual or Witt in this meeting. We have therefore restricted the range of possible values of the extinction coefficients to: $0.00< DH<0.35$; $0.44< DV<0.68$; $0.54< DB<0.79$ and $0.64< DU<0.89$.

The optical thickness of spiral galaxies can be determined only if the relative distribution of the emitting stars and of the absorbing dust are known. To this aim we can adopt a sandwich model (Disney et al 1989). In this assumption, the internal extinction in a galaxy depends on the inclination, the relative thickness of the dust and of the star layers and on the optical thickness of the absorbing dust. We can then obtain the optical thickness of

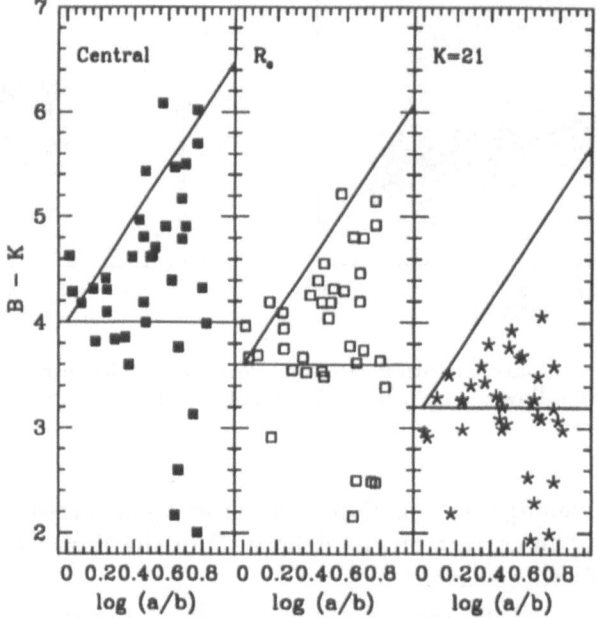

Figure 4. Radially averaged $B - K$ color at various positions in the galaxy. On the left is shown the extrapolated central disk color, in the middle the color at $r_{eff,K}$, and on the right at K=21 mag arcsec^{-2}

from the fact that not only is the color of the underlying stellar population unknown, but also the extinction law, since dust and stars are mixed, so that the extinction law depends on the quantity and on the distribution of the dust (see e.g. Evans 1994, Jansen *et al.* 1994).

In Fig. 4 we present disk colors at various galactocentric radii as a function of inclination. On the left we plot the extrapolated disk color $\mu_{0,B}$ - $\mu_{0,K}$, in the middle the color as measured at the effective radius in K, and at $\mu_K = 21$ in the right panel. Also drawn are two lines indicating the behavior of a very simple models of completely opaque and a completely transparent (horizontal line) galaxies.

Several effects are seen here:

- We find that in the center galaxies are much redder than in the outer parts, (see also the large color gradients in the individual galaxies (Peletier *et al.* 1994)).
- In the center there is a large spread in central color for edge-on galaxies.

The drinking water

THE OPACITY OF SPIRAL GALAXY DISKS

A Far Infrared Perspective

NICK DEVEREUX
New Mexico State University
Dept. of Astronomy, Box 30001/Dept 4500, Las Cruces NM 88003, USA

1. Introduction

The reasonable expectation that the diffuse interstellar dust responsible for extinction in galaxies will re-radiate the absorbed starlight at far infrared wavelengths has led a number of authors to use the far infrared luminosity measured by the Infrared Astronomical Satellite (IRAS) to probe the opacity of spiral galaxy disks. For example, Burstein & Lebofsky (1986) used the inclination dependence of IRAS detections to investigate the disk opacity. Additionally, decomposition of the IRAS fluxes into two or more dust temperature components has led to wide ranging estimates for the disk opacity (Crawford & Rowan-Robinson 1986; De Jong & Brink 1987, Lonsdale-Persson & Helou 1987). Still others have compared the radial distributions of far infrared and blue light emission to argue that spiral galaxy disks are both optically thick (Phillips et al. 1991) and thin (Bothun & Rogers 1992). Unfortunately, in addition to the lack of consensus, all of the above studies are based on the incorrect assumption that the far infrared luminosity measured by IRAS is related, either in part or whole, to the opacity of spiral galaxy disks. On the contrary, recent evidence, to be briefly reviewed below, indicates that the far infrared luminosity measured by IRAS is dominated by thermal radiation from dust heated primarily by O and B stars. Furthermore, evidence will be presented showing that the dust that *is* responsible for the opacity of spiral galaxy disks radiates longward of $100\mu m$ and would therefore not have been detected by IRAS.

2. The Origin of the Far Infrared Luminosity Measured by IRAS

Comparison of the *global* $40 - 120\mu m$ far infrared luminosities of spiral galaxies with their *global* Hα luminosities indicates that the O and B stars

J. I. Davies and D. Burstein (eds.), The Opacity of Spiral Disks, 269–280.
© *1995 Kluwer Academic Publishers.*

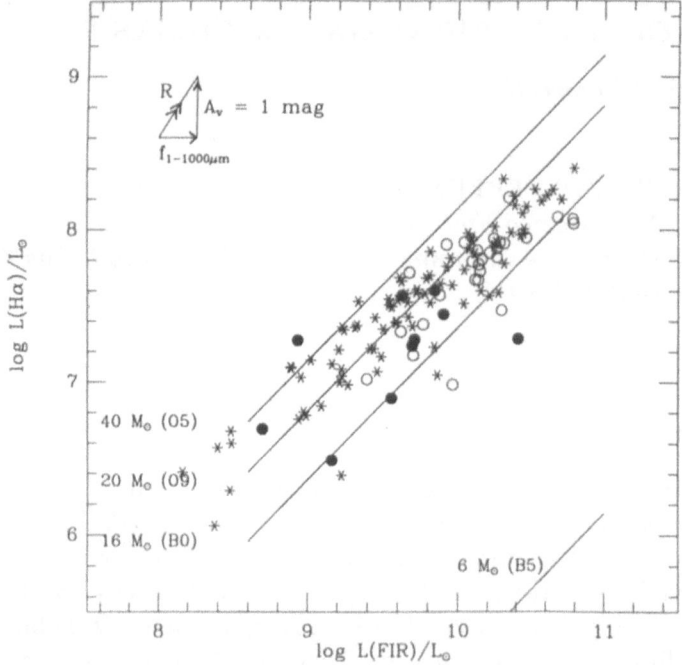

Figure 1. The correlation between the Hα and IRAS $40 - 120\mu m$ luminosity for 124 spiral galaxies. The points are coded by spiral type; filled circles identify early-types (Sa-Sab), open circles; intermediate types (Sb-Sbc) and stars; late-types (Sc and later).

which are required to ionize the hydrogen gas are also capable of powering the far infrared luminosity measured by IRAS (Devereux & Young 1990a). The evidence is presented in Figure 1 which illustrates the relationship between the far infrared and Hα luminosities for 124 spiral galaxies included in the Hα emission line survey of Kennicutt & Kent (1983).

The lines in Figure 1 identify the ratio of Hα to far infrared luminosity, $L(H\alpha)/L(FIR)$, that would be expected for HII regions powered by high mass stars of a variety of spectral types. The fact that the galaxy data points lie in between the lines labeled O5 and B0 indicates that such stars are capable of sustaining both the far infrared and Hα luminosities measured for spiral galaxies. The statement is true regardless of the extinction to the Hα emission line, the $S_{100\mu m}/S_{60\mu m}$ dust color temperature, and spiral morphological type as discussed in more detail by Devereux & Young (1991).

Additional evidence that O and B stars are responsible for generating the far infrared luminosity *within the disks* of spiral galaxies is provided by the comparison of high resolution far infrared *maps*, obtained with the Kuiper Airborne Observatory (KAO), and Hα images that have been convolved to the same resolution as the KAO data. Histograms illustrating the distribution of L(Hα)/L(FIR) ratios measured at multiple locations within the star forming disks of M51 and NGC 6946 point to O and B stars as the origin of both the far infrared and Hα luminosity (Devereux & Young 1992; 1993).

Unfortunately, with the exception of M51 and NGC 6946, the KAO has not yielded far infrared maps with sufficient resolution to permit the comparison with Hα images to be extended to other galaxies. On the other hand, the number of galaxies with high resolution far infrared *images* has recently increased significantly following application of a high resolution reconstruction algorithm, HiRes, to the IRAS data (Aumann, Fowler & Melnyck 1990). The HiRes algorithm improved the resolution by a factor of 2 or 3 over that of the original IRAS survey and generated useful high resolution (~ 105 arc sec) far infrared images for half a dozen or so nearby galaxies including the early-type spiral M31 (Devereux et al. 1994) and the late-type spiral M101 (Devereux & Scowen 1994).

M 31

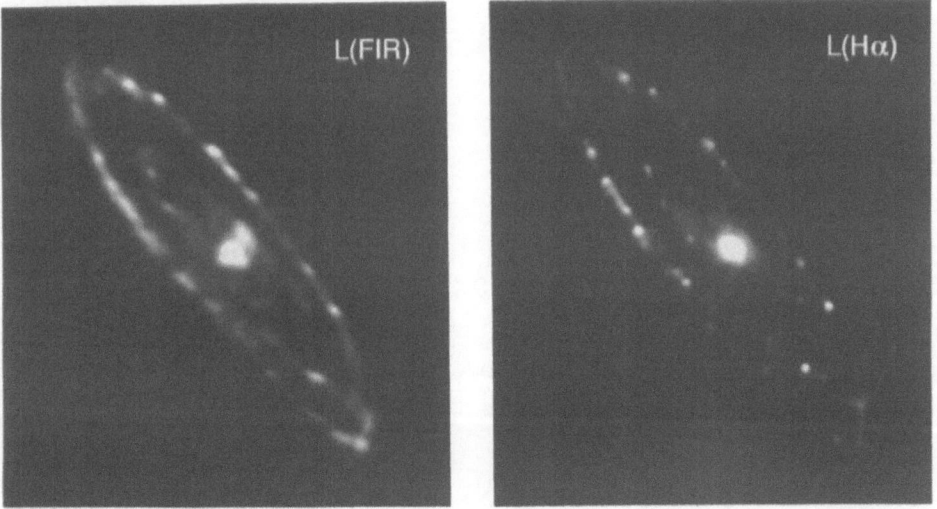

Figure 2. Two views of the Andromeda galaxy (M31). Far infrared (left panel) and Hα (right panel)

M101

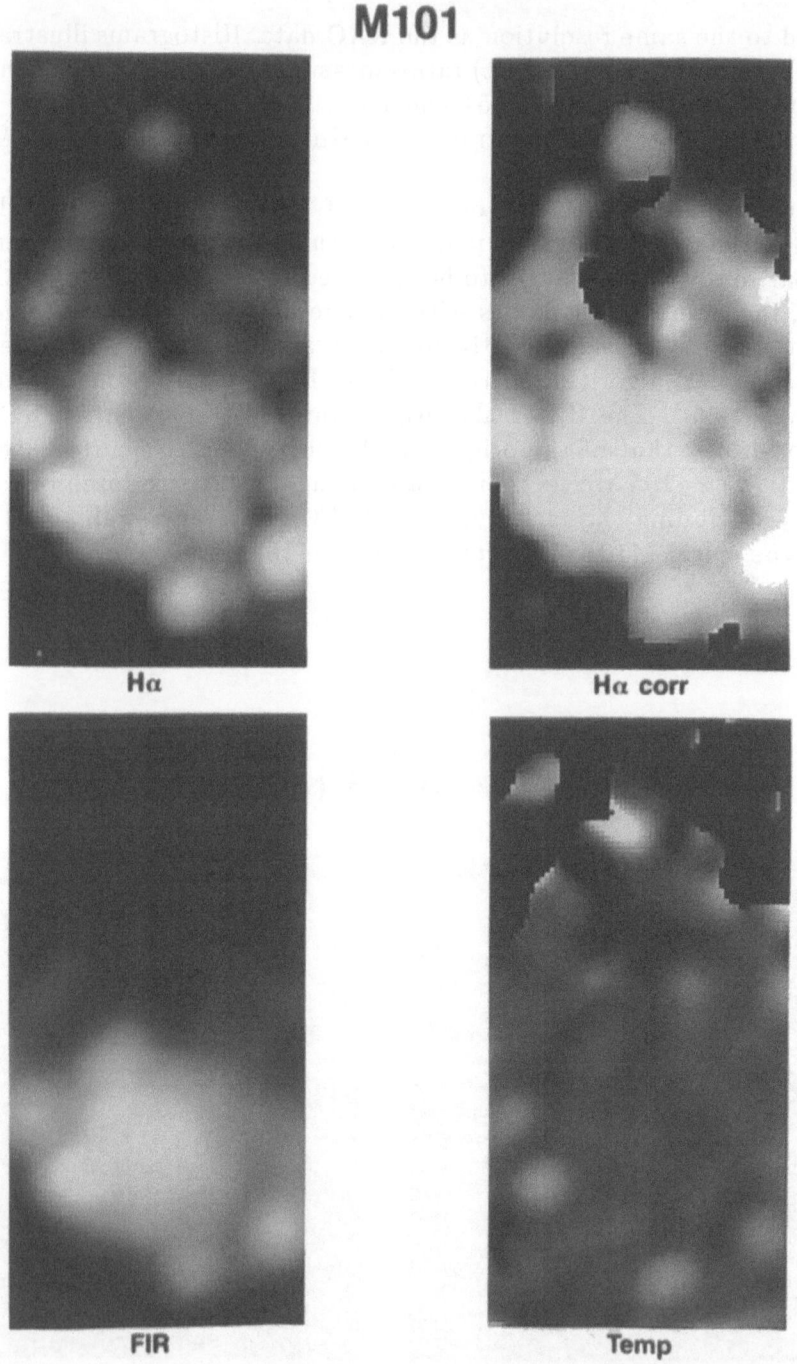

Figure 3. The distribution of the observed Hα luminosity (top left), extinction corrected Hα luminosity (top right), far infrared luminosity (lower left) and the $S_{100\mu m}/S_{60\mu m}$ dust color temperature (lower right) within the late-type spiral galaxy M101

Andromeda is the nearest and consequently the best resolved spiral galaxy. Images illustrating the far infrared and Hα morphology at a common resolution of 105 arc seconds are presented in Figure 2. The Andromeda galaxy has frequently been the focus of the debate surrounding the origin of the far infrared luminosity within spiral galaxies (Habing et al. 1984; Walterbos & Schwering 1987; Devereux et al. 1994). Since Walterbos & Schwering (1987), M31 has long been upheld as an example of a galaxy for which the far infrared luminosity is produced by diffuse dust heated by the general interstellar radiation field of primarily non-ionizing stars. However, the striking correspondence between the far infrared and Hα morphology, illustrated in Figure 2, indicates that it is the O and B stars which are required to ionize the hydrogen gas that are primarily responsible for the luminosity measured in the far infrared, particularly within the star forming ring. The association is further supported by the fact that the L(Hα)/L(FIR) ratio is also measured to be indicative of HII regions powered by O and B stars.

The morphology of the far infrared and Hα emission within the late-type spiral galaxy M101 is presented in Figure 3. Like M31, there is generally a good correspondence between the far infrared and Hα emission features. In particular, the dust temperature peaks coincide exactly with peaks in the Hα image pointing to O and B stars as the luminosity sources responsible for heating the dust. In addition, a histogram of the distribution of L(Hα)/L(FIR) ratios measured within the disk of M101 closely resembles that measured for Galactic HII regions as illustrated in Figure 4.

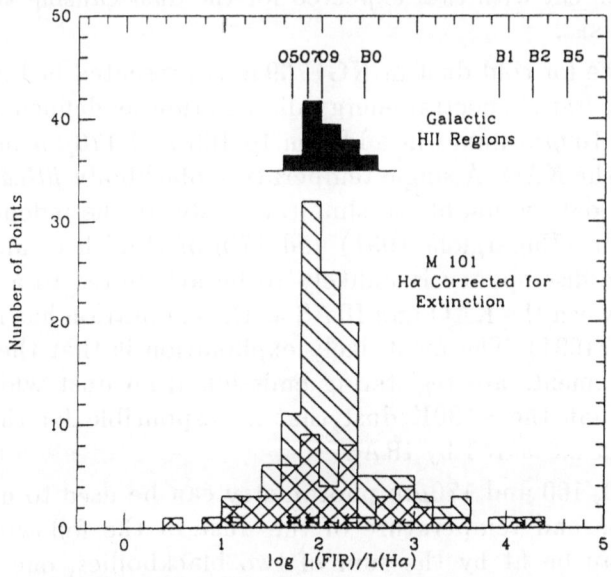

Figure 4. Histogram illustrating the similarity between the distribution of L(Hα)/L(FIR) ratios measured within M101 and Galactic HII regions

In summary, it is the close correspondence between the far infrared and Hα morphology, the positional coincidence between the dust temperature peaks and Hα peaks in addition to L(Hα)/L(FIR) ratios indicative of HII regions which suggest that the far infrared luminosity measured by IRAS is more an indicator of the high mass star formation rate than the disk opacity.

3. Measuring the Disk Opacity

There are several independent lines of evidence which suggest that the dust responsible for the opacity in spiral galaxy disks radiates longward of $100\mu m$ and would therefore have gone undetected by IRAS. First, only 10% of the total dust mass in galaxies is at the $\sim$ 30 - 40 K dust temperatures required to radiate the 60 and $100\mu m$ emission measured by IRAS (Devereux & Young 1990b). The result quite reasonably suggests that while O and B stars may be able to heat $\sim$ 10% of the dust to 30 - 40 K, they are unable to heat all of the dust in galaxies up to such high temperatures. The implication is that the bulk ($\sim$90%) of the dust mass in galaxies is much colder and radiating at wavelengths longward of $100\mu m$. The second piece of evidence is that 160 and $170\mu m$ photometry obtained with the KAO has yielded direct observational evidence for large masses of cold dust in the spiral galaxies M51 and NGC 6946 (Devereux & Young 1992; 1993). The third piece of evidence is that the temperature derived for the cold dust, $\sim$ 17K, is consistent with that expected for the dust causing the opacity in spiral galaxy disks.

The evidence for cold dust in NGC 6946 is presented in Figure 5 which shows the far infrared spectral energy distribution as defined by the IRAS 12, 25, 60 and $100\mu m$ fluxes in addition to 160 and $170\mu m$ measurements obtained with the KAO. A single temperature blackbody *fitted* to the IRAS 60 and $100\mu m$ data is unable to simultaneously fit the independently determined $160\mu m$ (Engargiola 1991) and $170\mu m$ (Smith et al. 1984) measurements. The discrepancy is unlikely to be attributed to a difference in calibration between the KAO and IRAS as the calibration has been checked by Engargiola (1991). The most likely explanation is that the $160\mu m$ and $170\mu m$ measurements are registering emission from dust which is significantly colder than the $\sim$ 30K dust that is responsible for the $60\mu m$ and $100\mu m$ emission measured by IRAS.

The 60, 100, 160 and $170\mu m$ photometry can be used to uniquely constrain the mass and temperature of the dust in the following way. The observations can be fit by the sum of two blackbodies, one *fitted* to the IRAS 60 and $100\mu m$ measurements (curve a) representing the *warm* dust mass and a second blackbody (curve b) representing the remainder of the

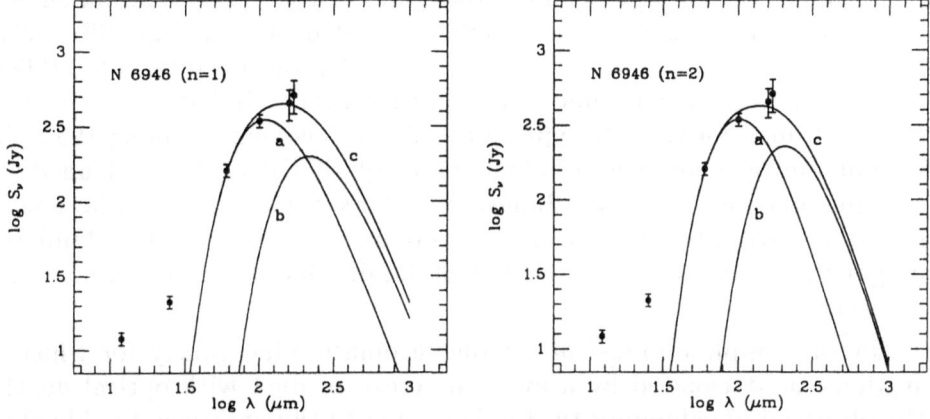

Figure 5. The global far infrared spectral energy distribution of NGC 6946. The independently measured 160μm and 170μm fluxes both exceed those expected if all the dust were radiating at the same temperature as that required to produce the IRAS 60 and 100μm emission (curve a) regardless of whether the dust emissivity index n=1 (left panel) or n=2 (right panel).

dust mass that is cold. The temperature and amplitude of the second black-body are varied to satisfy the *dual* requirements that ∼ 90% of the dust *mass* be in the cold component and the *sum* of the two blackbodies (curve c) provide a satisfactory fit to the observations. It is important to appreciate that the temperature and the amplitude of the two blackbodies are completely constrained by the observational data.

For an emissivity index n=1, curve c in Figure 5 represents the sum of the emission expected if 10% of the dust mass is at a temperature of 34 K (curve a) and 90% of the dust mass is at a temperature of 17 K (curve b). For an emissivity index n=2, curve c represents the sum of the emission expected if 4% of the dust mass is at a temperature of 29 K (curve a) and 96% of the dust mass is at 14 K (curve b). In both cases, the far infrared spectral energy distribution of NGC 6946 indicates that the bulk, 90 - 96%, of the *dust mass* is at a temperature of 14 - 17 K, regardless of the dust emissivity law, be it proportional to λ^{-1} or λ^{-2}.

The 14 - 17 K temperature range derived for the bulk of the dust is consistent with the temperature expected for large, ≥ 0.1μm, grains heated by the general interstellar radiation field within NGC 6946 (Devereux & Young 1993). Such large dust grains are expected to dominate the optical extinction. Consequently, it is the far infrared *luminosity* of the *cold dust*, not the warm dust measured by IRAS, that is to be identified with that responsible for the opacity in spiral galaxy disks. The distinction is an important one because even though the cold dust represents 90 - 96% of the dust *mass*, it produces only ~ 22 - 24% of the total 40 - 1000μm *luminosity*. NGC 6946 is not unique in this regard, similar results have also been obtained for the late-type spiral M51 (Devereux & Young 1992). The submillimetre common user bolometer array (SCUBA) that is planned to be commissioned for use on the James Clerk Maxwell Telescope, at Mauna Kea in Hawaii, will greatly facilitate measurement of the cool dust luminosity in galaxies. Measuring the cold dust luminosity is not the whole story, however.

In the simplest model of a monochromatic blue galaxy for which the starlight is attenuated by a uniform screen of dust with optical depth τ, the observed blue luminosity, L_B, is related to the unattenuated blue luminosity, $L_B^{o,i}$ according to the following relation

$$L_B = L_B^{o,i} e^{-\tau} \tag{1}$$

The luminosity of the cold dust measured in the far infrared, L_c, is equated to the luminosity absorbed by the dust such that

$$L_c = L_B^{o,i} - L_B \tag{2}$$

and substituting for $L_B^{o,i}$ from equation (1) leads to

$$L_c = L_B(e^\tau - 1) \tag{3}$$

Hence, substituting measurements of L_c and L_B in principle allows a determination of τ. The actual situation for galaxies, however, is much more complicated. First, galaxies are not monochromatic because stars radiate over a wide range of wavelengths from the ultraviolet to the near infrared. Second, the dust opacity, τ, has a wavelength dependence. Thus, equation (1) should be integrated over all wavelengths. The third, and perhaps greatest source of uncertainty however, is the 3 dimensional distribution assumed for the dust and stars. The uniform screen, assumed above, is unlikely to be a realistic representation for the distribution of dust in galaxies. Witt et al. (1992) have shown the opacity to be *highly dependent* on the star and dust geometry. Thus, while it may be possible to construct various models, their usefulness in predicting the disk opacity will always be compromised by the uncertainty surrounding the *actual* distribution of stars and dust within the galaxies with which the models are being compared.

Consequently, measuring the cold dust luminosity may not be the best way to determine the opacity of spiral galaxy disks.

4. Summary and Conclusions

Evidence has been presented showing that the far infrared luminosity measured by IRAS represents the thermal radiation from dust that has been heated by O and B stars. As such, the IRAS database constitutes an excellent resource for studying high mass star formation in galaxies, but regrettably, has little to do with the opacity of spiral galaxy disks. The dust that is responsible for the opacity in spiral galaxy disks has been detected in long wavelength, 160 μm observations made from the KAO. The dust has a temperature of $\sim$ 14 - 17K, and produces $\sim$ 22 - 24% of the total 40 - 1000μm *luminosity* in the two galaxies, M51 and NGC 6946, for which it has been measured. In principle, the far infrared luminosity radiated by the cold dust can be used in combination with the stellar luminosity to estimate the disk opacity. In practice the technique is fundamentally limited by our inability to determine the 3 dimensional distribution of stars and absorbing dust within individual galaxies.

References

1. Aumann, H.H., Fowler, J. W., & Melnick, M., 1990, AJ **99**, 1674.
2. Bothun, G.D., & Rogers, C., 1992, AJ **103**, 1484.
3. Burstein, D., & Lebofsky, M.J., 1986, ApJ **301**, 683.
4. Crawford, J., & Rowan-Robinson, M., 1986, MNRAS **221**, 923.
5. De Jong, T., & Brink, E., 1987, In *Star Formation in Galaxies*, p. 323. ed. C. Lonsdale (Washington D.C.: NASA).
6. Devereux, N.A., & Young, J.S., 1990a, ApJ Letters **350**, 25.
7. Devereux, N.A., & Young, J.S., 1990b, ApJ **359**, 42.
8. Devereux, N.A., & Young, J.S., 1991, ApJ **371**, 515.
9. Devereux, N.A., & Young, J.S., 1992, AJ **103**, 1536.
10. Devereux, N.A., & Young, J.S., 1993, AJ **106**, 948.
11. Devereux, N.A., Price, R., Wells, L.A., & Duric, N., 1994, AJ (in press).
12. Devereux, N.A., & Scowen, P.A., 1994, AJ (in press).
13. Engargiola, G., 1991, ApJ Supp. **76**, 875.
14. Habing, H.J., *et al.*, 1984, ApJ Letters **278**, 59.
15. Kennicutt, R.C., Jr., & Kent, S. M.,1983, AJ **88**, 1094.
16. Lonsdale-Persson, C.J., & Helou, G., 1987, ApJ **314**, 513.
17. Phillips, S., Evans, Rh., Davies, J.I., & Disney, M.J., 1991, MNRAS **253**, 496.
18. Smith, J., Harper, D.A., & Loewenstein, R.F., 1984, Airborne Astronomy Symposium Proceedings. ed. H. A. Thronson, Jr., & E.F. Erickson (NASA Conf. Pub. 2353) p. 277.
19. Walterbos, R.A.M., & Schwering, P.B.W., 1987, A&A **180**, 27.
20. Witt, A.N., Thronson, H.A., & Capuano, J.M., Jr., 1992, ApJ **393**, 611

Question
Valentijn

In N4565 KAO maps by Engarziola and Harper reveal a larger extend (or radial scale length) for the "cool dust" than found for the "warm dust". Did you find this in your maps?

Answer
Devereux

That is an interesting result and to be honest, I have not looked into it for the galaxies I have been studying.

Question
Valentijn

I believe the answer to this question is critical for understanding the opacity in the outer regions.

Answer
Devereux

Maybe, although it will probably tell us more about the dust temperature gradient in spiral galaxies.

Question
Wang

Surprising to see your H_α map of M31 showing a strong ring but lacking emission inside (except for the very central region). What happens to spiral structures? Any selection bias?

Answer
Devereux

What you see is what you get !

Question
Ryder

I am concerned you may be seriously underestimating the extinction at H_α. First the Kennicutt and Kent data are based on spectra, not aperture photometry; second, heavily extinguished HII regions may not be seen in H_β, or even H_α, third as my poster on radio continuum to H_α ratios in M83 show, previous estimates of the mean extinction at H_α may be biased to the lowest values. Thus, O and B stars may not be capable of fully accounting for the $L_{FIR}/L_{H\alpha}$ ratios, if the extinction at H_α has been underestimated.

Answer
Devereux

You are wrong, the K+K (1983) data are in fact based on aperture photometry not spectra. With regard to your second point, I agree, obviously if on HII region is completely obscured, no H_α emission will be detected. I should add, however that if the extinction has been underestimated, then correcting for that will cause the $L(FIR)/L(H_\alpha)$ ratio to be even lower, indicative of even higher mass stars than presently indicated by the histograms.

Question
Evans

As I showed in my talk on Wed, we can fit the FIR spectrum of M51 with very different fits, and have ~ 60% of the FIR radiation (1-1500µm) being due to <u>cool</u> dust, the restriction on needing 90% (no more or less) of the mass in cool dust arises because the calculated warm dust mass suggests we are only seeing 10% of the dust mass. This is a circular argument, a different initial warm dust temperature would lead to a different % of warm/cool dust, the answer is not unique.

Answer
Devereux

It is unfair to call the argument circular but I concede that it is not unique.

Comment
Wang

I agree with you there could be more cold dust in galaxies. In fact, COBE has seen ~20K dust in our Galaxy. The question of whether ~0.1µm dust can be responsible for IRAS band emission is more complicated, I think. If you put them nearby OB stars it's possible to heat them up to ~30K.

Question
Thronson

I have two worries. First, when you incorporate the 160μm flux from the KAO, it certainly looks like there is an "excess" over a spectrum fit through the 60+100μm IRAS data. However, it is widely believed that small grains contribute to at least some of the 60μm emission. What are your conclusions about cold dust if you fit the 100μm and 160μm data under the assumption that these two bands are dominated by the same grains? Next, you showed a plot of $L_{FIR}/L_{H\alpha}$ for the Milky Way and two other galaxies. There must be some selection effects which contribute to the narrowness of the histogram: declining numbers of massive stars, detectability, sensitivity of the KAO, and so on. Please comment.

Answer
Devereux

The whole point about fitting the IRAS 60+100 points with a single temperature is to elucidate blackbody emission from dust. Dust that is colder than required to radiate the IRAS 60+100μm emission may be caused by small grains large grains or any kind of grains. The excess emission at 160μ over a single temperature blackbody fitted to the IRAS 60+100μm emission simply proved the existence of dust with temperatures colder than that required to radiate the IRAS 60+100μm emission.

The histograms L(FIR/L(Hα) are narrow for the Milky Way and other spiral galaxies because there are two properties of high mas stars that prevent very high and very low values of L(FIR)/L(Hα). On the high end, where the L(FIR)/L(Hα) ratio indicates B stars, there is a cut-off because BO and later stars do not produce HII regions, hence no H_α emission, and apparently no far infrared emission either. On the low end, where the L(FIR)/L(H_α) ratio is indicative of O stars, there is a cut-off because the L(FIR)/L(H_α) ratio reaches an asymptotic value of ~ 1.8 due to the high mass stars and the fact that there are not many stars in any galaxy with masses ≥40M☉.

Comment
R Arendt

COBE/DIRBE results reported recently by Sodroski et al (1994, ApJ, 428,638) find an ~18K component to the 140-240μm emission within a few degrees of the galactic plane. Its optical depth is well correlated with the extinction observed in the near-IR (1.25-2.2μm) as long as the near-IR is optically thin.

DIRBE would probably not be very sensitive to much colder dust..

THE FAR INFRARED/STELLAR ENERGY BALANCE

Cydbwysedd yr egni is-goch pell/serol

RHODRI EVANS
Swarthmore College,
Department of Physics and Astronomy, Swarthmore, PA 19081
USA

Abstract.
In this talk I will cover the following topics
(a) the different measurements of the B–band luminosity used in Astronomy, and which is the correct one to use when comparing the stellar output of a galaxy to its far infrared (FIR) output.
(b) The relative contributions of the warm and cool dust components in galaxies, in particular I will concentrate on M51.
(c) What does a particular value of L_{IRAS}/L_B imply about the opacity and optical depth of a particular galaxy and
(d) Specific calculations of the central B–band optical depth $\tau_B(0)$ and the opacity A_B for M51.
I will show that, contrary to other claims, there is a large uncertainty in what is the dominant heating source of the dust seen by IRAS in M51, which in turn leads to a broad range in the possible opacities for this galaxy. Finally, I will briefly discuss the data which can help us narrow this uncertainty.

1. Introduction

One of the ways we might hope to solve the vexing question of how opaque (or transparent) normal spiral galaxies are is to use the radiation produced by the obscuring material – the dust. When IRAS was launched in the early 1980s it came as quite a surprise to most astronomers just how much of a galaxy's flux was being radiated in the FIR part of the spectrum. For normal galaxies this radiation arises from the dust grains re–radiating the stellar flux they have absorbed.

281

J. I. Davies and D. Burstein (eds.), The Opacity of Spiral Disks, 281–290.
© *1995 Kluwer Academic Publishers.*

That it is dust which is responsible for the FIR flux seen by IRAS is not in dispute. What is in dispute, however, is how much of that flux is due to warm dust associated with massive star formation, and how much is due to dust heated by the general interstellar radiation field (ISRF). We would expect the warm dust (30–50 K) to be concentrated in the sites of star formation, whereas the cool dust (13–20 K) would be more diffuse, and it is this cool component which may provide enough attenuation of starlight to render galaxies optically thicker than has historically been thought.

Hence, observations in the FIR part of the spectrum may be a way to resolve the present lack of consensus on the opacity of spiral galaxies, but if we are to use this method we must do so carefully and correctly. If we are going to compare the FIR output of galaxies to some measurement of their observed stellar output, we need to make sure that we are comparing like with like, i.e. that our values for the FIR flux and the stellar flux are based on the same physical units. For example, what does

$$\frac{L_{IRAS}}{L_B} = 1.0$$

tell us about the optical depth and opacity of a galaxy? Does a value less than unity imply galaxies are optically thin, does a value greater than unity imply it is optically thick? As I will show, things are much more complicated than they initially appear to be.

2. Using the correct measurement of the Blue luminosity

The "blue luminosity" L_B The "blue luminosity" L_B, measured in Wm^{-2} which is often compared to the IRAS luminosity L_{IRAS} (also measured in Wm^{-2} and defined as the flux between 42.5 and 122.5 μm , Helou etal. 1984) is based on the total B–band magnitude of a galaxy B_T and the absolute blue magnitude of the Sun $(M_B)_\odot = (5.48^m$, Allen 1973). For M51 $B_T = 8.96$ (RC3) and the distance D of the galaxy is 9.6 Mpc (Devereux & Young 1990) which leads to a blue luminosity L_B of $L_B = 1.296 \times 10^{-11} \, Wm^{-2}$.

The "in–band blue luminosity" L_b A more correct measurement of the flux coming from a galaxy in the part of the spectrum covered by the Johnson B filter is the "in–band" blue luminosity, which we shall call L_b. To calculate L_b we start with the same B–band magnitude B_T for the galaxy, but then convert this magnitude to a flux in the B–band using the absolute calibration of Johnson (1966). We finally multiply by the effective bandwidth $\Delta\nu_B = c\Delta\lambda_B/\lambda_B^2$, where $\lambda_B = 0.435$ μm and $\Delta\lambda_B = 0.1$ μm (Johnson 1966). For $B_T = 8.96$ we get $1.83 \times 10^{-12} \, Wm^{-2}$.

As we can see, L_b, the flux in the B–band, is more than a factor of 7 less than L_B, which is the quantity usually used. So, we can write for M51 that

$$\frac{L_{IRAS}}{L_B} = 0.38 \quad \text{but} \quad \frac{L_{IRAS}}{L_b} = 2.68$$

3. What is the main dust heating source in M51

It was realised quite early on in the study of IRAS data (e.g. de Jong etal. 1984, Helou 1986) that the 60 and 100 μm data points were probably heated by a combination of a warm dust component (T_d = 30–50 K) associated with massive star formation and a cool dust component (T_d = 13–20 K) heated by the ISRF. What is still controversial is which one dominates the flux at 100 μm, or to put it another way, how much is the flux measured by IRAS dominated by warm dust associated with massive star formation. Because the total flux of a dust grain scales as $T^{4+\beta}$, where β is the emissivity index (usually thought to lie in the range 1 to 2 – see Draine 1989), a small amount of warm dust can easily dominate over a large amount of cool dust. This is one of the things which makes the cool dust so hard to find. With only IRAS 60 and 100 μm data points we do not have enough information to constrain our possible fits, the 60 and 100 μm points can be fitted by a single temperature dust grain, or by any two components.

In the case of M51 we have an additional data point at 160 μm, measured by the Kuiper Airborne Observatory (KAO) by Smith (1982). It is clear from this data point that a single fit to the 60 and 100 μm data (which yields a component at 33 K if $\beta = 1$) requires an additional, cool component to account for the measured 160 μm flux (see figure 1a).

This additional data point reduces our flexibility of possible fits considerably, but not completely. Devereux has claimed (Devereux & Young 1990, – henceforth DY90, 1992 – henceforth DY92, Devereux, these proceedings) that we can use the FIR spectrum together with the observed gas to dust ratio in M51 to determine uniquely the correct fit to the FIR spectrum, and hence determine the relative contributions of the two components to the FIR energy. The logic of their argument (which I will show is flawed) is the following:

- If we fit the 60 and 100 μm IRAS data points with a single component we see a need for a cool component to account for the observed flux at 160 μm . In figure 1(a) I have reproduced DY92's fit to the data for a β of 1. The temperature of the warm component is 33 K.
- Using the formula of Hildebrand (1983)

$$M_d = C S_{100} D^2 (\exp 144/T - 1) M_\odot$$

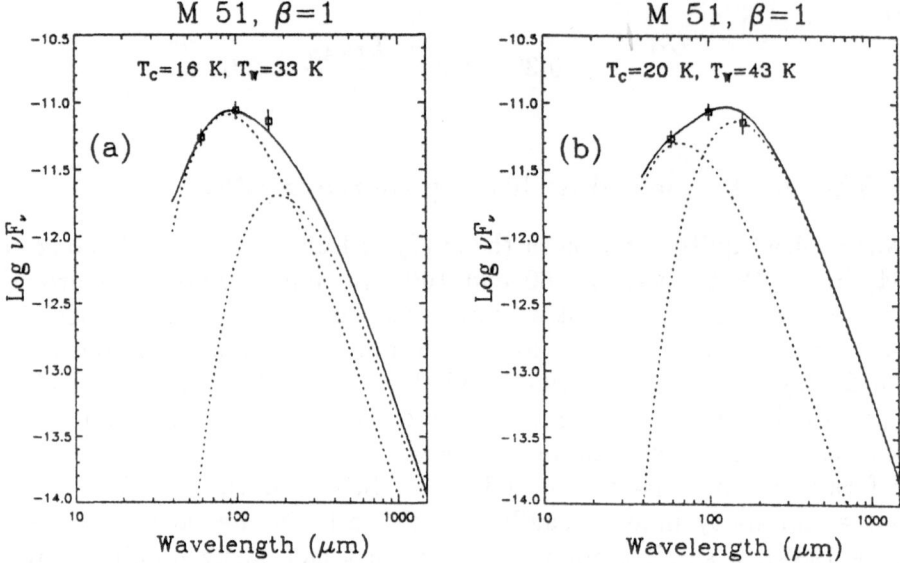

Figure 1. Two possible fits to the FIR spectrum of M51, fit (a) is the fit of Devereux & Young (1992), fit (b) is an alternative, equally correct fit. The relative contributions of the warm and cool dust components in the two fits is very different, as shown in table 3.

where C is a constant which depends on the grain opacity, S_{100} is the 100 μm flux in Janskys, D is the distance of the galaxy in Mpc, and T is the temperature of the dust grains, DY90 go on to calculate the mass of warm (33 K) dust, assuming a C of 4.58 and a D of 9.6 Mpc. They find $\log M_d = 6.99 M_\odot$ (DY90).

- Using the H_2 measurements of Young etal. (see DY90) and the HI measurements from Weliachew & Gottesman (1973), they find, only considering the HI within the inner half of the optical disk, that the mass of gas is $\log M_g = 10.17 M_\odot$. Thus, the observed gas–to–dust ratio for M51 is found to be $M_g/M_d = 1500$.

- They then conclude that if the canonical gas–to–dust ratio for our Galaxy of 100–150 is true for M51, then we are only detecting 10% of the dust at IRAS wavelengths. When the FIR spectrum is fitted with 10% of the dust at 33 K it is found that the remaining 90% needs to be at 16 K (if $\beta = 1$) to fit the 60, 100 and 160 μm data points. We have shown this fit in figure 1(a).

However, their claim that this is an unique fit, that the percentages of warm and cool dust are known, is <u>wrong</u>. It is a circular argument to say that the fit is constrained uniquely by the 60, 100 and 160 μm fluxes and the "known" percentage of warm dust, when this percentage of warm dust is determined from a fit to the data in the first place.

To illustrate this, we have tried fitting the data with a different warm dust component. We find that if we let T_W be 43 K, this produces a flux at 100 μm of $S_{100} = 127.31$ Jy (note, this is the flux due to the 43 K component only, not the total flux at 100 μm which is measured to be 292.08 Jy – Rice etal. 1988). Keeping C and D at the same values that DY90 used, we find now that $\log M_d = 6.17 M_\odot$ so that the observed gas–to–dust ratio is $M_g/M_d = 10020$, very different from the 150 found when T_W was 33 K. If we assume the true value to be 150, this means that the warm component constitutes 1.5% of the dust mass, and the cool component 98.5%. If we use these percentages of warm and cool dust to fit the observed spectrum, we find we can fit it with a cool component at 20 K, as shown in figure 1(b).

Although the two fits both pass through the data points equally as well, the energetics of the two fits are very different. In table 1 we summarise the percentage of energy due to each component in the 40–1000 μm range. Fit (a) (DY92) has 78% of the energy in this range as being due to the warm dust component, whereas fit (b) has only 38% of the energy in the 40–1000 μm range being due to the warm dust component, and 61% being due to the cool dust component. We also include in table 1 the data for the $\beta = 2$ case, although we do not show these fits in figure 1.

Thus, the conclusion that DY92 came to that, based on the FIR spectrum of M51 and the "known" observed gas–to–dust ratio, the dominant heating source of the dust in this galaxy is massive star forming regions, is not proven. It may be correct, but we have shown an equally valid case, based on the same procedure, which has the cool dust dominating over the warm dust. As we shall see, if we are to use the FIR flux of galaxies to determine their optical depths, we need to lessen this uncertainty. The argument made by DY90 and DY92 is thus a fallacious one.

TABLE 1:THE ENERGETICS OF VARIOUS FITS TO THE FIR SPECTRUM OF M51

Fit	β	T_W	T_C	40 μm –1000 μm	
				E_W	E_C
a	1	33 K	16 K	78%	20%
b	1	43 K	20 K	38%	61%
c	2	28 K	14 K	79%	20%
d	2	38 K	18 K	34 %	64%

4. Determining the opacity from the FIR/stellar energy balance

How much of the energy radiating in the FIR is due to the cool dust component is important for several reasons. If the flux measured by IRAS is dominated by warm dust, as claimed by e.g. DY90, DY92, then we can reliably use the IRAS flux as a measure of current massive star formation. If, however, the cool dust contributes a substantial fraction of the IRAS flux, then we cannot do this.

In addition, if we are to try and determine the opacity of a galaxy by comparing the observed stellar flux to the observed FIR flux, we need to be sure what fraction of this observed FIR flux is due to the cool dust which is going to provide the general obscuration in a galaxy and hence, possibly, higher opacities than historically thought.

Disney, Davies & Phillipps (1989 – henceforth DDP89) and Engargiola (1991), & Crawford & Rowan–Robinson (1986), among others, have compared the FIR flux of galaxies to their stellar output, in an attempt to determine their opacities. All of these studies have made the incorrect assumption that the stellar photons emerge at a single wavelength. To do the calculation properly (see Evans 1992) we need to consider the non–ionizing radiation at all wavelengths from UV to the near infrared. The optical depth due to a given amount of dust changes as a function of wavelength, and it is thus a much more complicated problem than simply comparing the FIR flux to L_B or L_b.

We first need to determine the total magnitude of a galaxy at all these different wavebands, to do this we can use the B_T from e.g. RC3, and, for most galaxies, the multi–wavelength aperture photometry published by Longo & de Vaucouleurs (1983) and de Vaucouleurs & Longo (1988) to get the total magnitude in each waveband from U to the K–band.

Then, using the absolute flux calibrations of Johnson (1966) for U to I–band, and Campins etal. (1985) for J, H and K–bands, we can convert these magnitudes to fluxes, and hence produce an observed spectral energy distribution (SED) for the starlight of the galaxy in question. Figure 2 shows the observed SED for M51 calculated in this way. We have fitted the 8 data points with the sum of two blackbody curves, and find our range of possible fits to be quite tightly constrained. In figure 2(a) we show one of the fits, where we have one stellar component with $T = 3000K$ and the other with $T = 6000K$. The curves have been normalised to pass through the B–band and J–band data points, and by summing under the entire curve from 0.1 to 30 μm we get the total observed stellar flux from non–ionizing photons. We have explored the full range of possible fits, and get that the observed stellar flux of M51 lies in the range

$$1.00 \times 10^{-11}\, \text{Wm}^{-2} \quad \text{to} \quad 1.24 \times 10^{-11}\, \text{Wm}^{-2} .$$

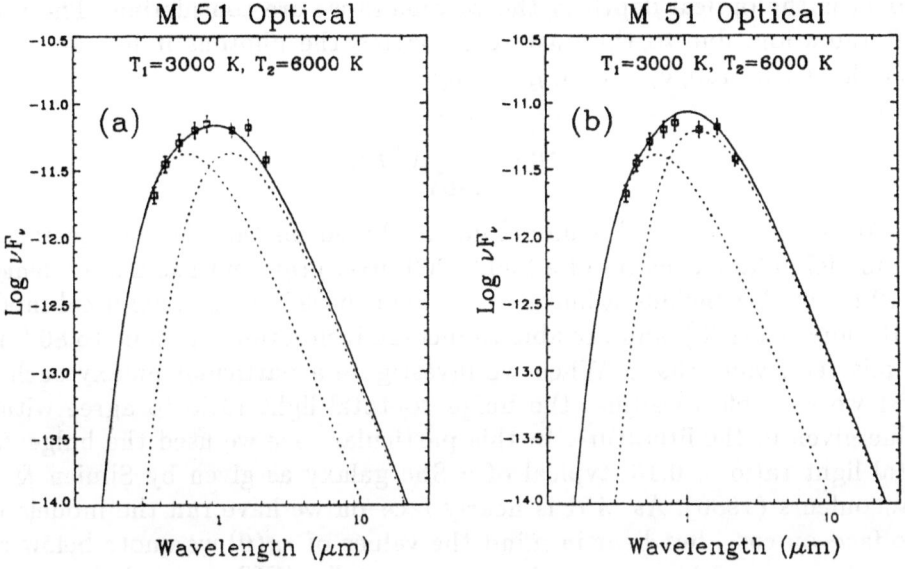

Figure 2. Two possible fits to the Stellar spectrum of M51. Because of the 8 points our fits are tightly constrained, here we show the two most extreme fits to the data, fit (a) -1.00×10^{-11} Wm^{-2} and fit (b) -1.24×10^{-11} Wm^{-2}.

This small range is because of the tight constraints having 8 data points puts on our fitting.

The next step is to calculate the flux in each waveband, this is simply found by summing under the curve between the lower and upper wavelength limits for each filter. This gives us $(L_\star)_\lambda$, where

$$L_\star = \sum_{\lambda=UV}^{NIR} (L_\star)_\lambda.$$

The third step is to assume a model for the stars/dust geometry. We have adopted the Triplex model of DDP89, where the dust and stars have the same exponential scale length, and the relative scale height of the dust to stars is given by ζ, the layering parameter. In a particular waveband we can write (see Evans 1992)

$$\frac{(L_{FIR})_\lambda}{(L_\star)_\lambda} = \frac{1 - \exp(-r/\alpha)\Theta_\lambda}{\exp(-r/\alpha)\Theta_\lambda}$$

where

$$\Theta_\lambda = \exp(-\tau_\lambda)\left[1 + \frac{\tau_\lambda^2}{(\zeta+1)(\zeta+2)} + \frac{\tau_\lambda^4}{(\zeta+1)(\zeta+2)(\zeta+3)(\zeta+4)} + \cdots\right]$$

and τ_λ is the optical depth in the waveband we are considering. The total FIR radiation, due to the dust re–radiating the photons it has absorbed over this entire range, is then given by

$$L_{FIR} = \sum_{\lambda=UV}^{NIR} (L_{FIR})_\lambda.$$

We have constructed a model galaxy based on the Triplex geometry, the model galaxy consists of a 200×200 pixel grid, and has a scale length of 10 pixels. We include a bulge component, based on the analytical model of Hernquist (1990), and are able to include inclinations from 0 °to 80 °(for details see Evans 1992). When we investigate a particular galaxy such as M51 we are able to adjust the bulge–to–total light ratio to agree with a value given in the literature. In this particular case we used the bulge–to–total light ratio of 0.16, typical of a Sbc galaxy as given by Simien & de Vaucouleurs (1986). As M51 is nearly face–on we have run the models for the face–on case, but bear in mind the values of $\tau_B(0)$ we quote below are the values required to give us the observed stellar/FIR energy balance, and so would be the observed $\tau_B(0)$, whether the galaxy is face–on or inclined.

For a particular value of ζ and a particular central optical depth $\tau_B(0)$ the model calculates for each waveband the ratio

$$\frac{(L_{FIR})_\lambda}{(L_\star)_\lambda} = \frac{1 - \exp(-r/\alpha)\Theta_\lambda}{\exp(-r/\alpha)\Theta_\lambda}$$

in each pixel in the grid, to give us a total L_{FIR} for the whole galaxy.

The final step is to compare the total L_{FIR} of the model to the observed (or inferred) $(L_{FIR})_{cool}$, the FIR luminosity of the <u>cool</u> dust component. This allows us to determine the opacity A_B of the galaxy. For a given layering parameter ζ this will also provide us with an optical depth at the centre of the galaxy, $\tau_B(0)$. However, $\tau_B(0)$ is dependent on the chosen value of ζ, whereas A_B is only dependent on the ratio of $(L_{FIR})_{cool}$ to $L_\star$, the total stellar flux from non–ionizing stars. For M51 this flux ranges from

$$1.52 \times 10^{-12}\,\mathrm{Wm}^{-2} \quad \text{(fit c)} \quad \text{to} \quad 5.90 \times 10^{-12}\,\mathrm{Wm}^{-2} \quad \text{(fit d)} .$$

The large uncertainty is due to how much of the FIR flux is due to the cool dust component, as we discussed in section 3. The calculated value for the central B–band optical depth $\tau_B(0)$ is

$$\tau_B(0) = 2.0 \ \text{(fit c)} \quad \text{to} \quad \tau_B(0) = 11 \ \text{(fit d)}.$$

In terms of opacities this translates to

$$A_B = 0.27^m \quad \text{(22\% of the B–band light lost)} \quad \text{to}$$
$$0.74^m \quad \text{(50\% of the B–band light lost)}.$$

In terms of an average optical depth, these figures translate to

$$\tau_B(0) = 2.0 \Rightarrow <\tau_B> = 0.55$$
$$= 11.0 \Rightarrow <\tau_B> = 1.95.$$

For those interested in what fraction of M51 is optically thick, assuming τ_B decreases as $\tau_B = \tau_B(0)\exp(-r/\alpha)$ as assumed in the Triplex model, we find

if $\tau_B(0) = 2.0$ $\tau_B > 1$ to 0.7 of a scale length

if $\tau_B(0) = 11.0$ $\tau_B > 1$ to 2.4 scale lengths.

5. Reducing the uncertainty

The range of possible values of L_* is small because we are fitting two blackbody curves to 8 data points, and so the range of possible fits is small. The same cannot be said for the FIR spectrum, for most galaxies all we have is the IRAS 60 and 100 μm data, and even for M51 with its additional 160 μm data point the value of $(L_{FIR})_{cool}$ has a range of nearly a factor of 4. In addition to not knowing the temperatures of the warm and cool dust components, we at present do not know at what point the emissivity index β becomes 2 (see Draine 1989).

However, additional data can solve both of these unknowns. Studies at 800 μm and longer, from the ground, have often failed to find much evidence for cool dust in galaxies (see e.g. Eales etal. 1989), but all studies in the 100 to 350 μm range have shown a need for a cool dust component. As illustrated by Evans (1992), observations in the sub–mm region of the spectrum are not that sensitive to cool dust, we need to look where this dust would emit most of its energy, in the 100 to 300 μm range. Observations at sub–mm wavelengths, can, however, resolve the issue of what the value of β is in the FIR/sub–mm part of the spectrum.

Fortunately the coming 5 to 10 years should see a number of telescopes and instruments capable of making these crucial measurements. The 100 to 300 μm part of the spectrum will be explored by the Infrared Space Observatory (ISO) and, hopefully, by the Space Infrared Telescope (SIRTF) and the Stratospheric Observatory For Infrared Astronomy (SOFIA). The sub–mm part of the spectrum is soon to be explored using the Sub–mm Common User Bolometer Array (SCUBA), which will improve our sensitivity in this part of the spectrum spectacularly (Cunningham & Gear 1991).

6. Summary

We have shown that it is possible to use the FIR output of galaxies, in principle, to determine their opacities and optical depths. In this paper we have presented what we feel is the most detailed approach so far to doing this, considering the entire spectral range of non–ionizing photons, not just the photons emerging in a particular waveband. We have shown that the origin of the dominant dust heating source is still not resolved, even for M51. This uncertainty leads to a broad range of possible opacities and optical depths for M51, it can be optically thin if the dust is mainly heated by ionizing photons in star forming regions, or it can be optically thick out to about 2.4 scale lengths if the dominant dust heating source is the ISRF. Resolving which is the case should be possible in the next few years with ISO, SIRTF, SOFIA and SCUBA.

References

1. Allen, C.W., 1973, Astrophysical Quantities.
2. Campins, H., Rieke, G.H. & Lebofsky, M.J., 1985, AJ **90**, 896.
3. Crawford, J. & Rowan–Robinson, M., 1986, MNRAS **221**, 53.
4. Cunningham, C.R. & Gear, W., 1991, in 'The SPIE Symposium on Astronomical Telescopes and Instrumentation for the 21st Century'.
5. Devereux, N.A. & Young, J.S., 1990, ApJ **359**, 42.
6. Devereux, N.A. & Young, J.S., 1992, AJ **103**, 1536.
7. Disney, M.J., Davies, J.I. & Phillipps, S., 1989, MNRAS **239**, 939.
8. Draine, B.T., 1989, in "The Interstellar Medium in Galaxies, eds. H.A. Thronson and J.M. Shull, 483.
9. Eales, S.A., Wynn–Williams, C.G. & Duncan, W.D., 1989, ApJ **339**, 859.
10. Engargiola, G., 1991, ApJ Supp. **76**, 875.
11. Evans, Rh., 1992, PhD Thesis, University of Wales.
12. Helou, G., Soifer, B.T. & Rowan–Robinson, M., 1984, Bull. A.A.S. **16**, 471.
13. Helou, G., 1986, ApJ **311**, L33.
14. Hernquist, L., 1990, ApJ **356**, 359.
15. Hildebrand, R.H., 1983, Qu. J. R. astr. Soc. **24**, 267.
16. Johnson, H.L., 1966, Ann. Rev. Astr. Ap. 4, 193.
17. de Jong, T., Clegg, P.E., Soifer, B.J., Rowan-Robinson, M., Habing, H.J., Houck, J.R., Aumann, H.H. & Raimond, E., 1984, ApJ **278**, L67.
18. Longo, G. & de Vaucouleurs, A., 1983, A General Catalogue of Photometric Magnitudes and Colours in the U,B,V System... , Texas University Press.
19. Rice, W.J., Lonsdale, C.J., Soifer, B.T., Neugebauer, H.J., Kopan, E.L., Lloyd, L.A., de Jong, T. & Habing, H.J., 1988, ApJ Supp. **68**, 91.
20. Simien, F. & de Vaucouleurs, G., 1986, ApJ **302**, 564.
21. Smith, E.P., 1982, ApJ **261**, 463.
22. de Vaucouleurs, A. & Longo, G., 1988, Catalogue of visual and infrared photometry from 0.5 μm to 10 μm (1961–1985), Texas University Press.
23. Weliachew, L. & Gottesman, S.T., 1973, A&A **24**, 59.

The pose II

OPACITY FROM LUMINOSITY FUNCTIONS

The IRAS Survey Does Not Show That Galaxies Are Optically Thin

M. TREWHELLA, J. DAVIES, M. DISNEY AND H. JONES

University of Wales College of Cardiff
Department of Physics and Astronomy
PO Box 913
Cardiff CF2 3YB

Abstract.

We have compared the 60 micron IRAS luminosity function [1] with that in the optical [2] to obtain the average blue band optical depth in late-type galaxies. Contrary to previous work [1] we find that late-type galaxies have an average optical depth near to unity.

1. Background

Saunders *et al.* [1] created and integrated an IRAS $60\mu m$ luminosity function to get the local luminosity density. Combining this with $100\mu m$ data enabled them to calculate the local far infrared (FIR) luminosity density. Assuming the FIR energy comes from absorbed starlight they compared this to a similar blue band local luminosity density to infer an optical depth of $\tau_B = 0.26$. We feel that the technique and the model used are too simplistic for this type of study. In this paper we present what we believe to be a more realistic interpretation of the observational data.

2. Modifications to Saunders *et al.* model

In this section we describe the four modifications to the Saunders *et al.* model. At the end of each subsection we summarise the effect of that modification alone (only in section 3 are they combined).

J. I. Davies and D. Burstein (eds.), The Opacity of Spiral Disks, 293–297.
© *1995 Kluwer Academic Publishers.*

2.1. ELLIPTICALS

This correction was suggested by Saunders *et al.* to account for the discrepancy between their value of $\tau_B = 0.26$ and that quoted in the RC2 [4] of $\tau_B \sim 0.39$. Generally elliptical galaxies have less gas and dust, emit weakly in the FIR and therfore contribute little to the $60\mu m$ luminosity function while still contributing to the optical luminosity function. Loveday *et al.* [2] splits the luminosity function up into early and late type galaxies allowing us to remove the elliptical contribution to the luminosity function. This is shown below.

Saunders *et al.* ....$\tau_B = 0.26$
Ellipticals removed ....$\tau_B = 0.52$

2.2. RADIATIVE TRANSFER MODEL

The radiative transfer equation Saunders *et al.* [1] used to get the optical depth from the observations corresponds to a screen geometry [5]. This model is unrealistic but has simple mathematics and has been used historically. For instance, it applies to studies of the extinction in our own galaxy, where the dust *does* lie as a screen between us and individual stars. The problem with using a a screen model is as follows, for a dust layer with a given optical depth a screen geometry produces the maximum amount of extinction possible. This is because *all* the starlight must pass through the layer. This means that if you measure a certain extinction (drop in optical output), it is therefore only necessary to invoke a relatively small optical depth to explain a given ratio of blue to FIR energy.

If one looks at a disk galaxy edge on, one sees the dust confined to a layer along the centre of the galaxy. This behaviour is mimicked by the sandwich model [5]. The ratio of the scale heights of dust to stars is defined as $\zeta = \alpha_s/\alpha_d$. Using a sandwich model (with $\zeta = 0.5$) gives the following result.

Saunders *et al.* [1] ....$\tau_B = 0.26$
Sandwich model ....$\tau_B = 0.57$

2.3. "IN BAND" BLUE LUMINOSITY

When you integrate a Schechter luminosity function [8] to get the local luminosity density, you get an expression containing L^* (the characteristic luminosity). Most Schechter functions are parametrised in terms of their characteristic magnitude (M^*). There are three quite different ways of converting from M^* to L^* [6,7]. Saunders *et al.* use a method that gives "the blue luminosity" [6,7], this has no real physical meaning except under the assumption that all galaxies have the same SED as our sun. The quantity required for this work is the "in band" blue luminosity [6,7]. To convert from "the blue luminosity" to the "in band" blue luminosity it is necessary to multiply by a factor of 0.14 [6,7], this has the following effect.

Saunders *et al.* ....$\tau_B = 0.26$
With "in band" ....$\tau_B = 1.13$

2.4. OPTICAL SED

Saunders *et al.* assumed that only absorption of blue light is an important contributor to the FIR energy output. The U.V. starlight is readily absorbed by dust but there is a relatively small amount given off by the galaxy, $\Longrightarrow$ low contribution to FIR output.

Although there is a large amount of energy in the red region of the spectrum, the dust absorbs it to a much lesser degree, $\Longrightarrow$ low contribution to FIR output.

These arguments are reasonable, the blue band is the largest contributor to the FIR output *but* the other regions still contribute a significant amount. We therefore need to account for this. It is not practical to construct a luminosity function in every band so we have used a typical spiral galaxy SED [7] normalised to the blue band to calculate the energy densities in the other bands. To take into account the differing degree of absorption from these bands we have scaled all the optical depths relative to the blue band ie. we have $\tau_\lambda = k_\lambda \tau_B$ (we have taken k_λ to be proportional to $1/\lambda$ [9]).

$$\frac{<L_B>}{<L_{FIR}>} = \sum_{\lambda=U.V.}^{NIR} x_\lambda \left(\frac{1 - R(k_\lambda \tau_B, \zeta)}{R(k_\lambda \tau_B, \zeta)} \right) \tag{1}$$

Here R is just a geometry dependant function that relates the amount of light lost through a galaxy to the optical depth and x_λ is the relative SED. We have solved this equation iteratively to find τ_B and obtained the following result.

Saunders *et al.* ....$\tau_B = 0.26$
SED model ....$\tau_B = 0.06$

3. Conclusions

If we put in all the corrections described above, we get:

Saunders *et al.* ....$\tau_B = 0.26$
Trewhella *et al.* ....$\tau_B \sim 1$

We feel this is the most realistic solution possible at this time with existing models and data. Better galaxy models, involving patchiness and different stellar populations and a more detailed knowledge of the FIR spectral energy distribution will enable us to do a more rigorous determination using this method.

4. References

1. Saunders, W., Rowan-Robinson M., Lawrence A., Efstathiou G., Kaiser N., Ellis R. and Frenk C. (1990), *MNRAS* **242**,318-337.
2. Loveday J., Peterson B., Efstathiou G. and Maddox S. (1992), *ApJ* **390**, 388-344.
3. Efstathiou G., Ellis R. and Peterson B. (1988), *MNRAS* **232**, 431-461.
4. de Vaucoulers G., de Vaucoulers A. and Corwin H. (1976) *Second Reference Catalogue of Bright Galaxies (RC2)*
5. Disney M., Davies J. and Phillipps S. (1989), *MNRAS* **239**, 939-976.
6. Witt A., Thronson H. and Capuano J. (1992), *ApJ* **393**, 611
7. Evans Rh. (1992) Opacity in Spiral Galaxies, *Ph.D Thesis University of Wales.*
8. Shechter P. (1976), *ApJ* **203**, 297-306.
9. Mathis J. (1990), *Ann. Rev. Astr. Astrophys.* **28**,37

Comment
Burstein

Aside from correction for E's to SO's, there are also corrections for bulges! I see an inherent uncertainty in this kind of comparison at the factor of two level. No wonder Rowan-Robinson did not want to attend this meeting!

Answer
Trewhella

When I spoke to him I told him that I was working on this but unfortunately did not have time to go through the details but he assured me that many people have redone this work and they all came to the same answer as he did.

Question
Giovanelli

Can you define τ_B? Is it such that you'd detect only $e^{-\tau_B}$ of the total flux of the "transparent" galaxy?

Answer
Trewhella

No. It is defined as in a sandwich model.

$$\frac{L(dust)}{L(no\ dust)} = \%\ of\ galaxy\ light\ that\ escapes$$

$$= \frac{\langle L_B \rangle}{\langle L_B \rangle + \langle L_{FIR} \rangle}$$

$$= \left(\frac{1-\zeta}{2}\right)(1+e^{-\tau_B}) + \frac{\zeta}{\tau_B}(1-e^{-\tau_B})$$

If you were referring to a background light source then YES τ_B would be defined as you said.

Question
Devereux

I do not believe that it is so simple to obtain the optical depth from the $L_{FIR}/L_{H\alpha}$ ratio because Witt et al. (1992) showed that the optical depth inferred from the $L_{FIR}/L_{H\alpha}$ ratio depends very sensitively on the dust/star geometry and you can basically get any value of τ, i.e. it is a completely unconstrained problem.

Answer
Trewhella

Yes, we have used our best guess at the geometry factor.

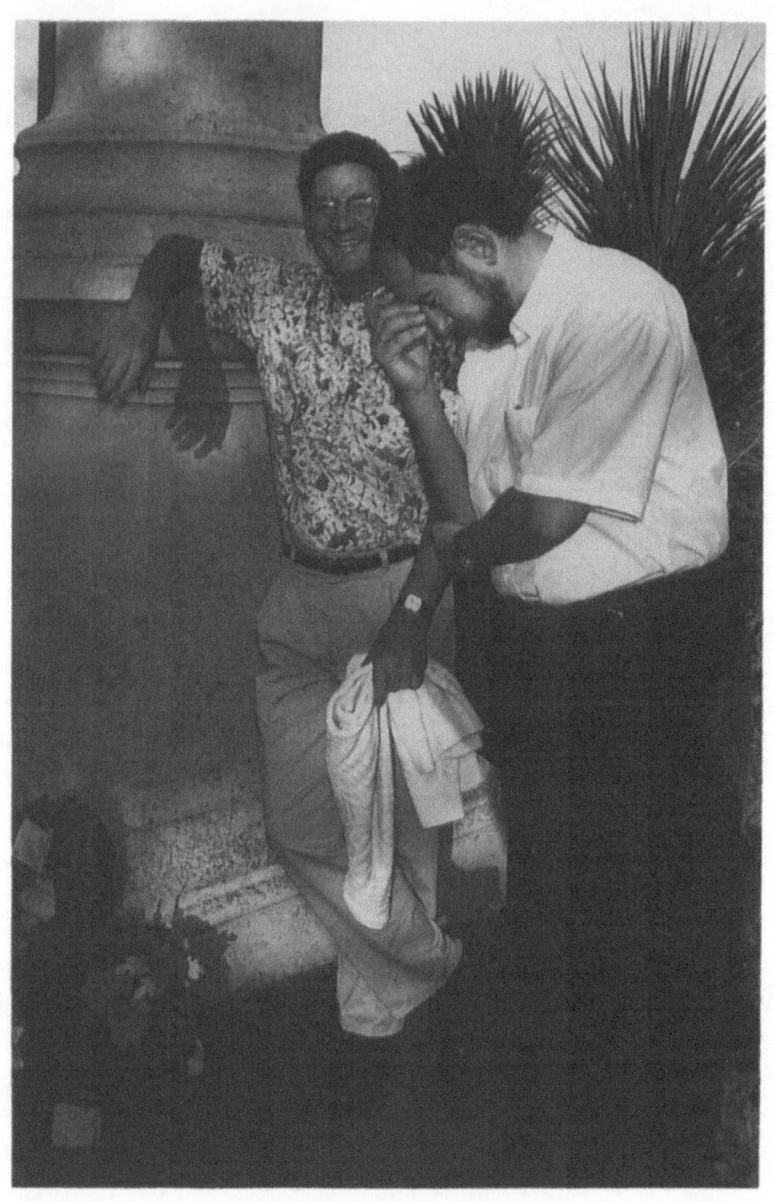

The thinker

ESTIMATING DISK OPACITIES USING INFRARED IMAGES

WIM VAN DRIEL

Kiso Observatory
University of Tokyo, Mitake, Kiso, Nagano 397-01, Japan

AND

Nançay Radio Observatory
Observatoire de Paris-Meudon, 92195 Meudon Cedex, France

Abstract. A study of galactic opacities was made, comparing disk scale-lengths at optical, near-infrared (K band), and far-infrared (50 and 100 μm) wavelengths. The far-infared values were measured for all 55 nearby spirals spatially resolved with the $\sim$90" circular beam of the CPC instrument on board IRAS, and compared to optical scalelengths. The CPC data are completely independent of the IRAS Survey instrument data, which have a much larger beam size. Reliable far-infrared scalelengths could thus be determined for 27 galaxies, generally out to a radius of three scalelengths; for most of them the Survey instrument's beam size is too large for this purpose. The scalelengths at 100 μm are on average 1.2 times larger than at 50 μm, and 1.1 times smaller than in the optical.

For the derivation of disk opacities through a comparison with optical and near-infrared scalelengths, new dust distribution models needed to be explored, as the one-component model of Bothun & Rogers (1992) was found to be inappropriate for this kind of study. Since it turned out to be very difficult to match all available observational data on far-infrared, visual, and near-infrared disk scalelengths with single temperature or single scalelength dust models, two-component dust models have to be considered instead – the first results indicate a total central opacity in the B band of about 4.5, and an opacity at the de Vaucouleurs' radius R_{25} of about 0 and 0.6 for the warmer and cooler dust component, respectively, while the scalelength of the cooler dust is about 3–4 times that of the stars (as measured in the K band, virtually unimpeded by dust).

J. I. Davies and D. Burstein (eds.), The Opacity of Spiral Disks, 299–311.

1. Introduction – The IRAS CPC Instrument

Galactic disk opacities can, in principle, be determined through a comparison of disk scalelengths at optical, near-infrared, and far-infrared wavelengths, due to the strongly wavelength-dependent influence of dust, which generally impedes the starlight strongly in the optical but negligeably in the near-infrared, while it dominates the emission in the far-infrared.

For our galactic opacity studies we use the far-infrared observations of nearby spiral galaxies obtained with the Chopped Photometric Channel (or CPC) instrument on board the IRAS satellite. The CPC images are independent of, and in fact complementary to, the IRAS Survey instrument data; see van Driel *et al.* (1993) for further details. The CPC mapped selected objects at 50 and 100 μm wavelength with a resolution of about 90″, matched to the diffraction limit of the IRAS telescope at 100 μm wavelength, significantly better than that of the IRAS Survey instrument (180″×300″ at 100 μm) – compare, e.g., the clearly separated images of the M51 (NGC 5194/95) pair (Fig. 1) to those in Rice *et al.* (1988). The CPC image size is about 12′×9′, which limits our measurements to about 3 disk scalelengths, compared to the 5 reached in Survey instrument images (see Bothun & Rogers 1992, hereafter BR92).

The strength of the CPC lies in its superior resolution, but its weak points are a lower sensitivity and a less accurate absolute calibration than the Survey instrument. However, the lower sensitivity is not so important for the infrared-bright, large galaxies discussed here, and the calibration could be adjusted satisfactorily through a comparison with Survey data of the same objects. Furthermore, a scaling of the flux density has no effect on measuring exponential scalelengths.

The analysis of the CPC galaxy images is presented in a series of papers (van Driel *et al.* 1993, 1994; van Driel & Valentijn 1994), while the interpretation of the data in terms of galactic disk opacity is linked to the observational and theoretical studies of Valentijn and Peletier *et al.* (see Peletier, this Volume, and Valentijn, this Volume, and references therein).

2. Far-infrared and Optical Scalelengths

The CPC spatially resolved 55 galaxies, which are practically all nearby spirals of type Sb or later (see van Driel *et al.* 1993). For the 31 objects with an inclination i<65° we obtained radial luminosity profiles by averaging the CPC images in 20″ wide rings which are circular in the plane of the galactic disk, while for the 24 more inclined galaxies we obtained luminosity profiles in a ±10° wide wedge around the major axis.

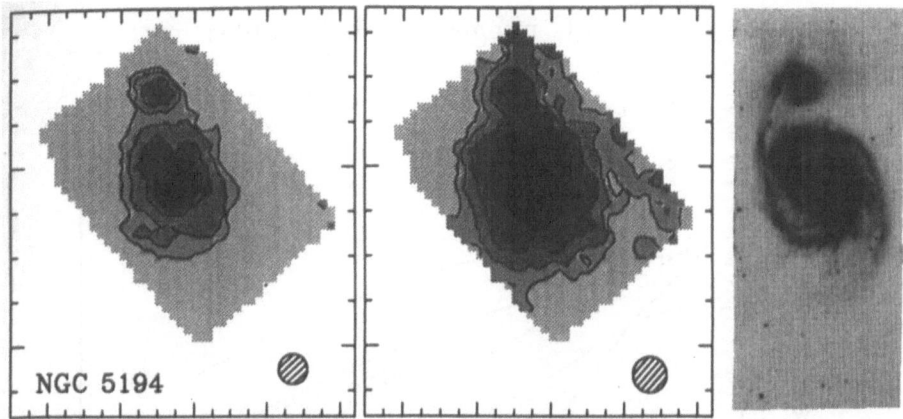

Figure 1. M51 - the galaxy pair NGC 5194/95: averaged far-infrared CPC images at 50 and 100 μm wavelength (left and middle, respectively), together with an optical image on the same scale (right). The lowest contour level is at the 3 σ level, and the hatched circles indicate the FWHM resolution (about 90″); see van Driel *et al.* (1993) for details.

The radial profiles of 43 galaxies (see Fig. 2 for some examples) could be fitted by an exponential disk, though sometimes at radii exceeding 75–125″ only if the central region shows a peak or plateau (see Fig. 1). We examined each CPC image and profile and divided the galaxies in a number of categories (see Figs. 3 and 4), depending on the estimated accuracy of the infrared scalelength determination. Profiles could be determined out to a radius of on average 260″, i.e. about 3 scalelengths.

A comparison of the 50 and 100 μm far-infrared and optical scalelengths of the nearby spirals for which CPC far-infrared and reliable optical disk scalelengths are available is shown in Figure 4. The optical scalelengths used were measured preferably in the B band, or in the V band, if not available. They were taken from the literature or measured from wide-field Kiso Observatory Schmidt CCD images.

Using only the 27 CPC profiles from which accurate infrared scalelengths could be derived, we find the following relations, with their 1σ deviation, between the various scalelengths:

$r_{100} = 1.21 \pm 0.32\ r_{50}$, $r_{opt} = 1.36 \pm 0.57\ r_{50}$, and $r_{opt} = 1.12 \pm 0.33\ r_{100}$.

3. Estimating Disk Opacities Using Far-infrared Images

A simple single-temperature dust model for the local conditions of the far-infrared (60 μm) and optical (B-band) disk emission was developed in

CPC images – radial profiles

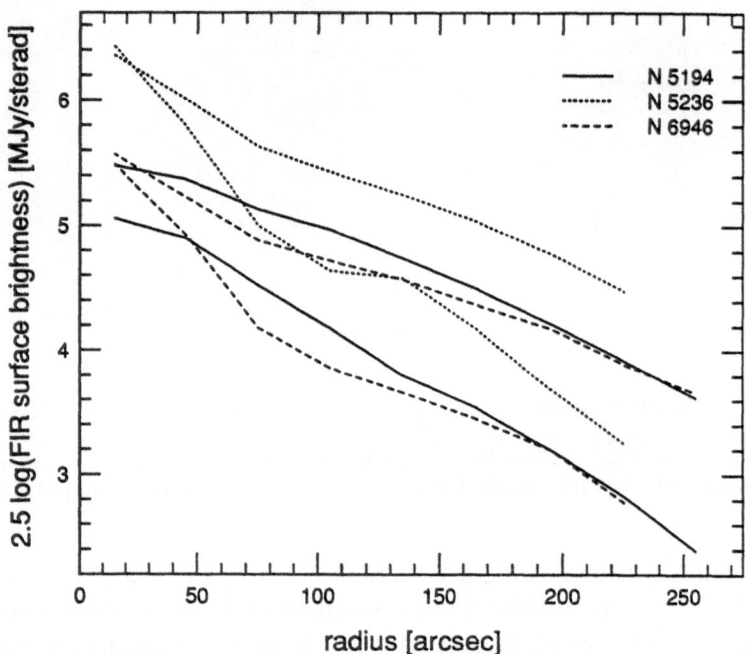

Figure 2. Radial CPC profiles at 50 and 100 μm wavelength of three well-resolved spiral galaxies: NGC 5194, 5236, and 6946. The 100 μm profile is the higher curve, and the 50 μm one the lower for each object. The curves extend to about three disk scalelengths.

BR92. Here, the far-infrared emission originates from an isothermal slab of dust at temperature T_d, transparent to emission, where the dust grains are in local radiative equilibrium with the interstellar radiation field. Stars and dust are assumed to be well-mixed and the optical emission is a mixture of directly seen, attenuated as well as dust-scattered starlight.

For disks which are everywhere optically thick the model predicts a far-infrared scalelength of half the (blue) optical value, if the stars and dust follow similar density distributions. In case of a low optical depth, similar far-infrared and optical scalelengths are predicted when there is no significant variation across the disk of the infrared temperature gradient, dust-to-stellar mass ratio, and stellar population mix. The dust temperature gradient is found to be small in the inner regions (out to 3 scalelengths) where we can use the CPC images, and its influence is noticeable at large radii only (BR92); a small gradient may explain the difference in scalelengths at 100 and 50 μm, though. Using the BR92 model to fit our data we would conclude that galactic disks are optically thin, confirming their

Scalelength comparison – 50/100 μm

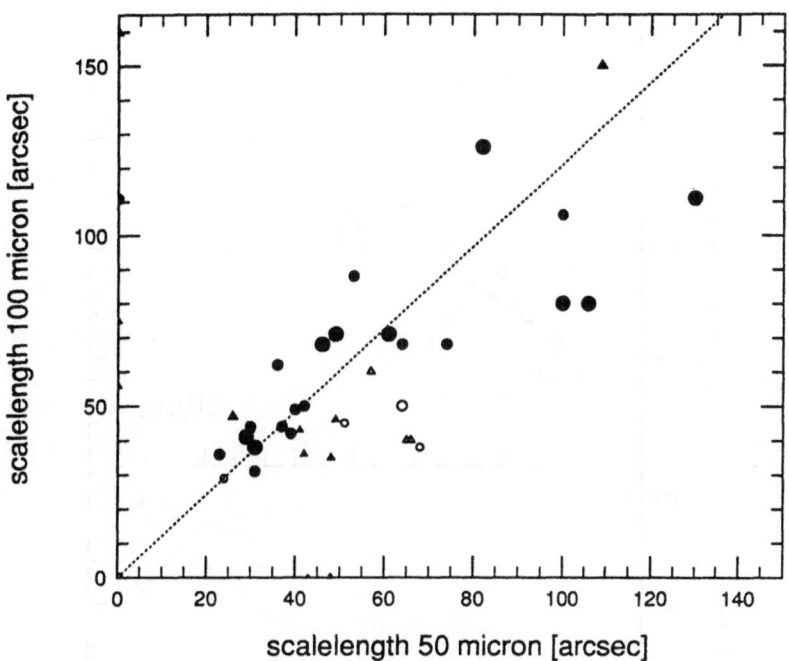

Figure 3. Comparison of exponential disk scalelengths derived from the CPC images at 50 and 100 μm wavelength. The dotted line represents a fit to the best quality scalelengths only, with a slope of 1.21. The symbols indicate the method by which the scalelengths were derived (see text): • – i<65°, △ – i>65°. Closed symbols indicate an accurate infrared scalelength, open symbols less reliable values.

conclusion, which was based on an analysis of independent IRAS Survey instrument images of large galaxies.

However, upon closer investigation (see below), we have found this model to be unsuitable for the determination of optical depths from the comparison of infrared and optical scalelengths:

In Eq. (5) of BR92 for the relation between the emissivity and the mean emissivity due to stars at a dust grain,

$$J_\nu = (32\pi/3)^{2/3} n_\star^{-1/3} \eta_\nu$$

the implicit assumption is that $n_\nu \propto n_\star$, where $n_\star$ is the number density of stars. Thus, the mean blue intensity of the mean interstellar radiation field $J_B \propto n_\star^{2/3}$. This implies, if we ignore the diffuse component, that the blue flux per pixel is

$$f_B = (1 - e^{-\tau_B}) J_B n_\star^{1/3} (cn_d)^{-1}$$

Scalelength comparison – far-infrared/optical

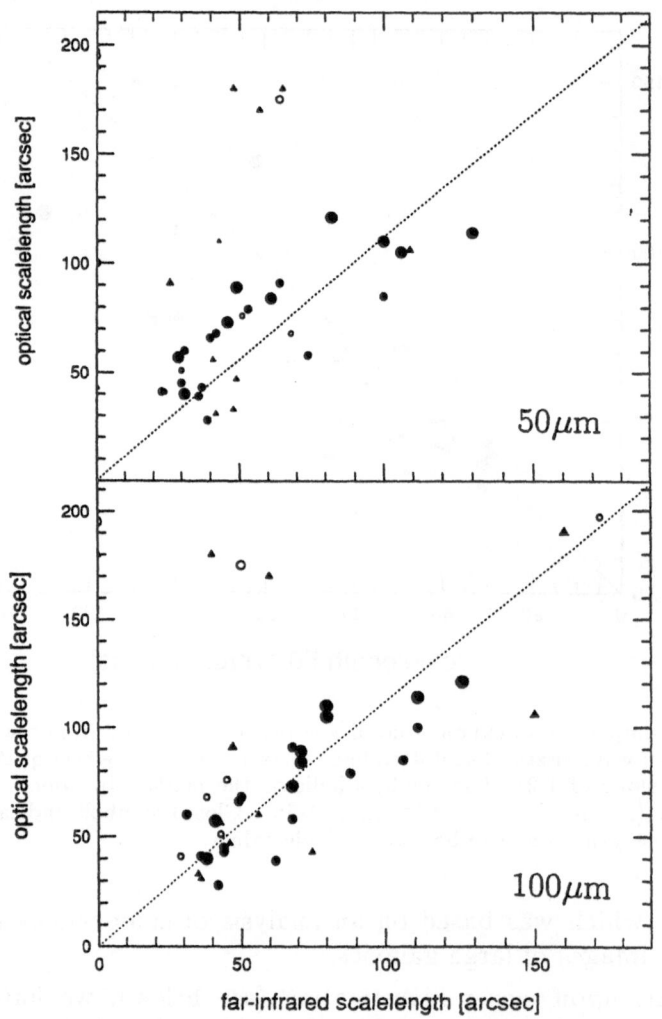

Figure 4. Comparison of exponential disk scalelengths measured in the optical and at 50 μm (top) and 100 μm wavelength (bottom). The dotted lines are a fit to the best quality CPC data only. For an explanation of the symbols used, see Fig. 3.

where n_d is the dust number density, and c is the cross section of dust particles. In the optically thick case, $\tau_B \gg 1$, this becomes

$$f_B = J_B n_\star^{1/3}(cn_d)^{-1} \ , so \ \ f_B \propto n_\star(cn_d)^{-1}$$

Thus, we can compare various disk scalelengths using the latter expression, $f_B(r) = n_\star(r) \ (c \ n_d(r))^{-1}$. If we assume that the radial number density distribution of the stars and dust can be expressed as $n_*(r)=n_*(0)\,e^{(-r/r_*)}$ and $n_d(r)=n_d(0)\,e^{(-r/r_d)}$, respectively, this impies that $f_B(r)$ does not vary as function of r, i.e., the luminosity profile is flat. This problem has been discussed in Valentijn (1991); one needs instead to make an exact solution for $f_B(r)$ using realistic values for τ, see Valentijn (1994).

In BR92 it is found, however, that for optically opaque disks the blue scalelength is twice that at 60 μm, $\alpha_B=2\alpha_{60}$, but this can only be derived if one assumes that the blue light and dust distribution scalelengths are equal, i.e., if $\alpha_B=\alpha_d$. This assumption can not be valid, though, since the scalelength of the blue light is severly affected if $\tau \gg 1$; even for optically non-thick cases α_B can easily be 1.4 times the true stellar scalelength, as measured in the K-band. Therefore Eq. (8) in BR92 can not be used to derive disk opacities through an optical and far-infrared scalelength comparison.

3.1. NEW DUST DISTRIBUTION MODELS

Thus, alternative dust distribution models were developed to interpret the various scalelengths observed in terms of disk opacity; for further details we refer to Peletier *et al.* (1994b). The most simple model is a one-component dust distribution with a dust scalelength equal to that of the stars, i.e. as measured in the K-band, $\alpha_*=\alpha_K$. In the model, equations are evaluated at radii of 1 and 3 scalelengths, the region where most exponential fits are made. This yields:

$$1 = \frac{\alpha_{obs}}{\alpha_*} - \frac{\alpha_{obs}}{\alpha_d} - 0.5\,ln\,\frac{(1 - e^{(-\tau_0 exp(-3\alpha_{obs}/\alpha_d))})}{(1 - e^{(-\tau_0 exp(-\alpha_{obs}/\alpha_d))})}$$

where τ_0 is the central opacity in the observed wavelength band, $\tau(r)=\tau_0\,e^{(-r/\alpha_d)}$. Note, that when both K and B-band photometry are available a unique solution for τ can be found, if we assume a certain $\alpha_d=\alpha_K$ ratio, and that in the optically thin case $\alpha_{obs}=\alpha_*$, like in BR92.

To derive the far-infrared scalelength we follow BR92 (see above): $f_{60} \propto F_{60}(T_d)\,\tau_j$ J, with $\tau_j(r)=\tau_j(r=0)\,e^{(-r/r_d)}$, where J$\propto n_*^{(2/3)}$, giving

$$f_{60}(0)e^{(-r/\alpha_{60})} \propto F_{60}(T_d)\tau_j(0)exp(-r/\alpha_d)n_*(0)e^{(-2r/3\alpha_*)}$$

Evaluating this formula at 1 and 3 scalelengths, like we did for the previous formulae, yields

$$\alpha_{60}/r = 1/(r/\alpha_d + (2r/3\alpha_*) + 0.5\,ln\,F(T_{d,r=1})/F(T_{d,r=3}))$$

where r can be any scalelength, like α_*, α_B, or α_K. Assuming, for instance, equal scalelengths for dust and stars, and assuming there is no temperature gradient, we get $\alpha_{60}/\alpha_\star=0.6$ This would imply that one would always find a scalelength at 60 μm of 0.6 times the true stellar value, independent of the optical depth τ_0. We checked this conclusion using Disney's slab model, assuming all optical radiation is re-radiated in the far-infrared, which yields a similar ratio of 0.6–0.7.

The mean observed scalelength ratios are: $\alpha_{100}/\alpha_{50}=1.21$, $\alpha_B/\alpha_{50}=1.36$, $\alpha_B/\alpha_{100}=1.12$, and $\alpha_B/\alpha_K=1.4$–1.9.

Now, we face these observational data with single compenent dust models, and we first evaluate the simplest case with $\alpha_\star=\alpha_d=\alpha_K$ (but not equal to α_{obs}, as assumed in BR92).

According our model this configuration can be expressed with:

$$\alpha_{50}/\alpha_\star = 1/(5/3 + 0.5 ln F(T_{d,r=1})/F(T_{d,r=3})) \tag{1}$$

According to this model (see also Peletier et al. 1994), an α_B/α_K ratio of 1.4–1.9 corresponds to a central opacity of $\tau_0=10$–100 and a value at the de Vaucouler's radius of $\tau_{25}=0.04$–0.025, i.e., a disk which is opaque near the centre and transparant in the outer parts, a characteristic of models with $\alpha_\star=\alpha_d$.

The observed α_B/α_K ratio, then predicts, according to eq. (1), that

$$\alpha_B/\alpha_{50} = (2.3 - 3.1) + (0.7 - 0.9) ln F(T_{d,r=1})/F(T_{d,r=3})$$

which is observed to be 1.36 on average, so $F(T_{d,r=1})/F(T_{d,r=3}))=0.25$–0.15. Thus, this configuration implies a higher dust temperature in the outer disk. Unless we adopt a dust emissivity law $\propto \lambda^{-2}$, in which case we a natural situation with $T_d >60$ and 40 K in the inner and outer disk, respectively (see Fig. 3 in BR92). We conclude that the configration with $\alpha_\star=\alpha_d=\alpha_K$ requires a very large temperature gradient over the disk.

Alternatively, we can use the single-component model to determine which α_B/α_K ratio fits the observed α_B/α_{50} ratio. The latter observed ratio of 1.36 implies an α_B/α_K ratio of 0.85, if $F(T_{d,r=1})=F(T_{d,r=3})$). This means an optically transparent disk with equal scalelengths in blue and near-infrared, but it is in clear contradiction with the observed ratio of 1.4–1.9.

In order to further pin down the solution to the opacity problem in connection with far-infrared data, one in fact needs a good determination of α_B/α_K ratios for a fair number of galaxies observed with the CPC, but such K-band observations are not yet available. However, the failure of single component models to match all observational optical and (near- and far-) infrared data strongly suggest the need for two component dust models, in which the second component should have a larger scalelength and/or a lower dust temperature.

Such models are still under development (see also Peletier or Valentijn, this Volume), but the first results for models with a similar scalelength of the stellar distribution and that of the warmer dust, for which IRAS was most sensitive, and a free scalelength ratio for the cooler dust component and the stars, indicate a total central opacity in the B band of about 4.5, with an opacity of the warmer dust component of about 1–3 in the centre and about 0 at the de Vaucouleurs' radius R_{25}, and an opacity of about 0.6 at R_{25} for the cooler component, which has a scalelength of about 3–4 times that of the stellar distribution (as measured in the K band) and the warmer dust.

The work presented here was done together with E.A. Valentijn, D. Kussendrager, R. Peletier, and P.R. Wesselius of the Kapteyn Astronomical Institute in Groningen, The Netherlands, whom I wish to thank for their continued collaboration. We would like to thank G. Bothun for making his optical data available to us, G. Bothun and C. Rogers for discussions of their model, and staff and visitors of the Kiso Observatory of the University of Tokyo for their help in obtaining CCD images.

References

1. Bothun, G.D. and Rogers, C. (1992), Surface Photometry of Disk Galaxies at 60 and 100 μm: Radial Gradients in Dust Temperature in Optically Thin Disks, *AJ*, **103**, 1484 (BR92)
2. Peletier, R.F., Valentijn, E.A., Moorwood, A.F.M. and Freudling, W. (1994a), The Distribution of Dust in Sb's and Sc's – K-band Imaging of a Diameter Limited Sample of 37 Galaxies, *A&AS*, in press
3. Peletier, R.F., *et al.* (1994b), *A&A (Letters)*, submitted
4. Rice, W., *et al.* (1988), A Catalog of IRAS Observations of Large Galaxies, *ApJS*, **65**, 1
5. Valentijn, E.A. (1991), in E. Bloemen (ed.), *IAU Symp.* **144**, pg. 245
6. Valentijn, E.A. (1994), The Opacity of Galaxies using Volume Representative Samples, *MNRAS*, **266**, 614
7. van Driel, W., de Graauw, Th., de Jong, T. and Wesselius, P.R. (1993), IRAS CPC Observations of Galaxies: I. Catalog and Atlas, *A&AS*, **101**, 207
8. van Driel, W., Valentijn, E.A., Kussendrager, D. and Wesselius, P.R., (1994), IRAS CPC Observations of Galaxies: II. Scalelengths and Disc Opacity, *A&A (Letters)*, in press
9. van Driel, W. and Valentijn, E.A. (1994), IRAS CPC Observations of Galaxies: III. Radial Distributions and Dust Models, *A&A*, in prep.

Question
<u>Burstein</u>

1.   In your plot of α_B vs $\alpha 50\mu$, there are 4 points that deviate strongly and systematically - any comment?

2.   How many beams are there over the 250" radius you have sampled? I would estimate 3-4 beams. Also, what are the errors on these measurements, especially at low IR surface brightnesses?

Answer
<u>Van Driel</u>

1.   I looked if I could find any significant difference between these 4 galaxies that deviate from the average relation, but, as I remember; did not find any deviating optical and/or infrared property. These are mainly rather edge-on galaxies with radial profiles from which we could not determine accurate scalelengths; I do not have the raw data for these objects at hand, however.

2.   With a beam size of about 90", there are indeed about 3 independent points in radius out to a ~260" radius. Each point is an azimuthal average covering a number of independent beams per ring, though. I do not have the data at hand here to give you the errors in the points on the radial profiles, unfortunately.

Question
<u>Kylafis</u>

Just a clarification. You said you assumed an <u>exponential</u> radial distribution, but on the other hand you said you used a <u>sandwich</u> model. Usually, when we talk about sandwich models we mean slabs of <u>uniform thickness</u>. So, do the dust <u>and</u> stars have exponential radial distributions?

Answer
<u>Van Driel</u>

Yes, in our sandwich model both the absorbing layer and the stars have exponential radial distributions.

Question
Disney

1.   The Cardiff group also showed there is something seriously wrong with the Bothuns and Rogers one-component dust distribution model.

2.   Didn't Rice et al (19??) also compare the 60/100μm and optical scalelengths directly from their IRAS Survey Instrument Data, showing results similar to those of Bothun & Rogers?

Answer
Van Driel

1.   We agree, there are implausible assumptions embedded in it. They were not apparent at first to us, since the authors considered only optically opaque or transparent disks.

2.   I can't remember at the moment. At least our independent CPC data show a similar far-infrared/B scalelength ratio. As those derived from the survey data, as presented by Bothun & Rogers.

Question
Braun

Are you sure the 50 and 100μm CPC beams are really identical? My recollection was that the aperture size was the same, but after convolving with the diffraction limited point spread function these were significantly different beam sizes.

Answer
Van Driel

I forgot the actual physical aperture size(s) of the CPC, and if it has one or two apertures, but actually measured FWHMs of calibration sources and many other point (like) sources indicate a mean FWHM resolution of 20" and 95" at 50 and 100μm, respectively.

Question
R Giovanelli

1.   Did you say that you find the cold dust distribution should be 3-4 times that of the stellar population?

2.   Since you map the FIR out to ~3 stellar scale lengths, doesn't the influence of an unseen cold dust population with such a huge scale length bother you a little? In our galaxy that's about 20 kpc. How's that different from a flat distribution?

Answer
Van Driel

1.   Yes, from the observed change in the B/K-band scalelength ratio as function of radius, assuming that all galaxies have a similar dust/K-band scalelength ratio (independent of inclination), and requiring that edge-on galaxies have an optical depth of about 5 times that of face-on objects at similar radii, we do find that the cooler dust component has a scalelength of 3-4 times that of the intrinsic stellar distribution.

2.   Actually, the far-infrared IRAS CPC images led us to assume a warm dust component with a scalelength equal to that of the stars. The scalelength estimate of the cool dust component, for which IRAS is not really sensitive, at this point is derived from the B/K-band comparison only; we have not yet calculated the emitted far-infrared flux densities of the two dust components. So the long scalelength of a component we do not really detect in the IRAS images does not disturb us; the scalelength estimate comes from the B/K comparison only.

Question
Huizinga

Since you use galaxies which have been detected with a large S/N at 50 and 100μ, is there not a selection effect towards dustier FIR-bright galaxies?

Answer
Van Driel

Actually, not all of the 55 galaxies we considered to be resolved (actually, meaning their measured FWHM≥1.6 the beam FWHM) have a high far-infrared surface brightness, especially at 50μm Of these 55, 27 are large and bright enough in the FIR to allow the determination of a, we think, reliable scalelength at 50 and 100μm. So we do not think this introduces a bias towards dustier galaxies, as we tested with histograms of various infrared-related properties.

Question
Ryder

In a recent ApJ paper we compared V and I scale lengths for 34 nearby spirals with the scale lengths of their H_α distributions, and found the H_α scale lengths to be typically 75% longer than the broad-band scale lengths. Although we interpreted this in terms of evolution in the large-scale star formation pattern, we could not rule out a systematic extinction gradient. Since the H_α is clearly more intimately associated with the dust than the B-band light, perhaps a comparison of H_α scalelengths with K-band scalelengths would be more appropriate?

Answer
Van Driel

In our model we assume that the general interstellar radiation field heats the dust grains, a generally accepted situation for such models. It is not clear to us at the moment that taking the scalelength of the H_α emission would be a better choice.

DIRBE OBSERVATIONS OF GALACTIC EXTINCTION

R. G. Arendt, T. J. Sodroski,
Applied Research Corp.
Code 685.3, NASA GSFC, Greenbelt MD 20771
M. G. Hauser,
NASA Goddard Space Flight Center
Code 680, NASA GSFC, Greenbelt MD 20771
N. Boggess, E. Dwek, T. Kelsall, S. H. Moseley, R. F. Silverberg,
NASA Goddard Space Flight Center
Code 685, NASA GSFC, Greenbelt MD 20771
T. L. Murdock,
General Research Corp.
Technology Dept., 5 Cherry Hill Dr., Suite 220, Danvers, MA 01923
G. B. Berriman, N. Odegard, and J. L. Weiland
General Sciences Corp.
Code 685.3, NASA GSFC, Greenbelt MD 20771

Abstract
The short wavelength bands of the Diffuse Infrared Background Experiment (DIRBE) instrument on NASA's Cosmic Background Explorer (COBE)* mission allow us to examine the large-scale distribution and character of the extinction in the Galactic plane. An analysis of the near-IR emission of the Galaxy at 1.25, 2.2, 3.5, and 4.9 µm yields the line-of-sight extinction in these bands and a mean near-IR interstellar reddening law.

1. Introduction

The DIRBE instrument has mapped the full sky with a $0°.7$ square beam at ten wavelengths from 1 to 240 µm (Hauser et al. 1990, Boggess et al. 1992), and has measured polarization from 1 to 3.5 µm (Berriman et al., 1994). DIRBE maps of the Galaxy at 1.25, 2.2, 3.5, and 4.9 µm clearly show a relatively thick, smooth Galactic disk of starlight surrounding an elliptical bulge. Extinction still significantly influences the appearance of the Galaxy at 1.25 µm (Weiland et al. 1994, Arendt et al. 1994, Dwek et al. 1994), but it quickly becomes less significant with increasing wavelength.

This report summarizes the characteristics of the interstellar extinction as observed in the short wavelength DIRBE bands. Comparisons are made with the optical depths derived from the long wavelength emission (140 and 240 µm bands) observed by DIRBE. More detail on much of this work can be found in Arendt et al. (1994) and Sodroski et al (1994).

2. Data

The near-IR data referred to here include observations spanning the six month interval from Dec 1989 to May 1990. All observations were interpolated to 90° elongation, so

313

J. I. Davies and D. Burstein (eds.), The Opacity of Spiral Disks, 313–315.
© *1995 U.S. Government.*

that the signal contributed by interplanetary dust could be uniformly removed using an empirical fitting function (Hauser, 1993). The estimated inaccuracies in our removal of this local emission are <5% of the median brightness of the inner Galaxy. The far-IR data have had no model of the zodiacal emission removed as the zodiacal emission is very faint at wavelengths longer than 100 μm.

3. Color–Color Diagrams

Color-color diagrams of $I(1.25 \ \mu m)/I(2.2 \ \mu m)$ vs. $I(2.2 \ \mu m)/I(3.5 \ \mu m)$ and $I(2.2 \ \mu m)/I(3.5 \ \mu m)$ vs. $I(3.5 \ \mu m)/I(4.9 \ \mu m)$ are the guides to deriving the properties of the Galactic extinction and stellar emission (see Arendt et al 1994).

From such diagrams we find: 1) the colors in the inner Galaxy are well correlated along a linear trend, 2) the dispersion of points along the linear trend decreases towards the outer Galaxy and at higher latitudes, 3) the range of colors is greatest for the shortest wavelengths, 4) trends in the 3rd and 4th Galactic quadrants are very similar to those in the 2nd and 1st quadrants and, 5) the Cygnus region and other areas of widespread star formation show a greater range of colors than suggested by the general trend with longitude, and in the shorter wavelengths the trend has a significantly steeper slope.

4. Extinction

These trends in the color-color plots are strongly suggestive of extinction effects. The strong linear trend in the inner Galaxy suggests reddening of background stellar sources by foreground extinction. A linear least squares fit to the data with the assumption of only foreground extinction gives a slope for this reddening line that is quite close to that predicted using the Rieke–Lebofsky reddening law (Rieke & Lebofsky 1985). Other tabulated reddening laws (e.g. Mathis 1990, Koornneef 1983) do not fit the DIRBE data as well, particularly in the redder colors.

In the direction of the inner Galaxy and bulge, the reddening is too large and too linear to be caused by a uniform mixture of dust and stars along the line of sight. The light is dominated by that of the inner regions of the disk and the bulge as seen through several magnitudes of extinction (at 1.25 μm). The optical depth reaches 1.0 within ~2° of the Galactic plane at 1.25 and 2.2 μm. The maximum optical depth inferred is ~4 at 1.25 μm towards the Galactic center. This means that the different wavelengths used here will sample different path lengths in the Galactic plane, complicating any detailed analysis. At high latitudes (above $|b| > 10°–20°$) and in regions towards the outer Galaxy, the color-color plots do not show enough correlation to extract meaningful values for the extinction.

If we assume that the intrinsic colors of the Galactic disk population are constant throughout the Galaxy and ignore the possible slight color difference between the disk and bulge, then we can use the ratio of $I(1.25 \ \mu m) / I(2.2 \ \mu m)$ to construct an optical depth $(\tau_{1.25})$ map of the Galaxy. The $\tau_{1.25}$ map has been compared with the optical depth map at 240 μm constructed from the dust emission in the 140 and 240 μm DIRBE bands assuming a dust grain emissivity $\sim \lambda^{-2}$ (Sodroski et al. 1994).

Where $\tau_{1.25} < 1.0$, it is well correlated with the DIRBE 240 μm optical depth. This implies that where the line of sight is optically thin at 1.25 μm, we observe the same dust in both absorption and emission. Those areas where excess 240 μm optical depth is observed will generally be lines of sight along which dust is too "distant" to be detected through its near-IR extinction. The long wavelength emission in the 140 and 240 μm bands is characterized by a dust temperature of ~18 K locally. Sodroski et al

(1994) reports in detail about this emission and the large-scale temperature gradient, which is observed at these wavelengths but is undetectable at shorter wavelengths.

5. Cygnus and Other Star–Forming Regions

The different character of the color-color diagrams in regions of intense star formation activity can be characterized by various mixtures of stellar light and emission from very hot dust ($T_{dust} \approx 900$ K). The high temperature poses some difficulty. Such dust will be found only very close to stars ($\lesssim 1$ pc). Emission at these wavelengths from compact H II regions and high latitude clouds is characterized by dust temperatures in the 400 – 800 K range. The trends cannot be characterized by stellar light mixed with line or continuum emission from the ionized gas within H II regions. More detailed examination of the Orion region (Wall et al 1994) finds that a fraction of the 3.5 µm emission can be attributed to PAHs. Giard et al. (1988, 1989, 1994) have observed and modeled the correlation of excess PAH emission in a narrow 3.3 µm band with 12 and 100 µm emission.

6. Conclusions

It is interesting that the DIRBE observations of our own Galaxy can be fit with a simple screen model of the extinction when we view the bright inner Galaxy. It may also surprise some that the effects of extinction are still significant at 2.2 and even 3.5 µm. The DIRBE data also directly reveal the equivalence of the dust which causes the optical and near-IR extinction and that which emits in the far-IR. Models of radiative transfer in galaxies will ultimately need to balance the energy absorbed and emitted by dust grains.

7. References

Arendt, R. G., et al. 1994, ApJ, 425, L85
Berriman, G. B., et al. 1994, ApJ, 431, L63
Boggess, N., et al. 1992, ApJ, 397, 420
Dwek, E., et al. 1994, ApJ, submitted
Giard, M., et al. 1988, A&A, 201, L1
Giard, M., et al. 1989, A&A, 215, 92
Giard, M., et al. 1994, A&A, 286, 203
Hauser, M. G. 1993, in "Back to the Galaxy," eds. S. S. Holt, and F. Verter, (AIP: New York), p. 201
Hauser, M. G., et al. 1990, in "After the First Three Minutes," eds. S. S. Holt, C. L. Bennett, V, Trimble, (AIP: New York), p. 161
Koornneef, J. 1983, A&A, 128, 84
Mathis, J. S. 1990, ARAA, 28, 37
Rieke, G. H. & Lebofsky, M. J. 1985, ApJ, 288, 618
Sodroski, T. J., et al. 1994, ApJ, 428, 638
Weiland, J. L., et al. 1994, ApJ, 425, L81
Wall, W., et al. 1994, in preparation

* COBE is supported by NASA's Astrophysics Division. Goddard Space Flight Center (GSFC), under the scientific guidance of the COBE Science Working Group, is responsible for the development and operation of COBE.

KINEMATICS OF EDGE-ON GALAXIES AND THE OPACITY
OF SPIRAL DISKS

A. BOSMA
Observatoire de Marseille
2 Place Le Verrier, 13248 Marseille Cedex 4, FRANCE

Abstract. Following up on the paper by Bosma, Byun, Freeman and
Athanassoula (1992), I will present further HI, CO and Hα observations
of a number of edge-on galaxies of various types. These observations cor-
roborate the hypothesis that is the molecular gas as traced by the CO which
is responsable for the opacity of the inner parts of large spiral disks.

1. Introduction

As already shown in Bosma et al. (1992) and discussed by Ken Freeman
(this meeting), it is possible to infer some information about the optical
depth in edge-on galaxies from comparative studies of the kinematics of
several tracers (e.g. Hα, CO and HI). In particular, if galaxies are really
optically thick at the outer edge in the optical wavelengths region, one
expects to see only Hα-emission from the outer edge, and thus a position-
velocity curve with a shallow gradient. On the other hand, since we see the
HI line emission and the CO emission from the whole range of radii, even in
the edge-on situation, we can expect to recover fully the true rotation curve
from such data if resolution effects are unimportant. Any optical thickness
in the line emission will only diminish the measured intensities, but not
affect the radial velocity range of the emission along the line of sight. If a
galaxy is transparent, we should see Hα emission from the inner regions,
and thus be able to recover the rotation curve from Hα data as well. For a
good description of the expected shape of the position velocity diagram in
the HI, see the classic study of Sancisi and Allen (1979) on NGC 891. Goad
and Roberts (1981) were the first to take spectra of "superthin" edge-on
galaxies in Hα, and to discuss the effects of extinction on them.

J. I. Davies and D. Burstein (eds.), The Opacity of Spiral Disks, 317–324.
© 1995 *Kluwer Academic Publishers.*

The data by Bosma et al. (1992) for the small Sc galaxy NGC 100 show that we can see the Hα-emission from the inner regions since we measure radial velocities corresponding to the tangent point (i.e. on the line of nodes). For the Sb galaxy NGC 891, however, we see the Hα-emission only as far in as the outer rim of the CO-distribution. One possibility is that this reflects a ring-like distribution of the HII-regions : they should then only occur beyond a radius of 4' (or $\sim$ 0.6 R_{opt}, where R_{opt} is the optical radius at the 25th B magn. arcsec^{-2} isopohote level). The other possibility is that the region inside 4' radius is sufficiently optically thick at Hα so that we cannot see the emission from further in. In any case, the outer regions of NGC 891 seem transparent at Hα.

2. Comments on the method

Several criticisms can be raised against the method, some of which are more serious than others. It is clear that the method only works for galaxies seen nearly edge-on, since for more face-on situations one can see through a galaxy entirely. This is partly due to the thinness of the absorbing layers themselves. Typical slitwidths employed are 1''- 2'', corresponding to $\sim$ 50 - 100 pc for a galaxy at 10 Mpc distance. Furthermore, the method does not give directly a value for the optical depth : only an indication is obtained about the visibility of HII-regions inside a spiral galaxy seen edge-on. Another issue sometimes raised is the question of resolution effects (in general not too important, since the galaxies under study have a large angular size).

Clearly the method is only indicative, and our lack of knowledge about the actual value of the inclination (88° or 90° ?), the presence of optical warps which are hard to detect, and the filling factor and the scaleheight of the absorbing material is hampering interpretation. Nevertheless, the indications sofar can be used to guide further discussions about the statistical studies, and some of the proposals made, e.g. spiral galaxies are optically thick at the optical radius, seem to be ruled out by these observations.

3. Further results for several Sb and Sc galaxies

I will present here new data for a number of other galaxies, of various types and mean rotational velocities, in an effort to strengthen the conclusions reached in Bosma et al. (1992). Images and Hα-spectra have been collected using respectively the 120-cm and the 193-cm telescopes at the Haute Provence Observatory, CO-data in the ^{12}CO(1-0) line have been obtained with the IRAM 30-m telescope at Pico Veleta at an angular resolution of 23'', and HI data either with the VLA or with Westerbork. I will concentrate here on the comparison of the Hα and CO-data.

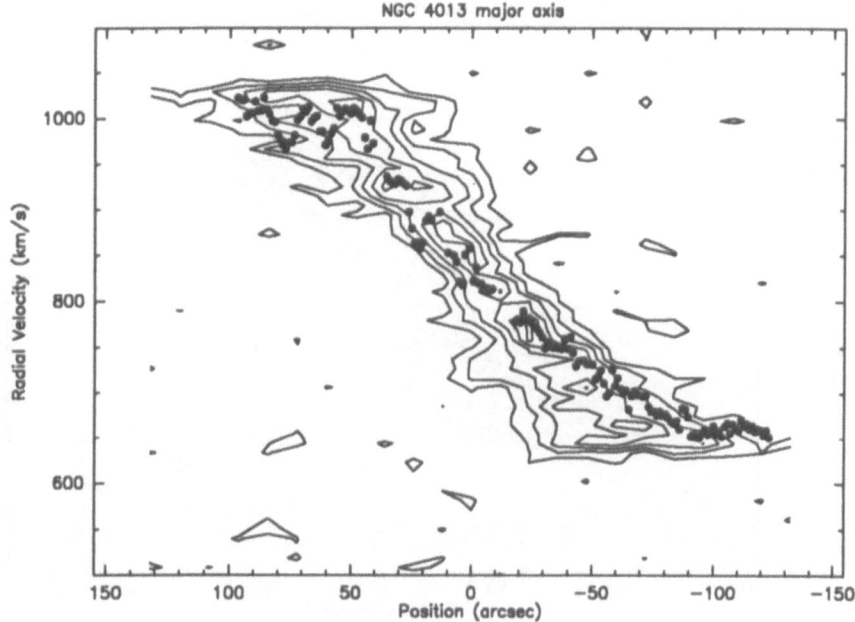

Figure 1. Position-velocity diagram for NGC 4013. Dots are Hα-velocities, and contours are from the CO data.

3.1. NGC 4013

This Sb galaxy is seen exactly edge-on, as the dust lane coincides with the midplane of the light distribution (cf. Van der Kruit & Searle 1982). Figure 1 shows on the same scale the data for Hα and CO. The situation here is similar to that in NGC 891 : the Hα is seen as far in as the outer rim of the CO distribution, in particular at the low velocity side. Note that there is an asymmetry in that we see Hα at the low velocity side at a radius of 90″, while at the high velocity side it is at a radius of 70″. Since the optical radius is 157″, we conclude that the galaxy is transparent beyond ∼ 0.6 R_{opt} or about 2 disk scalelengths.

3.2. NGC 4217

This Sb galaxy is not entirely edge-on, and the dust lane is seen off to one side. Spectra were taken along the major axis just above the dust lane, and in the dust lane itself (which is offset by ∼ 5″ with respect to the major axis). The results are shown in Figure 2. The Hα in the dust lane is seen up till the edge of the CO distribution, but the Hα above the dust lane comes

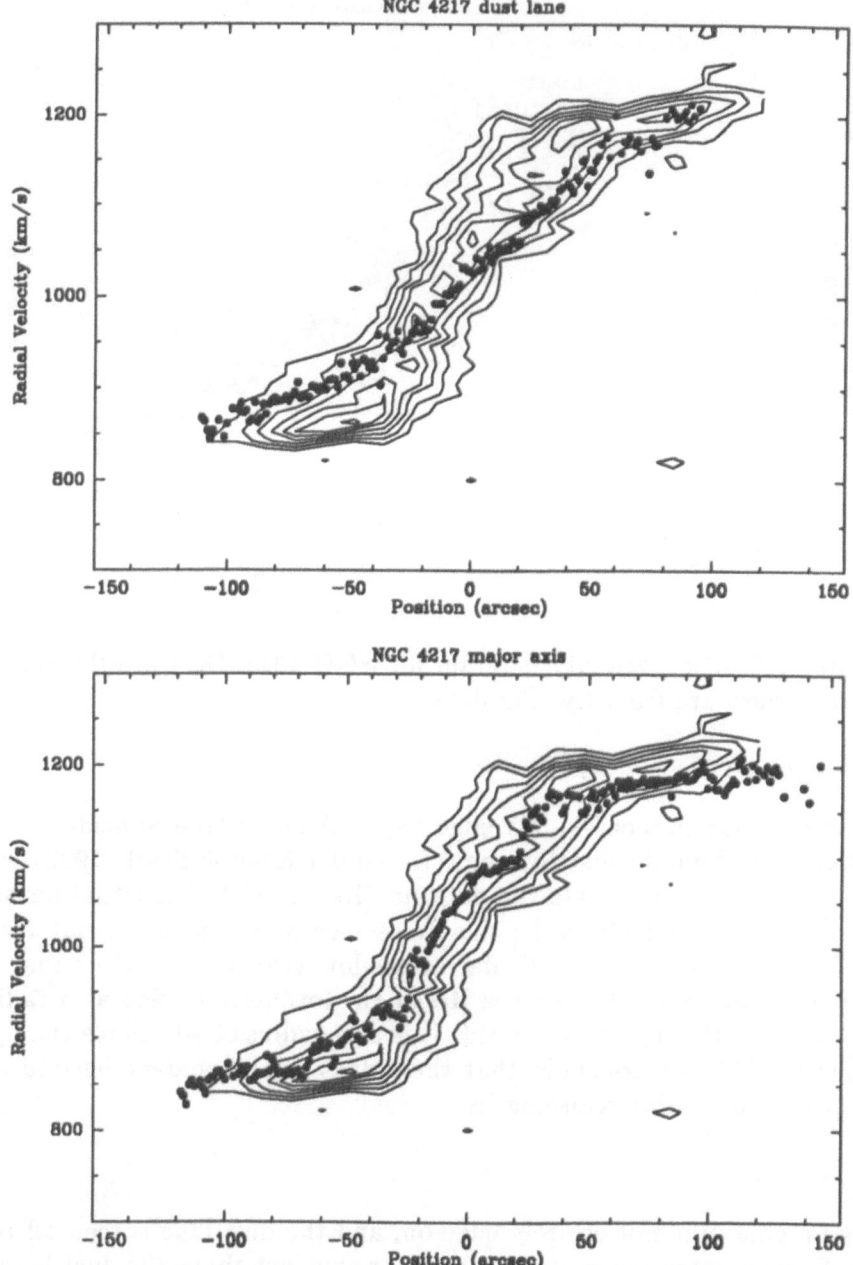

Figure 2. Position-velocity diagrams for NGC 4217. Dots are Hα-velocitie
and contours are from the CO data. The top one refers to the Hα spectru
in the dust lane, and the bottom one to the spectrum at the major axis.

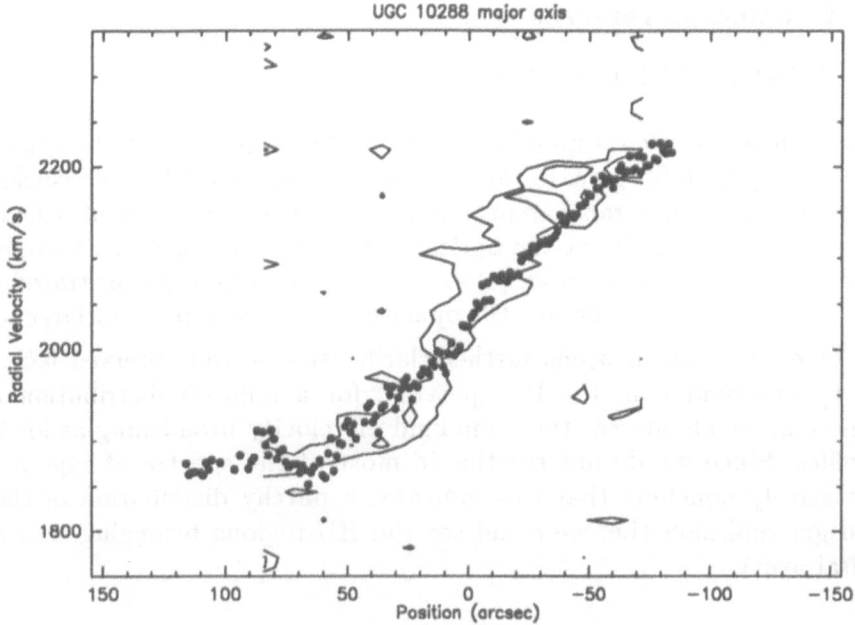

Figure 3. Position-velocity diagram for UGC 10288. Dots are Hα-velocities, and contours are from the CO data.

from further in, since it coincides with the ridge line of the CO emission. Again we can conclude that the dust lane is transparent beyond a radius of ∼ 0.6 R_{opt} , or about 2 disk scalelengths.

3.3. UGC 10288

Again this galaxy, of type Sc, is not seen entirely edge-on, but the spectrum was taken in the dust lane. The CO emission is much weaker here than in the previous two cases, but even so the Hα emission is seen as far in as the outer edge of the CO emission, that is out to ∼ 0.4 R_{opt} .

3.4. NGC 4183 AND OTHER SMALL LATE TYPE GALAXIES

A comparison between Hα and HI indicates that NGC 4183 is transparent. However, it may be that this galaxy is not entirely edge-on. No CO was detected in the central position of this galaxy. For NGC 5023 a similar situation occurs, and no CO emission has been detected with a 12-m telescope on the central position (Sage, 1993). For two other late type galaxies in Goad and Roberts (1981), more recent HI data can be combined with their results, and again the conclusion is that those galaxies are transparent.

4. Concluding remarks

4.1. SUMMARY OF RESULTS

Our results for several individual case studies can probably be generalized by stating that Sb galaxies are opaque out to about 2 disk scalelengths, large Sc's are semi-transparent, and small Sc's are transparent. Clearly the outer parts of spirals, at the optical radius, are transparent. Moreover, in agreement with Bosma et al. (1992), it is the molecular gas as traced by the CO which is responsable for the opacity of the inner parts of large spirals.

One issue which needs further clarification is the observed lack of velocity broadening in the Hα spectra : for a uniform distribution of the emission, we should see the same kind of velocity broadening as for the HI profiles. Since we do not see this in most of the spectra of edge-on's, we tentatively conclude that this indicates a patchy distribution of the dust and gas emission (i.e. we could see the HII-regions belonging to a strong spiral arm).

4.2. MAPPING THE COLD DUST

The cold dust which does most of the absorbing can be mapped with millimeter and sub-millimeter telescopes equipped with bolometers. The current trend is to use bolometer arrays. A good example of the potential of such instrumentation is given by the mapping of NGC 891 by Guélin et al. (1993) at 1.3 mm with the 30-m IRAM telescope and the MPIfR 7-channel bolometer array. A breakdown of the 1.3mm flux for this galaxy shows that 34% of the emission comes from cold dust in diffuse HI clouds, 42% from cold dust in the cores of molecular clouds, 13% from the ^{12}CO(2-1) line emission in the observed wavelength band, and 11% from warm dust as seen by IRAS. The dust emission corresponds remarkably well to the CO distribution. Preliminary work on M51 shows lots of interesting detail (Guélin et al. 1994). Hence, in a few years, it will be possible to map a number of spirals this way, and thus directly adress the question of the amount of dust and its distribution.

Acknowledgements

The Hα-observations have been collected at the Observatoire de Haute Provence. I thank the IRAM staff for their efficient help with the 30-m observations. Michel Guélin kindly allowed me to show the new 1.3mm bolometer results at the meeting. I thank Ken Freeman and Lia Athanassoula for discussions and encouragement.

References :

Bosma, A., Byun, Y.I., Freeman, K.C., Athanassoula, E. 1992, ApJ, 400, L21.

Goad J.W., Roberts, M.S., 1981. ApJ, 250, 79

Guélin, M., Zylka, R., Mezger, P.G., Haslam, C.G.T., Kreysa, E., Lemke, R., Sievers, A.W., 1993. A&A 279, L37

Guélin, M., Zylka, R., Mezger, P.G., Haslam, C.G.T., Kreysa, E., Lemke, R., Sievers, A.W., 1994. A&A (submitted)

Sage, L.J., 1993. A&AS, 100, 537

Sancisi, R., Allen, R.J., 1979. A&A, 74, 73

Van der Kruit, P.C., Searle, L., 1982. A&A, 105, 61

Discussion :

JAMES : Would you care to speculate further on the lack of Hα velocity broadening in your position velocity curves of edge-on galaxies, given that the broadening is apparent in the HI position - velocity diagram ?

BOSMA : Well, either we lack the spectral resolution to see it, or we lack the sensitivity to detect the diffuse Hα-emission, or , what I think is most likely, we see a manifestation of the patchiness in the dust and ionized gas distribution.

BURSTEIN : Albert, I would find it helpful if you would draw the "optically thick" line in the l-v diagram. As we think there are "holes" in the dust layer, such a line can be used as a convenient reference point, against which those HII regions at the edge of the galaxy can be easily separated from those in the interior.

BOSMA : Well, I think it's even better to present the raw Hα data super-imposed directly on the CO position velocity maps, so that is clear what emission corresponds to where, rather then just give the mean Hα velocity. However, this involves a bit more effort in manipulating the data across different software systems.

ZARITSKY : I just want to clarify your response to Dave Burstein's question. The Hα-data that scatter toward the solid body rotation curve in the position - velocity diagram need not be interpreted as evidence for clumpy dust. Could they not just be reflective of a clumpy gas distribution, that is you are not sampling uniformly to the tangent point along the line sight ?

BOSMA : Yes, this can be correct in principle.

DISNEY : Now that you have shown us the raw data, I am much less convinced that rotation curves tell us galaxies are transparent. From your photos it's quite clear that, because dust-discs are so narrow, that if the galaxy is not exactly edge-on, to within a degree or so, then you will see in across the top of the disc, and then of course you will see the Hα from way in.

BOSMA : As I showed, NGC 4013 is exactly edge-on, and the outer parts are transparent. Likewise, the small galaxies are very near or exactly edge-on and are transparent. But for NGC 4217, which is not exactly edge-on, the situation is as you say. The rotation curves certainly tell us that galaxies are more transparent than what you would think if you believed some of the earlier claims.

VALENTIJN : Your results are summarized as : optical thick inner regions in Sb's, semi-transparent in "large Sc's", transparent in "small Sc's". This would match my results reasonably when "large Sc's" correspond to high surface brightness galaxies and "small" to the fainter ones. Do you think that is the case ? (larger = brighter, smaller is fainter surface brightness) ?

BOSMA : Indeed it seems that we are converging towards similar answers, and yes, to a certain extent you can say that smaller galaxies can have lower surface brightnesses. However, a closer look might leave some scope for disagreement.

HUIZINGA : One of your conclusions is that Sb galaxies are opaque inside 1 - 2 scalelenghts. Is this when seen edge-on ? If so, does this mean that when seen face-on these galaxies are transparent almost to their centres ?

BOSMA : I only look at edge-on and nearly edge-on galaxies, and my results pertain to those. The method does not give an answer for the face-on optical depth. However, for NGC 4217, which is not entirely edge-on, I can do the asymmetry test described at this conference by Byun and by Knapen, and estimate the total amount of absorption.

SPECTROSCOPIC STUDIES OF THE DISK AND HALO OF M82

C D McKeith[1], A Greve[2], D Downes[2] and F Prada[1]

(1) Physics Department, Queen's University, Belfast, Northern Ireland
(2) Institut de Radio Astronomie Millimetrique, Grenoble, France

INTRODUCTION

We have been studying Galactic HII regions and nearby, spatially resolved galaxies using long slit CCD spectroscopy over wavelengths 3600 Å to 1.1 microns. Near-UV, visible, and near-IR spectra are recorded, sometimes simultaneously, in dual arm spectrographs. We obtain the Balmer decrement, and ratios of Balmer/Paschen series members below Paschen gamma, as well as ratios of other IR and near-UV lines, like S[II] 10300/4078, that are emitted from a common upper level.

As part of this programme, we have made several studies with high spectral and spatial resolution from near-UV to near IR wavelengths of the kinematics along the major and minor axes of the galaxy M82. We report here some conclusions from our spectra of the disk and halo of M82.

OBSERVATIONS

Observations of M82 were made in 1990 and 1992 with the 4.2 m WHT telescope and the dual arm ISIS spectrograph at the Observatorio del Roque de los Muchachos, La Palma, Spain. The slit length was 4' and spectral resolving power was about 3000 over the wavelength ranges 3600-4600 Å, 6000-7000 Å, and 8300-9100 Å. Emission line centres were determined from Gaussian fits to an accuracy of 20 km/s. The observed lines are the permitted Balmer and Paschen series below Pa 9 to the Balmer limit, the forbidden lines of NII, SII, SIII, and OII, and the stellar absorption lines of the CaII infrared triplet around 8500 Å. Details of the ISIS spectrograph configuration and data reduction are given in McKeith et al. (1993, 1994).

WAVELENGTH DEPENDENCE OF THE POSITION-VELOCITY CURVES

The data provide new insights into the rotational kinematics and the velocity distribution of the blowout along the minor axis of M82. Figure 1 (from McKeith et al 1993) shows the position-velocity diagram for some of the prominent spectral lines. The Paschen lines and the S[III] 9069 Å and OI 8446 Å lines trace HII regions, while the CaII absorption at 8500 Å originates in stars later than A-F, unlike the complex

325

J. I. Davies and D. Burstein (eds.), The Opacity of Spiral Disks, 325–328.

mixture of interstellar and stellar absorption in the near-UV CaII H and K absorption lines. The CaII IR triplet thus traces uniquely the kinematics of the stellar component to large radii along the major axis of M82.

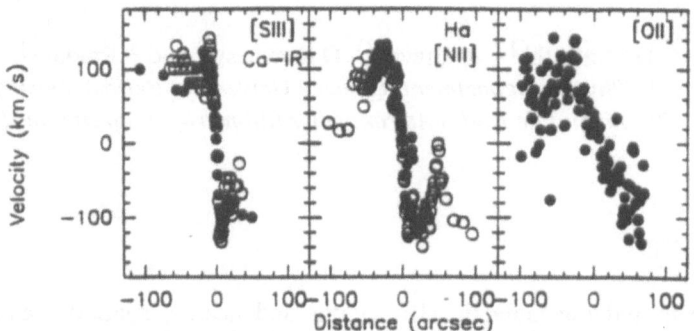

Fig 1   Position-velocity diagrams of M82's central region in different spectral lines (from McKeith et al 1993).

The most significant feature of the position-velocity curves of Fig 1 is the steepening of the velocity gradient in the central region with observing wavelength, increasing from 2.2 km/s per arcsec for the near-UV line of O[II] 3708 Å to 14.7 km/s per arcsec for the near-IR lines of S[III], Pa 10 and OI. The velocity gradient 14.7 km/s per arcsec at IR wavelengths is close to the 19.2 km/s per arcsec derived from the 12.8 micron NeII line by Beck et al (see Table 1 of McKeith et al, 1993). This is interpreted as an opacity effect: the nuclear region of M82 is optically thick at H alpha.

The gradients we measure in the near-IR lines also disagree with the gradients of 6-9 km/s obtained in older radio maps of the 1980's in HI and CO. Although the radio lines do not suffer extinction by dust, the apparent velocity gradients derived from the radio maps do vary with angular resolution. Most of the radio beams used prior to 1992 were larger than the aperture widths in the optical spectroscopy. In contrast to the older maps, more recent radio data, such as the HCN map of M82 (Brouillet & Schilke 1993; and private communication), made with the IRAM interferometer with a 2" beam, similar to our WHT observations, show good agreement with our S[III] 9069 Å position-velocity curve while, of course totally disagreeing with our O[II] 3708 Å curve in Fig 1, which cannot be tracing the central region.

It is obvious from Fig 1 that the IR wavelengths probe more deeply the centre of M82 than the visible and near-UV; the longer the wavelength of observation, the smaller the region in which one sees an apparent solid-body rotation curve. In the near-IR lines of S[III], Pa 10 and CaII, this central region with a linear velocity gradient extends 10" along the major axis on either side of the nucleus. At a radius of 10" from the nucleus, the linear curve turns over, at a value of 120 km/s, the tangential velocity of the nuclear disk or bar.

Moreover, at these longer wavelengths, the identical position-velocity curve in the emission lines of S[III] and Pa 10 and the CaII IR triplet absorption lines indicate a common origin in the central disk and common kinematics for the gas and the stars. Because of their higher opacity, the visible and near-UV lines of OII 3727 Å and

Balmer absorption lines come from regions at greater distances from the centre of M82, and hence give a lower apparent velocity gradient.

Following Bosma et al, (1992), we modelled this change in apparent velocity gradient with wavelength with a corresponding radial variation of opacity. We used a model for an edge-on galaxy with exponential distributions along the major and minor axes, that were different from the stars, the gas, and the dust. For an assumed rotation curve, we used the Galactic extinction function to predict the observed position-radial velocity curve at U, B, V, R, I, or Br gamma, NeII, S[III], H alpha, and O[II]. Figure 2 shows the result: the "true" position velocity curve is flattened out by the extinction, yielding a wavelength-dependent family of "apparent" position-velocity curves that agree with our observations. Because of the increasing opacity, the curves turn over and become flat at increasing radii with a decreasing wavelength. It is rather obvious that the dust in M82 is opaque at H alpha, causing the apparent position-velocity curve to deviate significantly from the "true" one.

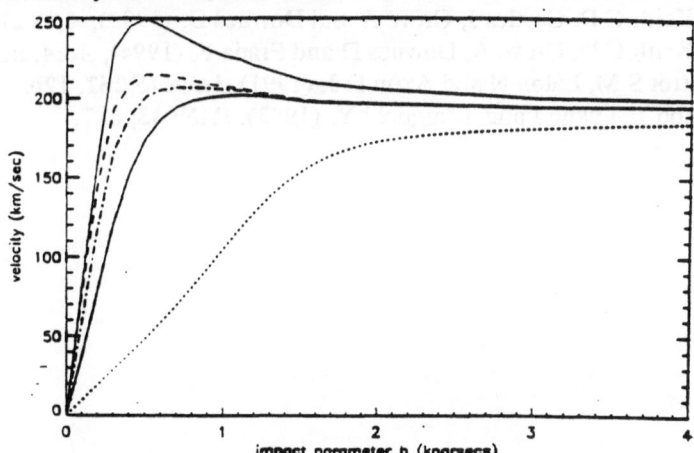

Fig 2 Model of the effect of extinction on the derived rotation curve of M82. The uppermost curve is the assumed true rotation curve. The 4 other curves represent the rotation curve as it would be derived from observations of O II, H alpha, S III and Ne II.

EMISSION IN THE HALO OF M82

We also took spectra along the minor axis of M82 to study the well-known split emission lines in the halo filaments (Axon & Taylor 1978; Amirkhanyan et al 1982, 1985, Bland & Tully 1999, Heckman et al 1990). In contrast to the split emission lines, stellar CaII IR triplet lines appear close to the 200 km/s systemic velocity of the galaxy with no velocity splitting or velocity gradient along the minor axis.

In McKeith et al (1994), we interpret these results according to the superwind blowout models for starburst galaxies (eg, Chevalier & Clegg 1985, Heckman et al 1990, Yokoo et al 1993).

Only a minor fraction of the emission from the halo is due to scattered radiation. It is clear that the larger-scale pattern of polarization vectors with radial symmetry observed by Scarrot et al (1991) represents scattering by dust which is in quiescent gas at much greater distances from the starburst zone than the regions we have studied in our spectroscopic observations.

REFERENCES

1. Amirkhanyan A S, Gagen-Torn V A, Reshetnikov V P, (1982), *Astrofizica* **18**, 31 and 1985, *Astrophizica* **22**, 239.
2. Axon D J and Taylor K, (1978), *Nature* **272**, 37.
3. Bland J and Tully R B, (1988), *Nature* **334**, 43.
4. Bosma A, Byun Y, Freeman K C, Athanassoula E, (1992), *ApJ* **400**, L21.
5. Brouillet N, Schilke P, (1993), *A&A* **277**, 381.
6. Chevalier R A, Clegg A W, (1985), *Nature* **317**, 44.
7. Heckman T M, Armus L, Miley G K, 1990), *ApJS* **74**, 833.
8. McKeith C D, Castles J, Greve A and Downes D, (1993), *A&S* **272**, 98.
9. McKeith C D, Greve A, Downes D and Prada F, (1994), *A&A*, in press.
10. Scarrot S M, Eaton N and Axon D J, (1991), *MNRAS* **252**, 12p.
11. Yokoo T, Fukue J and Taniguchi Y, (1993), *PASJ* **45**, 687.

The intimate conversation

DISK ORIGIN AND EVOLUTION

JOSEPH SILK
Departments of Astronomy and Physics, and Center for Particle Astrophysics
University of California, Berkeley, California 94720

1. Introduction

The theme of this workshop is the nature of galactic disks, and in particular, their opacity. Since my role on the program as a theorist and cosmologist is unique and unchallenged, I propose to use this minority power to describe recent ideas on three topics that are crucial to our understanding of galactic disks. I will first describe how galaxy halos may provide a reservoir of cold gas clouds, useful both for maintaining star formation activity in disks and for a possible opacity source when highly inclined disks are viewed at several disk scale lengths. I will then discuss the star formation rate in disks, and develop a model wherein the gas component can self-regulate itself via feedback from star formation. This leads to a derivation of the Tully-Fisher relation that is independent of cosmological initial conditions and depends only on the star formation prescription.

2. Historical Overview

Disk formation has a long and venerable history. The following remarks come largely from a recent review, where many references to the literature can be found (Silk and Wyse 1993). A dark halo is the first prerequisite, both to stabilize against global nonaxisymmetric instabilities and to provide an inert background against which protogalactic gas can torque and spin as it contracts. The initial angular momentum of a uniformly rotating sphere projects, if angular momentum is conserved as the sphere collapses into a disk, to give an exponential disk profile with flat rotation curve. How the initial angular momentum conspires to have the required uniform distribution is not understood, although the dimensionless spin parameter λ generated by tidal torques with nearby protogalaxies is found to have a

J. I. Davies and D. Burstein (eds.), The Opacity of Spiral Disks, 331–343.

mean value of 0.07, with a large dispersion. The collapse factor of the disk major axis is equal to λ. Centrifugal balance determines the final gas disk extent. Whether specific angular momentum is conserved is also unclear, although a more general argument suggests that baryon self-gravity first becomes important after collapse of a uniform sphere by a factor equal to the initial ratio of dark to baryon matter. Coincidentally, the calculated λ_i is approximately equal, within a factor of 2, to the ratio of disk to halo matter in luminous spirals. All of this stems from analytic theory.

Numerical simulations paint a far more confusing picture. Disks lose an inordinate amount of angular momentum to the dark matter, and excessive baryon accretion occurs. What is certainly true is that dominance of gas self-gravity is a necessary condition in order for stars to form. The forming disk may develop an exponential profile more generically by viscous effects, enhanced by gas motions induced once massive stars begin forming. Disk star formation occurs slowly, relative to the mean dynamical time. Energy input from forming stars is essential to counter the baryon "catastrophe," a consequence of overly efficient cooling. I shall show below that feedback effects from star formation result in an inefficiency factor ϵ, such that the star formation time-scale is of order $\epsilon\Omega^{-1}$, where $\epsilon \sim 0.01$. It is the star formation efficiency and rate that is crucial to deriving the detailed properties of present-day disks.

However, self-gravitating disks are unstable, and halos provide the necessary degree of stabilization against bar formation. Spheroids also have a stabilizing role, especially for early type, bulge-dominated spirals. In addition to their stabilizing influence, the dynamically hot components of spirals can provide a continuing source of gas. For spheroids, this occurs by accumulation of stellar ejecta in the disk. Early addition of enriched gas must have occurred in order to account for the metallicity distribution of disk stars. For late-type spirals, the observed spheroids are too low in mass to account for the initial yield of the disk. Infall from some other source is necessary. It has also been argued that gas infall is necessary to maintain the observed star formation rate in spirals. Any gas accretion which is occurring today onto the inner Milky Way disk must be at a modest rate, $\lesssim 0.1 M_\odot yr^{-1}$, to avoid conflict with observations of the diffuse soft x-ray background and of high velocity HI.

A natural solution to these and other problems would arise if the outer galaxy halo were to consist of dense cold molecular clouds. I shall commence by arguing that this rather radical possibility can by no means be excluded, either observationally or even in terms of a physical model.

3. Cold Gas Globules in Galaxy Halos

Why should any rational astronomer seriously consider the idea that massive dark halos might be largely gaseous? The following arguments have been presented by Pfenniger *et al.* (1994), who consider a disk-like extension of the dark halo consisting of a fractal distribution of dense, cold molecular cloudlets, and by Gerhard and Silk (1994), who develop a model of a spheroidal halo containing cold, dense gas globules. To begin, there is of course the cosmic coincidence that the amount of baryon dark matter (BDM) in the universe inferred from primordial nucleosynthesis ($\Omega_B = 0.0125h^{-2} > 0.025$ if $h < 0.75$) is similar to the mass fraction in dark halos, equivalent to $\Omega_{halo} \approx 0.03$ (for a halo mass-to-blue luminosity ratio of $50h$). Halos are an attractive solution for hiding the baryonic matter because of their spatial proximity to, and association with, the luminous baryon content of the universe. If the gas fraction in galaxy clusters and groups were representative of the universal baryon fraction, then intergalactic gas would be another reservoir for BDM. In galaxies outside of clusters, any substantial mass of gas is likely to be accreted into halos around galaxies unless there is very substantial intergalactic gas heating. Indeed, intergalactic hydrogen clouds, such as the damped Lyman alpha clouds, are most likely precursors to galactic disks that account for a considerable gas fraction at high redshift, comparable to that seen in nearby disks.

If halos indeed are baryonic, gas clouds seem the most appealing form for the BDM as a consequence of the MACHO and EROS searches for compact objects. These microlensing experiments are sensitive over the mass range $10^{-6} - 1M_\odot$, and have hitherto reported a halo candidate event rate towards the LMC that is apparently below the predicted value if the halo consists of MACHOs, while at the same time detecting an abundance of microlensing events towards the galactic bulge, where our disk and bulge are being probed. The strongest argument for halos consisting of hitherto undetected gas comes from studying the most extended flat rotation curves, as measured in terms of disk scale length, r_d. Consider one of the best examples, DDO 154, where the HI rotation curve is flat to $\sim 15r_d$. The HI column density, if the HI is assumed to be optically thin, decreases as r^{-1} over 15 scale lengths. This means that the HI surface density is proportional to, but only 10 percent of, the projected dark matter density. The stellar disk provides an exponentially declining contribution. Clearly, one would have a candidate for the halo dark matter if for some reason 90 percent of the gas was being missed. Moreover, the fact that one has a component of ordinary matter, HI, that tracks the dark matter over so wide a range of scale suggests that rather than invent exotic nonbaryonic matter to account for the rotation curve, a form of baryonic matter that is associated with the gas, but dark, is a possible resolution.

What could the BDM be? Again, if it is coextensive with the observed HI the BDM must necessarily be in a disk. One could plausibly imagine that the BDM is in a thick disk or flattened halo, having undergone some, but less, dissipation than the diffuse HI. There are few constraints on halo flattening. Flaring of the HI disk is consistent with flattening of up to E6, and minor axis halo kinematics studied in polar ring galaxies actually favour a degree of oblate flattening of around E6. Evidence for a flattened, low surface brightness halo around an edge-on disk galaxy has recently been reported (Sackett *et al.* 1994). Oblate halos are expected only if the halo dark matter is dissipative BDM: CDM produces generically oblate halos.

The cold clouds are strongly constrained by considerations of collisions, cooling and evaporation (Gerhard and Silk 1994). Typical parameters are as follows: masses $\gtrsim 10 M_{\odot}$, densities $\sim 10^7 \text{cm}^{-3}$, radii $\sim 0.02 \text{pc}$, temperatures $\sim 5 \text{K}$, and disk surface covering factor of about one percent. Such clouds are most likely enriched, to ~ 0.1 of the solar metallicity. There are two reasons for believing this to be the case. One needs pre-enrichment to this level prior to the formation of the disk. In galaxy groups, metallicity of this magnitude is measured in the intergalactic gas. It is likely that the gas in halos has a similar origin. One might expect that galaxy mergers, ram pressure stripping and evaporation would remove some of the halo gas to supply and enrich the intergalactic medium.

One can also make a case for the halo gas to contain grains, with about 10 percent of the interstellar ratio of dust to gas. My logic is based on the fact that intracluster gas absorbs soft x-rays from cooling flow cores. Searches for cold gas have been unsuccessful. It is primarily oxygen that absorbs the soft x-rays, and much of this element is most likely locked up in grains, if the absorbing matter is in the intracluster gas. Again, if the gas origin is similar, associated with the earliest stages of galaxy formation, one would expect grains in halo gas. Zaritsky (these proceedings) finds evidence for dust in halos. It is of interest to note that the large iron and oxygen abundances measured in clusters require an origin early in galactic evolution, before the disks formed, since standard chemical evolution of the observed disks would have produced insufficient heavy elements that could have been ejected via winds at earlier epochs. The intracluster gas has a heavy element abundance characteristic of Type II supernovae, and probably requires an early generation of massive stars that formed at the same time as the spheroids formed.

The relevance of dust in the cold halo clouds is the following. The dust enables molecules to form, and the clouds would be expected to be $\lesssim 10 \text{K}$ in temperature. If the dust grains are smaller than in the local interstellar medium, then dust extinction would be significant, especially if the halo is highly flattened and disks are viewed at high inclinations, within $\sim 10\,\mathrm{kpc}$, and possibly detectable to $\sim 100\,\mathrm{kpc}$. Optical depths, $\tau_B \sim 1$ magnitude, could be produced at a galactocentric distance of $\sim 10\,\mathrm{kpc}$ in massive galaxies. In these proceedings, Giovanelli presents evidence for mean extinctions of this order for luminous spirals. The surface covering fraction rises from 0.01 at $100\,\mathrm{kpc}$ to 0.1 at $10\,\mathrm{kpc}$, and would amount to $\tau_B \sim 0.1$ at $100\,\mathrm{kpc}$.

4. Disk Self-regulation

4.1. STAR FORMATION RATE

Halo clouds will dribble into the outer disk, as occasional cloud collisions and orbital decay must occur. The disk most likely grew inside-out, and indeed would still be growing as gas is accreted. Angular momentum conservation guarantees that it is the outermost disk that is presently accreting gas. The rare HII regions and CO clouds discovered at large radii, in the most extreme case at 28 kpc from the center of our galaxy (de Geus *et al.* 1993), may be examples of agglomerations of interacting cold halo clouds that have recently initiated star formation. Certainly one is well below the gravitational instability threshold of the Milky Way disk, which terminates at $\sim 13\mathrm{kpc}$.

Kennicutt (1989) has demonstrated that the star formation rate in disks appears to be suppressed below a critical surface density of gas, Σ_{cr}, that is close to the value given by the Toomre criterion for gravitational instability, derived for a cold gas disk by Goldreich and Lynden-Bell,

$$Q_g = \frac{\kappa v_g}{\pi G \Sigma_g} \equiv \frac{\Sigma_{cr}}{\Sigma_g}.$$

Here, κ is the epicyclic frequency, equal to $2^{1/2}\Omega$ on the flat portion of the rotation curve, v_g is the velocity dispersion of the gas and Σg is the total surface density of cold gas (HI and H_2). One may write $Q = \delta Q_g$, where the factor δ includes the stellar contribution to disk stability:

$$\delta^{-1} = 1 + \frac{\Sigma_* v_g}{\Sigma_g v_*}.$$

The maximum growth rate for the instability can be written as

$$\tau^{-1} \approx \kappa (1 - Q^2)^{1/2} Q^{-1}.$$

This leads to the following expression for disk star formation rate

$$SFR = \epsilon \Sigma_g \tau^{-1},$$

where ϵ is a parameter that represents the inefficiency of star formation. The large-scale disk instability controls the rate at which molecular clouds form and subsequently fragment into stars. Buried in the physics that determines ϵ is feedback from star formation that controls the destruction of molecular clouds. With a simple model for ϵ (discussed below), the star formation rate can be integrated to provide a plausible explanation of the gradients in disk star formation rate, metallicity distribution of both gas and stars, and gas fraction (Wang and Silk 1994).

4.2. STAR FORMATION EFFICIENCY

Only about 1 percent of a molecular cloud complex is converted into stars. Globally, this corresponds to the fact that the galactic star formation rate per unit surface area of the disk is equal to about 1 percent of the total gas surface density divided by the dynamical time, and evaluated, say, at the disk scale length, or for that matter, in the solar neighborhood. Evidently, $\epsilon \approx 0.01$.

I shall argue that self-regulation by feedback associated with the deaths of massive stars provides the keys to understanding why ϵ is so small, and more significantly, a relatively robust parameter. Supernova remnants as well as expanding HII regions and OB winds are responsible for stirring up gas motions and provide a source of momentum that is replenishes the velocity dispersion v_g, which also is driven by the development of gravitational instabilities in the molecular gas. At the same time, the molecular clouds dissipate energy by collisions. My conjecture is that a self-regulating balance arises, because $Q \propto v_g$ and $SFR \propto 1/Q \propto 1/v_g$, but however v_g is proportional to the momentum injected by, say, supernovae, and therefore in turn to the SFR. The details of the momentum coupling are likely to be complex, but a crude estimate (Silk 1992) suggests that

$$v_g \approx P_{SN} \cdot SFR \cdot \tau_d \cdot \Sigma_g^{-1},$$

where the dynamical time-scale, $\tau_d \approx \Omega^{-1}$, determines the rate at which gas clouds collide and reshuffle their momentum, and the specific momentum injected into the interstellar medium by supernovae is given by

$$P_{SN} = \frac{2E_{SN}}{V_c m_{SN}} = 1000 E_{51}^{13/14} m_{250}^{-1} n^{-1/2} \ \text{km s}^{-1}.$$

Here $E_{SN} \equiv 10^{51} E_{51}$ ergs is the initial kinetic energy of a supernova remnant, m_{SN} is the mass in stars formed per supernova, and

$$v_c = 413 E_{51}^{1/14} Z^{-3/14} n^{1/2} \ \text{km s}^{-1}$$

is the "cooling" velocity at which a spherical blast wave enters the approximately momentum-conserving snowplow phase in a uniform medium of density n and metallicity Z relative to solar. Insertion of the previous expression for the SFR into the estimate of v_g then leads to our derivation of star formation efficiency,

$$\epsilon \approx \frac{v_g^2 \Omega}{G \Sigma_g P_{SN}} = \frac{\Sigma_d}{\Sigma_g} \frac{2\pi v_g^2}{v_{rot} P_{SN}} \approx 0.01.$$

The efficiency parameter self-regulates to be about 1 percent: there is only an implicit dependence on star formation via m_{SN}, which depends on the IMF. For a power-law IMF, $dN/dM \propto M^{-1-x}$, and for $M \gtrsim M_L$, one has

$$m_{SN} \approx \frac{x}{x-1} \left(\frac{8M_\odot}{M_L}\right)^x M_L \approx 250 M_\odot \cdot \left(\frac{0.1M_\odot}{M_L}\right)^{x-1}.$$

Stars more massive than $8M_\odot$ are assumed to become Type II supernovae. If there is an equal number of Type I supernovae, as in our galaxy, then m_{SN} should be reduced by a factor of 2. The dependence on M_L is weak for a Salpeter IMF ($\mu = 1.35$), and is likely to reduce m_{SN} at most by a factor of a few if the IMF is truncated at $M_L \sim 3M_\odot$. The solar neighborhood IMF peaks at $M_L \sim 0.3M_\odot$.

Most interestingly, $\epsilon \propto v_g^2$ in a gas-rich disk ($\Sigma_g \sim \Sigma_d$). This provides a possible mechanism for understanding starbursts that are triggered by tidal interactions and mergers of galaxies. The gas velocity dispersion increases by up to an order of magnitude in a merger: hence the star formation rate will respond dramatically as ϵ increases from of order 1 percent to of order 100 percent star forming efficiency.

4.3. TULLY-FISHER REVISITED

I shall assume that the current star formation rate may be identified with the disk luminosity. This is certainly a good approximation in the blue: even in the red and near infrared, relatively short-lived red supergiants contribute thirty percent of the light in a constant star formation rate model (Bruzual and Charlot 1993). At 2μ, the old stellar contribution, from red giants, is only 30 percent of the total light in this model; at 1μ, the main sequence also contributes, and old stars make up about 20 percent of the light. At $5000A$, red giants only produce 5 percent of the light while the main sequence contributes 60 percent of the light, and red giants about 30 percent; there are mostly massive stars.

A global generalization of the disk star formation rate per unit area to the total star formation rate $(SFR)_{tot}$ enables us to write the luminosity of the galaxy in terms of $(SFR)_{tot}$ by introducing an appropriate conversion factor,

$$\alpha \equiv L/(SFR)_{tot}.$$

Of course, α depends on the bandwidth within which L is measured. The derived star formation rate prescription now gives

$$L = \epsilon\alpha M_g\Omega Q^{-1}.$$

If the condition of virial equilibrium $v_{rot}^2 = GM/R$ is applied, one obtains

$$L = \frac{\epsilon\alpha v_{rot}^3 f}{GQ_g}.$$

This is the Tully-Fisher relation. The function f is defined by

$$f = \left(1 + \frac{\Sigma_* \, v_g}{\Sigma_g \, v_*}\right)\frac{\Sigma_g}{\Sigma_* + \Sigma_g},$$

and varies between 1 for a gas-rich disk and $v_g/v_* \sim 0.1$ for a gas-poor disk. In other words,

$$(SFR)_{tot} = \left(\frac{\epsilon f}{Q_g}\right)\frac{v_{rot}^3}{G}.$$

We have already estimated the magnitudes of ϵ and f, and self-regulation requires $Q_g \sim 1$. Hence, inserting numerical values,

$$(SFR)_{tot} = 3.8 \left(v_{rot}/200\mathrm{kms}^{-1}\right)^3 (f/0.2)(\epsilon/0.1)Q_g^{-1}\mathrm{M}_\odot\mathrm{yr}^{-1}.$$

This results in the observed normalization to star formation rate for our own galaxy.

The conversion between star formation rate and luminosity, in the appropriate band, must depend on internal dust content, and is correspondingly uncertain. However even without attempting to refine the disk models, one has a Tully-Fisher relation that approximately matches the observed slope and normalization, near, for example, the R, I and H bands, where the reported slopes (Burstein, these proceedings) vary from 2.5 to 4.1, to the extent that a single power-law index can fit the data.

4.4. MASS-TO-LIGHT RATIO

One can also rewrite the above expression for L in the form

$$\frac{M}{L} = \frac{L^{1/6}}{\Sigma_L^{1/2}}G^{-1/3}\left(\frac{Q_g(SFR)_{tot}}{\epsilon f L}\right)^{2/3}$$

This demonstrates that if the surface brightness Σ_L is constant for disks, then the mass-to-light ratio should increase weakly with luminosity, as $L^{1/6}$. This is consistent with the observation that indeed M/L does increase towards early, more luminous Hubble types. There is a large overlap between Hubble types, but the correlation is apparent if one compares extreme types, for example Sa's and Sd's. This places spirals in the so-called fundamental plane, which for elliptical galaxies and bulges defines a correlation that is equivalent to $M/L \propto L^{0.2}$ (Bender, Burstein and Faber 1992). One might expect similar correlations to arise for the time-integrated star formation rate, and to therefore equally apply to single burst models of ellipticals.

5. Discussion

If dark halos are gaseous, they will surely be disrupted as galaxies fall into clusters. One might expect to see some impact of this on disk morphology: disks are unstable to global non-axisymmetric instabilities in the absence of dark halos. There are so few spirals in the cores of nearby clusters that this implication is not easily tested. However in a high redshift cluster recently studied with HST (Dressler *et al.* 1994), there is a much larger fraction of star-forming galaxies, many of which appear to be distorted disks. If fragility of halos were to explain this phenomenon, then a halo model of cold clouds would provide a simple explanation.

I have shown that the Tully-Fisher relation is likely to be independent of cosmological initial conditions. The derivation relies on self-regulation of disk star formation. Even dark halos are irrelevant, other than indirectly insofar as they may contribute to the disk rotational velocity. In fact, within 1 or 2 disk scale lengths, the rotation curve is dominated by disk matter in the form of ordinary stars.

References

Bender, R., Burstein, D. and Faber, S.M. (1992), *Astrophys. J.* **399**, 462-477.
Bruzual & Charlot, S. (1993), *Astrophys. J.* **405**, 538-553.
Dressler, A., Oemler, A., Sparks, W. B. and Lucas, R. A. (1994), preprint.
Gerhard, O. and Silk, J. (1994), preprint.
de Geus, E. J., Vogel, S. N., Digel, S. W. and Gruendl, R. A. (1993), *Astrophys. J.* **413**, L97-L100.
Kennicutt, R. (1989), *Astrophys. J.* **344**, 685-703.
Pfenniger, D., Combes, F. and Martinet, L. (1994), *Astron. Astrophys.* **285** 79-93.
Sackett, P. D, Morrison, H. L., Harding, P. and Boroson, T. A. (1994), *Nature* **370**, 441-443.
Silk, J. (1992), *Aust. J. Phys.* **45**, 437-450.
Silk, J. and Wyse, R. F. G. (1993), *Phys. Rep.* **231**, 293–365
Wang, B, & Silk, J. (1994) *Astrophys. J.* **427**, 759-769.

340

Question
Giovanelli

Since these objects must have dynamical properties similar to old halo stars and possibly be coeval with those, and since the latter are very metal poor, then: who made the metals at solar abundance that are necessary to keep the clouds cool?

Answer
Silk

The clouds can stay reasonably cold at abundances as low as one-thousandth of solar. Early massive star formation might well have occurred during the formation of the dark hole, without much formation of associated low mass stars. Examples of this are believed by some to occur in starbursts, and the process of structure formation plausibly involved similar phenomena induced by mergers. Only a small mass fraction forming massive stars would suffice to provide the initial metals.

Question
Giovanelli

You mentioned that one way of detecting the postulated population of cold clouds would be that of looking at the central parts of clusters of galaxies in their x-ray profiles. How would one discriminate this population from a cooling flow?

Answer
Silk

High spectral resolution data from ASCA can probably distinguish self absorption from the underlying cooling flow. In one or two extreme cases, the self-absorption is strong below 1 keV, and these examples are unambiguous even at lower spectral resolution.

Question
A N Witt

In estimating the optical depths attributed to the dark clouds in the halo, what ratio for gas/dust did you have to assume?

Answer
Silk

The gas/dust ratio could be anywhere from near solar at one extreme, to one-thousandth or even less, where a lower bound is required to allow molecule formation.

Question
Valentijn

Triggered by the 1990 opacity results, Gonzales and myself developed some models explaining flat rotation curves by the addition of compact molecular clouds in the disks and without the need for dark matter in the halo.

What is your motivation to assume the molecular clouds in the halo? (By the way, Lequeux's measurements are in the disk).

Answer
Silk

I would expect the cold clouds to be in an extremely flattened halo. The reasons are because of the physics, as there is likely to be some dissipation of orbital energy with collisions, and because of constraints, e.g. for modelling of polar ring galaxies, that probe the halo distribution along the galaxy minor axis.

Question
Wang

These compact Molecular Clouds are exposed to a UV radiation field. Won't you see them glowing at high latitude? Are there any limits on it?

Answer
Silk

The diffuse radiation field at high galactic latitude will not produce much ionisation of the cold gas globules. However traversal of the nearest giant HII regions by the rapidly moving globules will generate ionisation sheaths with emission measure contrast of unity and arc-second scale.

Question
Greenberg

If the hydrogen in the 10^{12} $M_\odot$ gas clouds in the halo had heavy elements shouldn't we be able to detect absorption lines <u>sometime</u>, <u>somewhere</u>?

Answer
Silk

High resolution observations of quasars with the Hubble telescope provide the best bet for seeing the clouds in our halo in absorption by metal lines, although the low surface covering factor means that at least 100 lines of sight would have to be studied.

Question
Huizinga

1. The temperatures you predict, 10 K, are significantly higher than those of Pfenniger, Combes & Martinet, 3 K. Since this may be the difference between being able to detect them and not, how hard is your 10 K estimate ?

2. Spirals are accompanied by dwarf galaxies, orbiting through the molecular clouds. What are the consequences of this for your model?

Answer
Silk

1. With Romff and des Forets, I have been computing the chemistry and temperature of dense halo clouds. The diffuse x-ray background provides a minimal heat source. I do not believe that the clouds will cool below about 5 K over the density range that allows them to be stable pressure-supported objects of stellar mass, that is in the mass range 0.1 to 100 $M_\odot$.

2. Some gas accretion into the dwarfs may occur, as clouds occasionally lose sufficient orbital energy, by collisions with ambient gas clouds in the dwarfs, or by dynamical friction, or by mutual collisions, to be trapped.

Question
Braun

The cold cloudlets were designed to be invisible. Are you aware of some plausible production mechanism which would make it easier to "believe" in the?

Answer
Silk

Pfenniger has argued that one finds evidence for a similar fractal structure culminating in the formation of very low mass cold molecular clumps resulting from turbulent cascade, even in nearby molecular clouds.

Question
White

Cold absorption in cooling flows is found in clusters with large accretion rates and such clusters tend to be very relaxed, as clusters go. Therefore, these clusters are the least likely to have their cold absorption due to infalling cloudlets associated with galaxy halos- the cloudlets will not survive the infall. If you attribute the absorption to galaxies in the outskirts which are falling in for the first time, such galaxies should present a screen-like geometry, not very centrally concentrated, since their distances are <u>much</u> greater than the 100-200 kpc scale over which cold absorption is seen. Finally, isn't the galaxy distribution too grainy (particularly in the outer parts) to provide the clean correspondence observed between the absorption region and the cooling flow centre?

Answer
Silk

One finds that up to 100 galaxy halos, from outside several cluster are radii, may be seen in projection against the cluster core. Optical depth of unity is possible, but how well this serves as an absorbing screen against the core has not yet been calculated in detail

The flight

THE LUMINOSITY AND OPACITY OF GALAXIES

BOQI WANG
Department of Physics and Astronomy
The Johns Hopkins University
Baltimore, MD 21218

1. Introduction

The question of whether the dust in disk galaxies is optically thick or thin remains a puzzle in astronomy. We wish to point out that the goal of determining categorically whether disks are optically thick or thin may not be feasible if there is a dependence of the optical depth on some galaxy internal characteristics. It is perhaps natural to expect such a dependence, since it is the internal properties of galaxies (e.g., star formation) that determine the properties of the dust, such as its formation and survival and the total column density through the disk (hence the optical depth). Here we shall investigate the dependence of the optical depth on one of the most important characteristics of galaxies, the luminosity; a more detailed description of this work can be found in Wang and Heckman (1994). That the luminosity should make a difference may not be surprising; luminous galaxies, having more heavy elements per unit mass to spare (Skillman 1989; Zaritsky et al. 1994), may have a higher dust abundance. Indeed, direct examinations of galaxy images reveal that there may be more dust with increasing luminosities (van den Bergh & Pierce 1991).

2. Flux Ratio and Luminosity

We investigate the correlation of the optical depth of dust in galactic disks with galaxy luminosity. We examine a sample of galaxies with measured far-ultraviolet ($\lambda \sim 2000$ Å) fluxes (Donas et al. 1987, DDLMH), and compile the corresponding fluxes in the far-infrared ($\lambda \sim 40 - 120\mu$m) as measured by IRAS. The UV to FIR flux ratio is found to decrease rapidly with increasing FIR (or FIR+UV) luminosities (figure 1). As both the UV and FIR radiation originate mostly from the young stellar population in late

345

J. I. Davies and D. Burstein (eds.), The Opacity of Spiral Disks, 345–347.
© 1995 *Kluwer Academic Publishers.*

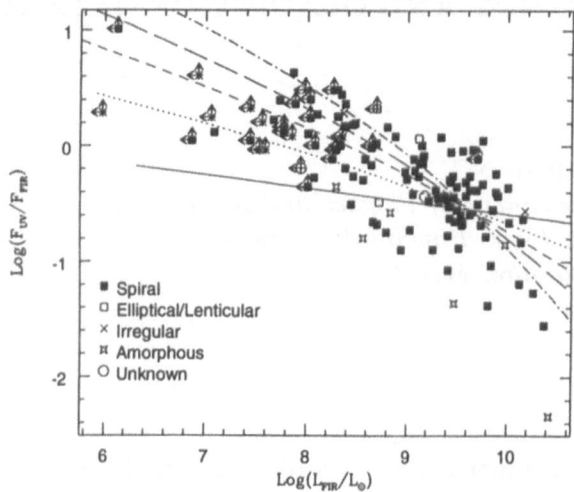

Figure 1. The ratio of the UV flux (λF_λ at $\lambda = 2000$Å) to the FIR flux (total flux from 42.5 to 122.5 μm) as a function of the FIR luminosity for the DDLMH sample. Those with only upper limits on the FIR fluxes are marked with the arrows. Model calculations are shown as the curves, with various values of β in equation (12): $\beta = 0.1$ (solid curve), 0.3 (dotted), 0.5 (dashed curve), 0.7 (long-dashed), and 1.0 (dash-dotted). We have taken the extinction optical depth $\tau_{B,1} = 0.8$ at $L_{B,1} \simeq 6 \times 10^9$ L$_\odot \simeq L_*$ in the blue.

type galaxies, the UV to FIR flux ratio is a measure of the fraction of the light produced by young stars escaping from galaxy disks. Thus, the strong correlations above imply that the dust opacity increases with the luminosity of the young stellar population. We also find that the ratio of the UV to FIR flux decreases with increasing galaxy blue luminosity (a tracer of the intermediate-age stellar population) and with galaxy rotation speed (an indicator of galaxy mass). We supplement the UV sample of galaxies with an optically-selected sample, and find that the blue to FIR flux ratio declines with both increasing FIR luminosity and galaxy rotation speed. We then consider an FIR-selected sample of galaxies and again find that the ratio of blue and FIR fluxes declines with increasing FIR luminosity. Finally, we examine a sample of galaxies for which the Hβ/Hα flux ratios can be obtained, and find that the Hβ/Hα ratio, which also measures the extinction, decreases with the increasing FIR luminosity. Thus, in all cases we find evidence that the opacity of galactic disks increases with increasing galaxy luminosity and/or mass.

We model the absorption and emission of radiation by dust in galactic disks with a simple model of a uniform plane-parallel slab in which the dust that radiates in the IRAS band is heated exclusively by UV light from relatively nearby hot stars. We then find that the relation between the various flux ratios and the observed luminosities can be explained by the face-on extinction optical depth τ varying with the intrinsic luminosity as a power-law in the intrinsic UV luminosity: $\tau = \tau_1 (L/L_1)^\beta$. Comparisons of the models to the observations show that, expressed in the blue band, the total extinction optical depth is $\tau_{B,1} = 0.8 \pm 0.3$ at the fiducial observed blue luminosity of a Schechter L_* galaxy and $\beta = 0.5 \pm 0.2$. Thus our models imply that most galaxies are optically thin to dust in the blue, and only those with $L \gg L_*$ are opaque. The increase in optical depth with luminosity can be attributed to the increase in both galaxy metallicity mentioned above and galaxy surface-mass-density (Wang 1991) with increasing luminosity. However, these model parameters fail to reproduce the observed properties of the most FIR-luminous galaxies in the FIR-selected sample, presumably because our assumed model is inappropriate to these extreme starburst systems (disturbed galaxies with strongly centrally-concentrated star-formation).

The inferred extinction-luminosity relation have significant implications in observations of galaxies. First, the relation should lead to a sharp drop in the numbers of galaxies observed at very large blue luminosities, providing a possible explanation for at least part of the difference in the shapes of the high luminosity portions of the galaxy luminosity functions observed in the blue and FIR. Second, as the total mass of dust in the interstellar medium of star-forming galaxies should have been higher in the past, the selectively greater extinction suffered by high-luminosity galaxies may also account in part for the "missing" high-luminosity galaxies in recent galaxy redshift surveys at intermediate and high redshifts. Third, an increase in opacity with galaxy luminosity would also result in a steepening of the slope of the Tully-Fisher relation from optical to near-infrared wavelengths.

References

Donas, J., Deharveng, J.M, Laget, M., Milliard, B., & Huguenin, 1987, A&A, 180, 12 (DDLMH).

Skillman, E.D., Kennicutt, R.J., Jr., & Hodge, P.W. 1989,

van den Bergh, S., & Pierce, M.J. 1991, ApJ, 364, 444.

Wang, B. 1991, ApJ, 383, L37.

Wang, B. & Heckman, T. 1994, submitted to ApJ.

Zaritsky, D., Kennicutt, R.C.Jr., & Huchra, J.P. 1994, ApJ, 420, 87.

The awful beer

DUST OBSCURATION IN STARBURST GALAXIES

D. CALZETTI

Space Telescope Science Institute
3700 San Martin Drive, Baltimore, MD 21218, USA

Abstract.

Understanding the role of dust obscuration in external galaxies is a key to our understanding of the intrinsic physical quantities, such as stellar content, ISM composition, nebular parameters, etc., which characterize the galaxies, and, ultimately, determine their evolution.

The analysis of the ultraviolet (IUE) and optical spectra of 39 star-forming galaxies shows a correlation between the UV spectral index β and the color excess $E(B - V)$. The spectral index β is derived from the fit $F(\lambda) \propto \lambda^{\beta}$ in the range 1250-2600 Å. The color excess $E(B - V)$ is derived from the Balmer decrement. The correlation between the two quantities reveals that the spectral shape of the UV spectrum in starburst galaxies is mainly sensitive to the dust obscuration.

The proportionality between β and $E(B - V)$ is used to derive an effective obscuration curve for starburst galaxies. The main characteristic of this curve is the absence of the 2200 Å dust feature.

1. Introduction

A reliable description of the intrinsic distribution of light in external galaxies, and its interpretation in terms of physical and chemical properties, such as stellar populations, Initial Mass Function, age, and metallicity, depends on how effectively the observed distribution of light can be corrected for the effects of dust obscuration.

For a long time galaxies have been thought to be essentially transparent systems. However, results from the IRAS satellite and other recent studies (Giovanelli et al. 1994; Valentijn 1990; see also various contributions in

J. I. Davies and D. Burstein (eds.), The Opacity of Spiral Disks, 349–354.

these Proceedings) indicate that at least the late Hubble type galaxies may contain a substantial amount of dust.

The difficulty of measuring dust effects on the radiation is in their elusiveness. Dust reddens the spectra and mimicks the aging of the stellar population. The lack of narrow features in the extinction curve makes a quantitative measurement of the obscuration difficult. Spectral and photometric information from external galaxies usually derive from extended regions, so the effect of dust on the emerging radiation is sensitive to the relative distribution of dust, gas and stars across the observed region. The differential dust extinction at different wavelengths may depend on global parameters such as the galaxy metallicity (Lequeux 1988).

Despite the difficulties listed above, correlations linking dust obscuration parameters have been found in the case of starburst galaxies (Calzetti, Kinney & Storchi-Bergmann 1994). Such correlations help to gain insight into the problem of dust extinction in external galaxies.

2. The Analysis.

The analysis is performed on a sample of 39 starburst and Blue Compact galaxies, for which both UV and optical spectra are available. The UV spectra come from the compilation of IUE spectra by Kinney et al. (1993) and span the wavelength range 1200-3200 Å. The optical spectra cover the wavelength range 3200-8000 Å and were observed in an IUE-matched aperture at the KPNO and CTIO observing facilities. The galaxies cover the metallicity range $8.3 \leq 12 + \log(O/H) \leq 9.2$.

Following standard techniques, the color excess E(B-V) is derived from the emission line ratio $H\alpha/H\beta$, adopting a Seaton (1979) extinction curve, after correction for the underlying stellar absorption. The color excess is a measure of the attenuation of the light due to the dust surrounding the ionizing regions.

The spectral index β is used here to characterize the dust obscuration at UV wavelengths. The spectral index is derived by fitting the observed UV spectra according to the power law $F(\lambda) \propto \lambda^{\beta}$, in the wavelength range 1250-2600 Å.

Calzetti et al. (1994) find that β is correlated with E(B-V) (see Fig. 1). The correlation between β and E(B-V) indicates that the shape of the UV spectra is mainly sensitive to the presence of dust obscuration. Variations in the spectral shape due to changes in the stellar populations contributing to the UV emission may still be present, e.g. they may be responsible for the spread in β at constant E(B-V), but they are not at the origin of the β-vs.-E(B-V) correlation. Another characteristic of the UV spectra in the sample is the weakness or absence of the 2175 Å bump. The absence of the

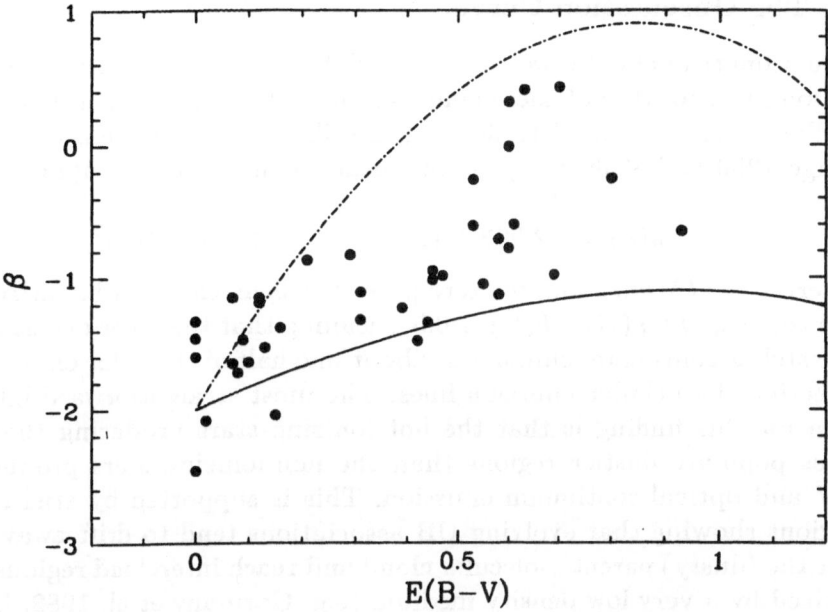

Figure 1. The observed β-vs.-E(B-V) correlation for the 39 starburst galaxies (filled dots) is compared with the correlation calculated for a model of clumpy dust in front of the emitting regions, using a Milky Way-type (continuous line) and a Large Magellanic Cloud-type (dot-dashed line) extinction curves.

bump indicates that the data do not support a model for the distribution of dust consisting of a foreground screen with a Milky Way-type extinction curve.

The predictions from various models for the geometrical distribution of stars and dust have been compared with the data. The most promising model for solving the problem of the 2175 Å bump is the screen with clumpy dust. This model predicts a fairly weak 2175 Å bump. However, it also predicts a trend β-vs.E(B-V) which is incompatible with the data (see the curves in Fig. 1).

The reason for the absence of the 2175 Å bump and for the correlation between β and E(B-V) is possibly a more complex (and more realistic) geometry for the stars-dust-gas distribution than those discussed here and in Calzetti et al. (1994). However, we cannot exclude that some of the peculiarities of the dust obscuration in starburst galaxies are due to a chemical composition of the dust grains different from that of the Milky Way.

3. The Obscuration Curve.

The proportionality between β and E(B-V) can be used to derive an obscuration curve for the galaxies in our sample in the same fashion that reddened stellar spectra are used to derive the Milky Way extinction curve. In the range $1250 \leq \lambda \leq 8000$ Å, the obscuration curve $k(x) = A(x)/E(B-V)$ is:

$$k(x) = -2.523 + 1.766x - 0.232x^2 + 0.013x^3, \qquad (1)$$

where $x = 1/\lambda(\mu m)$, and the zero point has been chosen arbitrarily $k(V) = 0$. From Eq.(1), $k(B) - k(V) \simeq 0.5$, meaning that the color excess affecting the stellar continuum emission is about one half of the color excess E(B-V) affecting the nebular emission lines. The most straightforward interpretation for this finding is that the hot ionizing stars producing the nebular lines populate dustier regions than the non-ionizing stars producing the UV and optical continuum emission. This is supported by studies of HII regions showing that evolving OB associations tend to drift away or blow out the (dusty) parent molecular cloud and reach intercloud regions characterized by a very low density medium (e.g. Garmany et al. 1982; Leisawitz & Hauser 1988).

The obscuration curve given in Eq.(1) is a reasonable tool to correct the emerging light from starburst galaxies for intrinsic reddening (e.g., Calzetti et al., 1995). Applications of Eq.(1) cover a variety of cases, including intermediate and high redshift galaxies, where the spectral and/or photometric information usually derive from the entire galaxy or a major fraction of it.

References

Calzetti, D., Kinney, A.L., & Storchi-Bergmann, T. (1994), *ApJ*, Vol. no. **429**, p. 582

Calzetti, D., Bohlin, R.C., Kinney, A.L., & Storchi-Bergmann,.T. (1995), *ApJ*, to appear on the April 1st issue

Garmany, C.D., Conti, P.S., & Chiosi, C. (1982), *ApJ*, Vol. no. **263**, p. 777

Giovanelli, R., Haynes, M.P., Salzer, J.J., Wegner, G., Da Costa, L.N., & Freudling, W. (1994), *AJ*, Vol. no. **107**, p. 2036

Kinney, A.L., Bohlin, R.C., Calzetti, D., Panagia, N., & Wyse, R.F.G. (1993), *ApJS*, Vol. no. **86**, p. 5

Leisawitz, D. & Hauser, M.G. (1988), *ApJ*, Vol. no. **332**, p. 954

Lequeux, J. (1988), in *Dust in the Universe*, eds. M.E. Bailey & D.A. Williams, Cambridge University Press, Cambridge, p. 449.

Seaton, M.J. (1979), *MNRAS*, Vol. no. **187**, p. 73P

Valentijn, E.A. (1990), *Nature*, Vol. no. **346**, p. 153

Question
Burstein

In looking at your poster, I see differences in the UV spectra of the two objects shown, from 2500-3000Å: the lower fluxed object shows absorption lines at MgI, MgII and Fe, while the higher fluxed object shows only MgII emission. Please comment.

Answer
Calzetti

Both the objects shown have Balmer emission lines (H_α / H_β) compatible with $\tau_B = 0$. The differences in the spectra are due to differences in the intrinsic stellar population. These differences are shown in the plot β vs τ_B as a spread in the values of β for constant values of τ_B. For this reason, we have merged spectra of galaxies with similar values of τ_B to create "template spectra" of "average stellar populations".

Question
Prada

Did you take into account the population gradient in your results?

Answer
Calzetti

Models show that a burst of star formation or continuum star formation give relatively constant values of β, till there are ionising photons produced. The ratio H_α / H_β is relatively constant for a long range of electron temperatures and densities; at least, the intrinsic variations in the ratio are well within our uncertainties. We correct for the underlying stellar absorption using H_γ, where possible.

Question
Whitworth

How did you normalise your extinction curve?

Answer
Calzetti

I normalise it using the decrement of the emission lines H_α and H_β. The resulting observation curve for the stellar continuum is grey (more grey than any known stellar extinction curve) and it gives $B - V \sim 0.5$.

If we normalise the observation curve for the stellar continuum in the standard way $(B - V = 1)$, the curve will look similar to the Milky Way stellar extinction curve (except for the 2200Å hump). However, the colour excess to be used in this case is about 1/2 of the colour excess you derive form the H_α / H_β emission line ratio.

Question
Adolf Witt

For clarification, it should be stated that the Balmer decrement referred to in your paper concerns the decrement of H_α / H_β in emission lines measured in HII regions, not the Balmer decrement in the continuum of the stars.

Answer
Calzetti

Thank you for the clarification.

POLARIMETRY OF DUSTY EDGE-ON GALAXIES

RAMON D. WOLSTENCROFT
Royal Observatory
Edinburgh EH9 3HJ,Scotland.

AND

S.M.SCARROTT
Physics Department
University of Durham, Durham DH1 3LE, UK.

1. Introduction

The controversy concerning the location and amount of dust in the plane of spiral galaxies will be resolved with a precise understanding on the relative geometry of the stars and dust and on the various contributions of scattered and absorbed starlight, none of which are well known *a priori*.

Light from galaxies can be polarized by scattering and/or absorption: in this paper we present imaging polarimetry data for dusty galaxies and will show that absorption is the dominant polarizing process and hence that magnetic fields are prevalent in such systems.

2. Examples of Dusty Galaxies

We discuss data for two edge-on galaxies which have conspicuous dust lanes.

2.1. M104 - THE SOMBRERO GALAXY

The orientation of the polarization in the central areas is parallel to the dust lane (Scarrott, Rolph & Semple 1990). Jura(1982) and Matsamura & Seki(1989) have proposed that the polarization arises from the scattering of starlight by dust grains but their models are unable to predict the orientations and degrees of polarization that are observed.

The wavelength dependence of polarization in the central parts of the dust lane follows the Serkowski relationship with a value of λ_{max} similar to the one found for the supernova behind the dust lane of Cen A

J. I. Davies and D. Burstein (eds.), The Opacity of Spiral Disks, 355–357.
© 1995 *Kluwer Academic Publishers.*

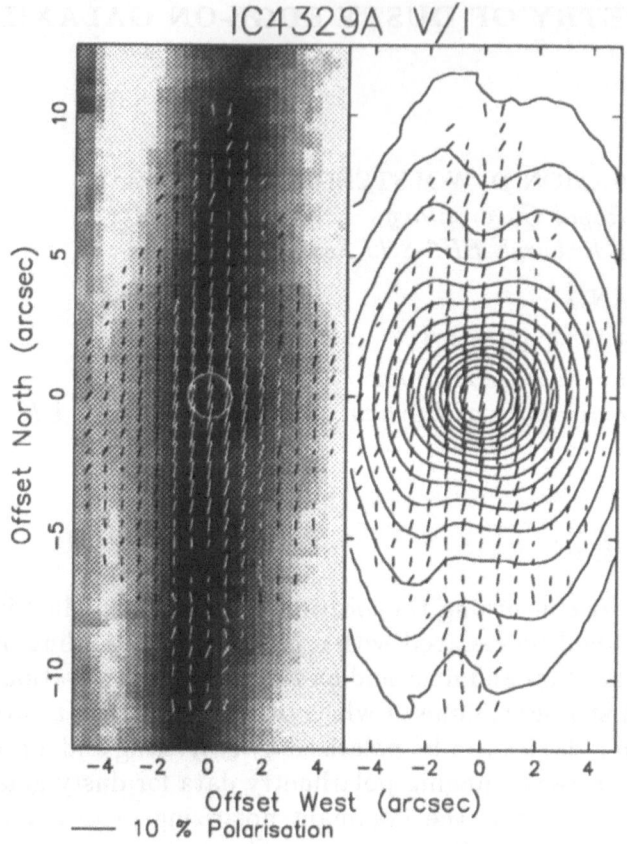

Figure 1. *Left* A greyscale map of the V-I colour with superposed polarization map for IC4329A. The darker regions are redder. *Right* V waveband polarization and intensity contour maps. The PA of the dust lane is 38° with the SW side at the top of the figure.

(0.43μm - Hough et al 1987). The Serkowski-like behaviour of the polarization in the dust lanes of these galaxies suggests that the polarization arises from dichroic extinction by magnetically aligned grains which in turn implies the presence of uniform galactic-scale magnetic fields in the regions of the dust lanes.

2.2. IC4329A

This spiral is a strong FIR & X-ray source with a heavily reddened Seyfert 1 nucleus. Our colour and polarimetry data are shown in figure 1. The V-I colour shows the dust lane to be much redder than elsewhere with a local maximum at the nucleus. The polarization is oriented parallel to the dust lane throughout the non-nuclear region of the galaxy and correlates with the reddening as defined by V-I. The similarity between the present polarization pattern and those in M104 and Cen A suggests that the mechanism giving rise to polarization in the dust lane of IC4329A is also dichroic extinction.

Maximum polarization (5.8% in V) occurs at the nucleus with a PA of 4° E of N relative to that in the dust lane. This offset is significant and it suggests to us that there is another polarizing mechanism in operation in addition to the dichroism found in the dust lane. The wavelength dependence of polarization in the nucleus of IC4329A (this paper, Martin et al,1982 & Brindle et al,1990) is very similar to that found by Webb et al(1993) for extreme AGN's such as 1Zw1 and Mkn 290. This leads us to believe that there is a synchrotron component to the polarization of the nucleus itself in addition to the dichroism induced by the intervening dust lane.

3. Conclusion

The polarization in the central regions of Cen A, M104 and IC4329a is due primarily to dichroic extinction by aligned grains which implies that there are uniform galactic-scale magnetic fields present in such systems. In the case of IC4329A there is an additional synchrotron component intrinsic to its active nucleus. A detailed discussion of the work on IC4329A is given in Wolstencroft, Scarrott & Scarrott (1994) where the relation between dust content and polarization in galaxies is also explored.

4. References

Brindle, C., Hough, J.H., Bailey, J.A., Axon, D.J., Ward, M.J., Sparks, W.B. & McLean, I.S. 1990, *Mon.Not.R.astr.Soc.* **244**, 577.

Hough, J.H., Bailey, J.A., Rouse, M.F. & Whittet, D.C.B. 1987, *Mon.Not.R.astr.Soc.* **227**, 1P.

Jura, M. 1982, *Astrophys.J.* **258**, 59.

Martin, P.G., Stockman, H.S., Angel, J.R.P., Maza, J. & Beaver, E.A. 1982, *Astrophys.J.* **255**, 65.

Matsumura, M. & Seki, M. 1989, *Astron.Astrophys.* **209**, 8.

Scarrott, S.M., Rolph, C.D. & Semple, D.P. 1990, *In Galactic and Intergalactic Magnetic Fields*, R. Beck et al eds, Kluwer, Dordrecht, 245.

Webb, W., Malkan, M., Schmidt,G. & Impey, C. 1993, *Astrophys.J.* **419**, 494.

Wolstencroft, R.D., Scarrott, R.M.J. & Scarrott, S.M. 1994, In prep.

The food

H II REGIONS AND EXTINCTION
IN THE SPIRAL GALAXY M83

S. RYDER
Dept of Physics & Astronomy, University of Alabama

A. HUNGERFORD
Dept of Physics, Western Washington University

M. DOPITA AND K. FREEMAN
Mt Stromlo & Siding Spring Observatories

Y.-I. BYUN
Institute for Astronomy, University of Hawaii

M. EHLE AND R. BECK
Max-Planck-Institut für Radioastronomie

R. HAYNES
Australia Telescope National Facility

AND

R. SUTHERLAND
Joint Institute for Laboratory Astrophysics

1. Introduction

M83 appears to be quite actively forming stars, having hosted a number of historical supernovae, and its radio continuum emission is dominated by non-thermal sources (Ondrechen 1985; Cowan & Branch 1985; Sukumar, Klein, & Gräve 1987). By comparing the optical Hα and radio continuum flux ratios of individual H II regions with those expected from theory, we wish to test whether or not M83 possesses a well-defined radial extinction gradient. Previous searches for such a gradient in M81 (Kaufman et al. 1987) and in M51 (van der Hulst et al. 1988) have been inconclusive, but the high H II region luminosities and significant dust content of M83 make it one of the best candidates for such a study. We present here the preliminary results of an analysis of this data.

J. I. Davies and D. Burstein (eds.), The Opacity of Spiral Disks, 359–361.

The Australia Telescope Compact Array (ATCA)[1] has been used in 3 of its highest resolution configurations to map M83 in both the 6 cm and the 13 cm continuum, so that non-thermal sources can be distinguished from H II regions. We produced a continuum–subtracted Hα image of M83 using narrow-band filters mounted in the beam of TAURUS II on the Anglo-Australian Telescope, and smoothed this image to the same resolution as the radio maps (6″). Aperture photometry of compact objects was carried out using the same aperture size and background annuli in each map.

2. Results

Of the 55 discrete sources identified on the 13 cm map, only 18 were found to have spectral indices ($S \propto \nu^{\alpha}$) in the range $\alpha = -0.1 \pm 0.3$. Many of the remaining sources have no 6 cm or Hα counterparts, and are often closely associated with the prominent dust lanes. This non-thermal emission most likely results from the compression of the large-scale magnetic fields with the passage of a spiral density-wave shock. For each of the 18 "thermal" sources, we have compared the ratios of 6 cm and 13 cm flux to Hα flux with those expected from theory (Caplan & Deharveng 1986), assuming a mean electron temperature of 6400 K (Dufour et al. 1980), in order to compute the mean equivalent extinction in the V band, A_V. Four of these objects (3 of which reside inside dust lanes) have no Hα counterpart, so only a lower limit to the extinction can be determined for these objects. Figure 1 shows a histogram of the derived extinctions.

3. Conclusions

From close inspection of these early results and comparison with previous studies, we draw the following (tentative) conclusions:

- The H II regions in M83 suffer at least 2 magnitudes of extinction, and on average, nearly 4 magnitudes of extinction in V. This is much greater than the mean extinction seen in either M81 (1.1 mag) or in M51 (2.2 mag). We point out however that previous studies tended to overlook thermal sources without Hα counterparts, thereby excluding the most heavily extinguished objects. While an error in flux calibration may account for a shift in the mean extinction, it could not explain the large and continuous *range* in A_V observed, right up to the point where the H II regions can no longer be seen optically.
- There is no well-defined radial extinction pattern, although a lack of pure thermal sources in the bar region limits our radial coverage.

[1]The ATCA is operated by CSIRO Australia Telescope National Facility.

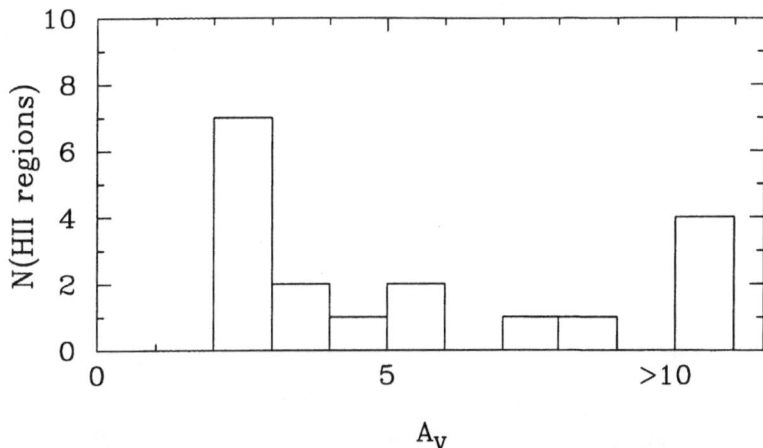

Figure 1. Histogram of H II region extinctions in M83

- If in fact only 5% of the Hα photons can escape the galaxy directly (as implied by our mean extinction), then this may go some way towards explaining the high FIR-to-blue light ratio (1.135; Rice et al. 1988), which is typical of rather more highly-inclined and more active galaxies.
- We plan to increase our sample of H II regions by adding more short-baseline observations, and perform a more effective separation of the thermal and non-thermal contributions to the radio fluxes.

We wish to acknowledge support from EPSCoR grant EHR-9108761 and NSF REU grant AST-9300413. We are grateful to the AAT and ATCA Time Allocation Committees for their support of this project.

4. References

Caplan, J., & Deharveng, L. 1986, A&A, 155, 297

Cowan, J. J., & Branch, D. 1985, ApJ, 293, 400

Dufour, R. J., Talbot, R. J., Jensen, E. B., & Shields, G. A. 1980, ApJ, 236, 119

Kaufman, M., Bash, F. N., Kennicutt, R. C., & Hodge, P. W. 1987, ApJ, 319, 61

Ondrechen, M. P. 1985, AJ, 90, 1474

Rice, W., Lonsdale, C. J., Soifer, B. T., Neugebauer, G., Kopan, E. L., Lloyd, L. A., de Jong, T., & Habing, H. J. 1988, ApJS, 68, 91

Sukumar, S., Klein, U., & Gräve, R. 1987, A&A, 184, 71

van der Hulst, J. M., Kennicutt, R. C., Crane, P. C., & Rots, A. H. 1988, A&A, 195, 38

A SEARCH FOR DUST IN GALACTIC HALOS

DENNIS ZARITSKY

Carnegie Observatories

813 Santa Barbara St., Pasadena, CA, 91101

1. Introduction

This meeting has summarized current understanding of dust within the optical disks of spiral galaxies, but does dust exist only within those disks? Because most of the mass of spiral galaxies is in their dark matter halos (cf. Zaritsky and White 1994), there may be a significant amount of dust in galactic halos even if only a minute fraction of that mass is in the form of dust. The question merits attention because the presence of dust would have serious repercussions on the study of galactic formation and evolution, dark matter, and the high redshift universe.

This paper outlines a recent search for dust in galactic halos, which is described in detail by Zaritsky (1994). The structure of the experiment is simple; to compare the colors of background galaxies observed through the halo of a foreground galaxy to the colors of background galaxies observed through unobscured reference fields.

2. The Observations

Observations of eight $21' \times 21'$ fields in B and I were made using a 2048^2 Tek CCD at the Las Campanas 40-inch telescope. Around each of two nearby galaxies, two inner fields (at about 60 kpc projected separation along the disk major axis) and two outer fields (at about 230 kpc projected separation) were observed. Each field was observed for between 4500 and 13500 seconds. Galaxies were identified using FOCAS (Jarvis and Tyson 1981) and a supplementary surface brightness criterium. Objects must be identified in both B and I to enter the catalog (note that this leads to a lower limit on reddening since highly reddened objects are excluded). The magnitude limit of the catalog is about I = 21 mag, although the I images

J. I. Davies and D. Burstein (eds.), The Opacity of Spiral Disks, 363–367.
© *1995 Kluwer Academic Publishers.*

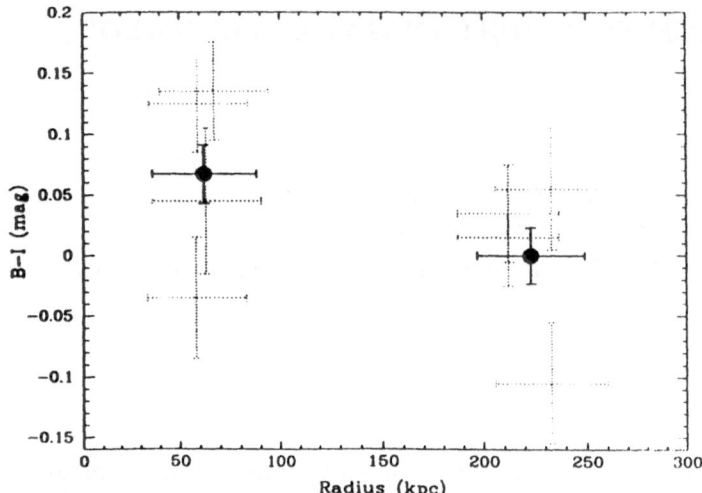

Figure 1. Differential reddening measurements for background galaxies vs. radius for all fields. The relative B-I color has been set to zero for the outer field mean. The dotted lines represent the measurements for each field, the solid lines represent the average of the values at either the inner or outer position.

themselves go much deeper (to about 24th mag). The I images were used for object classification.

Slight differences in the photometric zero point or local Galactic reddening can produce color offsets between fields. Therefore, the stellar color differences between fields, as measured using the technique described below, were set to zero for all images. Afterward, the relative colors of galaxies in the various fields were measured. The results for the four fields around each of the two galaxies (NGC 2835 and NGC 3521) are shown in Figure 1.

3. Analysis and Discussion

A robust estimator of color difference was constructed by using one frame as a reference from which to derive the "intrinsic" color distribution. Other fields were then compared by applying a range of color shifts and finding the shift for which χ^2 was minimized. The contribution to χ^2 from each object is the sum of Gaussians with exponents proportional to the square of the differences between its color and those of the reference objects within a small magnitude bin. This estimator proved to be more robust than either the mean or median color difference.

An important test of the data and color measuring procedure is to compare the mean reddening of the fields around NGC 2835 to that of the fields around NGC 3521. Because NGC 2835 is at a much lower Galactic latitude (18.5°) than NGC 3521 (52.8°), there should be differential reddening (estimated to be 0.18±0.05 mag by Burstein and Heiles (1984)). The measured value, 0.11±0.03, agrees with the predicted reddening difference, providing support to the contention that relatively small color difference can be measured accurately using these data.

As evident from Figure 1, there does appear to be a tendency of the inner fields to be *on average* redder than the outer fields. Since the color differences and the observational uncertainties are of the same order, not every inner field is redder than every outer field (although the reddest two are inner fields and the bluest one is an outer fields). Using the average reddening and uncertainties, the null hypothesis (that there is no excess reddening in the inner fields) can be discriminated against at the 98% confidence level.

The reddening difference in itself does not prove the existence of dust in galactic halos. It is possible that there are unaccounted for Galactic reddening differences or subtle variations in the intrinsic colors of background over large angular scales. The first is probably not a factor since the fields were chosen to have small 100μ flux differences and because the stellar colors were zeroed. The second is unlikely because the scale at which we know environment to affect galaxy colors (the scale of galaxy clusters) corresponds to a small fraction of a single image if the typical galaxy in the survey has a luminosity of L^*. The success of the test to recover the Galactic extinction also argues against this possibility. Nevertheless, before both of these alternatives can be completely ruled out and the results become unequivocal, additional observations over many other fields and galaxies are needed.

If confirmed, the presence of this dust has serious implications for the distribution of baryonic dark matter, for the evolution of galaxies, the processes that result in the presence of this material at such large radii, and for our view of the high redshift universe. The optical depths of dust calculated from these observations, assuming a standard extinction law, are quite similar to those postulated by Heisler and Ostriker (1986) to be necessary to obscure QSO's at $z > 2$. For sufficiently effective obscuration, this dust would have to be present in high redshift galaxies, but its presence in low redshift galaxies would indirectly support the conjecture.

References

Burstein, D., and Heiles, C. (1984) Reddening Estimates for Galaxies in the Second Reference Catalog and the Uppsala General Catalog *Ap. J. Supp.*, **54**, 33.

Heisler, J., and Ostriker, J.P. (1988) Models of the Quasar Population: II. The Effect of Dust Obscuration, *Ap. J*, **332**, 543.

Jarvis, J.F., and Tyson, J.A. (1981) FOCAS: Faint Object Classification and Analysis System, *A. J.*, **86**, 476.

Zaritsky, D. (1994) Preliminary Evidence for Dust in Galactic Halos, *A.J.*, in press.

Zaritsky, D., and White, S.D.M. (1994) The Massive Halos of Spiral Galaxies, *Ap. J.*, in press.

Question
Braun

You mentioned using foreground stars to test for systematic color errors and corrections. Could you tell me what the magnitude and accuracy of those corrections were?

Answer
Zaritsky

The magnitude of the corrections are similar to that of the results. However, because there are typically 10 times as many stars per frame as galaxies, the corrective color shifts are much better determined. Therefore, they contribute little to the final uncertainties.

Question
Ryder

As with many of the studies presented at this meeting, I am concerned that you may be biased towards the lowest reddenings/extinctions. For instance, if a galaxy was so heavily reddened that it only showed up ??? the I-band images, it would not be included. How many galaxies identified in the I-band images were not seen in the B-band images?

Answer
Zaritsky

It is possible that some galaxies are completely observed and not included in the sample. This would imply higher opt depth than inferred. Given that the current estimate is larger than expected, I hesitate to add to it, but you are correct. I have not done a counts analysis.

Question
Thronson

Yours calculated "dust halo" parameters are consistent with Silk's proposed "molecular halo" in that you may be detecting much less mass that he proposes, as you are presumably viewing between his clouds.

Answer
Zaritsky

The dust mass itself is low compared to the total mass. Simple estimates are $M_d = 10^8 M_O$ (as opposed to the total mass of spirals of $10^{12} M_O$). For standard gas/dust ratios the inferred gas mass is too low, but of course assumptions about gas/dust are complete guesses (as are any assumptions about their relative distributions).

Question
de Vaucouleurs

1.   First let me congratulate you for introducing a new, clever test for extinction in halos. I hope you can apply this test to another ~ 100 galaxy pairs.

2.   If the positive excess of reddening at 60 kpc you detect is confirmed, it would go a long way toward reconciling the low (~0) extinction in the B-H module based on stars at Z< 1kpc and the high extinction (E~ 0.05) in the RC2 model based on galaxies at 2> 1Mpc.

Answer
Zaritsky

1.   Thank you. I hope so too!

2.   That's a very intersting point that I had not considered.

The pose III

CONCLUDING THOUGHTS AND REFLECTIONS: DUST IN GALAXIES

HARLEY A. THRONSON, JR.

Wyoming Infrared Observatory, University of Wyoming

Abstract. Several themes ran through this conference. Above all, it is clear that dust plays an important role in significantly modifying stellar radiation. But, annoyingly, the dust affects light in such an ambiguous way that it is very difficult to determine whether or not it is present in a significant amount. Despite this, there were several techniques proposed to search for this elusive component of the ISM: surface brightness variations, absorption of background galaxies, far-infrared/sub-millimeter emission, and so on.

In this summary, I concentrate on some of the sociology of the meeting, identify a few topics which I believe should be given further attention, and pass on a handful of warnings about some of the most popular analytic tools discussed at this workshop.

Key words: Cardiff, Wales - Dust - Galaxies - Interstellar Medium

1. Dust in Galaxies: There is no Deader Hand than that of the Unchallenged Theory

Before I summarize those topics which I found most interesting at this meeting, I cannot let pass the opportunity to note how indebted we are to those iconoclasts — deliberate or otherwise — who have wrestled with the annoying problem of dust in galaxies. They have proposed annoying ideas, faced some severe criticism, ably defended their points of view, and succeeded in, at the very least, raising the awareness of the astronomical community to the central role that dust plays in our interpretation of galaxies. Many colleagues have contributed to the success of this conference —

J. I. Davies and D. Burstein (eds.), The Opacity of Spiral Disks, 369–374.
© *1995 Kluwer Academic Publishers.*

some by attending, some by not attending — but let me acknowledge a few key individuals.

Participants at this conference are indebted to Mike Disney, Jon Davies, and Steve Phillipps for reminding us very effectively in print several years ago how deviously very large amounts of dust can hide within even the most heavily studied objects in the Universe. [I cannot be the only one at this meeting who, in discussing these issues with colleagues, have easily sidestepped a lengthy description of radiative transfer in galaxies by simply muttering, "Oh, you know, just like Disney-Davies-and-Phillipps."] Of course, we are doubly indebted to Jon Davies for his work in organizing this conference.

More recently, Ed Valentijn has been a most effective advocate for optically-thick (at least in part) disk galaxies. It is a pleasure to acknowledge Ed for this work, although I do not think that he is correct. This is frustrating, as I think that he presented some effective arguments at this workshop and certainly has shown us data which begs for interpretation superior to that which I can give.

Similarly, Nick Devereux and co-workers have been arguing for years — in the face of good sense — that very young OB stars dominate the heating of the far-infrared-emitting dust in many disk systems. For the purpose of this meeting, I think that Nick's most useful contribution has been to continue to remind the audience about the significant amount of information about dust which is available at wavelengths beyond a few micrometers. Indeed, as Nick, Jon Davies, Rhodri Evans, and a handful of others pointed out at this meeting, it may be the coldest dust in a galaxy which is effectively absorbing the bulk of the stellar light. If so, this material must be searched for and analyzed in more depth than has been attempted so far.

Dave Burstein deserves recognition for so consistently keeping the conference moving in directions which he considers most interesting. Along with our hosts in Cardiff, he also contributed much to the organization of this meeting, for which we are all very grateful.

Finally, let me, as a descendent of Welsh miners, express my appreciation to my very distant cousins of a century ago who built this University in part via subscription, so that their children could be educated. In a small way, this fine workshop was a testimony to their dreams.

2. The Value of Mistakes: "Progress in an intellectually active field may be measured by the number of failures."

Some years ago, I was impressed by an article on economics which discussed interpretation of a variety of indicators of business activity. Most striking was the *positive* role of business failures — bankruptcies of various kinds — in a dynamic economy. Although painful for the individuals involved, large numbers of business failures may also be interpreted as the natural product of increasing numbers of new companies, new enterprises, and the eagerness of people to enter a vibrant market. Just so with the active areas of research presented here: quite a number of the advocates for different ideas will turn out to be wrong, but alternative ideas stimulate debate and a deeper examination of popular notions.

3. Missing Topics

One strong impression which I take away from this meeting is the wide range of research specialties which are impacted by the effects of dust: the colors and surface brightness distributions in galaxies, which have been at the center of much of the meeting, but also participants have noted the important role of dust in interstellar chemistry and gas heating, number counts of distant galaxies, and as a tracer of massive star formation.

Since it was not noted often during the meeting itself, I feel obliged to remind the readers of some of the instruments which will be available in the near future to investigate the distribution of solid state material in galaxies:

ESA's *Infrared Space Observatory* (ISO) is expected to be launched not long after these proceedings will appear. Not only will the dust thermal continuum be mapped, as some speakers noted, but there are a variety of solid state emission features throughout the mid- and parts of the far-infrared. These features should be useful to trace the location of the dust, as well as its chemistry, excitation, and relative abundance.

The Royal Observatory, Edinburgh is nearing completion of the *Sub-millimeter Common User Bolometric Array* (SCUBA), which will allow rapid, accurate, large area mapping of galaxy disks, in which the oft expected "cold dust component" may be studied. This material was invoked often throughout this meeting — somewhat too often to my thinking — as perhaps some of the material which produces the optically thick dusty disks. However, because it is so cold, it is very faint, even though it must also be luminous if it absorbs a significant fraction of a galaxy's general interstellar radiation field. Among many other important projects which SCUBA will undertake, the sub-millimeter array will be turned on disk galaxies in an effort to unambiguously identify the extent, location, and abundance of the most cold component of the solid state ISM.

James, Keel and White at this conference discussed the technique of using background light to search for the effects of dust in disks. This can be effective, say, where two galaxies overlap. A variant on this technique might be applied with the *Hubble Space Telescope* (HST) by mapping the reddening of individual stars distributed through a spiral disk. Unfortunately, my guess is that this technique [1] is very time consuming, [2] will have to be applied statistically, as the exact location of a individual target within a disk will be difficult to determine, and [3] may be limited to only the brighter stars, which will predominantly be young massive objects or else highly evolved stars, both of which are often associated with local dust.

Finally, I note that either of NASA's airborne observatories, the *Kuiper Airborne Observatory* (KAO) or, perhaps in 5 years, the *Stratospheric Observatory for Infrared Astronomy* (SOFIA), could make a major impact in the study of interstellar dust. Neither of these facilities has the sensitivity of a space-based mission, but airborne observatories have larger areal and wavelength coverage, higher angular resolution, and a wider variety of instruments.

Aside from several new facilities in the coming decade which will open, especially, the infrared to regular investigation, I would also like to point out at least two general observational techniques appropriate to this conference which should be pursued more vigorously. First, although it is depressingly difficult to interpret, the polarization of light carries at least some information about the presence or absence of dust.

I think much more promising is detailed study of grain emission features throughout much of the infrared. Of course, the half-dozen or so mid-infrared "PAH" features are starting to yield to persistent analysis, as they should, as these are the brightest spectral features in the ISM for many objects. However, limited observational and laboratory data has kept this important field from making the contributions that it might to conferences such as this one. Work to date indicates that grain features in emission and in absorption — from ices, silicates and silicon carbide, and a variety of hydrocarbons — possess information about abundances, size distribution, chemical composition, and excitation. This detailed information is essential for a complete picture of dust grains in galaxies.

4. A Beauty Myth: "Every event . . . every observable quantity . . . is ultimately the product solely of 'selection effects' "

Given the important role that appearance (that is, morphology) has played historically in astronomy, I was impressed that many of the results presented here did not appear to be closely related to galaxy type. At semi-regular intervals during the conference, there were requests to estimate various parameters of the dust — say, optical depth or abundance — for different morphological classifications (Sa, Sb, Sc . . .). However, no one at this conference to my knowledge has demonstrated that any important dust

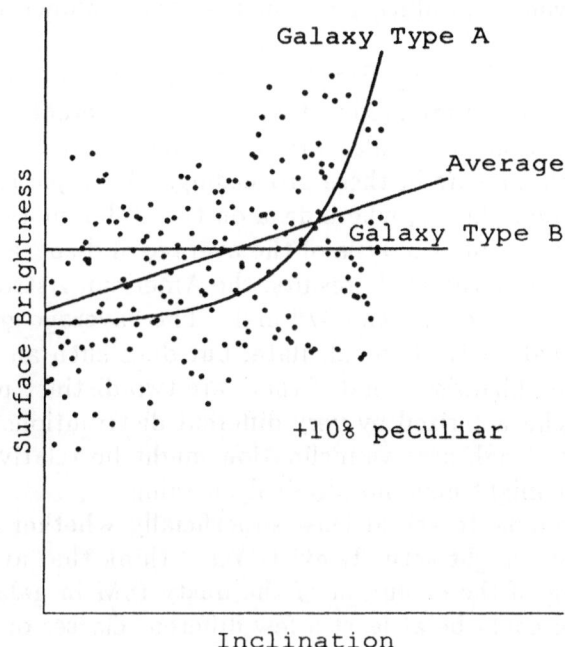

Figure 1. Large scatter can at least in part be the result of including more than a single populations. Here I show how two different models (plus a small amount of scatter due to peculiar systems) can produce a "surface brightness vs inclination" plot of the type shown elsewhere in these proceedings.

characteristic correlates well with Hubble type. I think that this was a pursuit which distracts many astronomers from more profitable tasks, although few of us argue against concentrating our interests on the most spectacular spirals simply because they are . . . well . . . intriguingly beautiful.

5. The Science is in the Scatter, not the Average

"Surface brightness vs inclination" tests were at the core of much of this meeting's discussions, where different models for the distribution of dust within disks predict different variations of a galaxy's visual surface brightness as viewed at different inclinations. Of course, as we cannot observe an individual galaxy at different orientations, large samples of disk galaxies are surveyed to produce plots such as those shown elsewhere in these proceedings.

As also pointed out by Ray White at the conference, a central danger of this technique is that mixing two or more different classes of galaxy in such a comparison might also produce a plot similar to the near-scatter diagram which appears in these proceedings. A simple example appears in the accompanying figure, where data on two different types of galaxies are combined in one plot. Analysis of the average of disparate populations can often lead to uninterpretable results: the American Astronomical Society is something like 1/5 female and 4/5 male. The "average gender" of the AAS is mathematically trivial to calculate, but does such an average make any physical sense? Likewise, what if there are two distinct populations of disk galaxy, each characterized by very different distributions of dust. A fit to a plot of surface brightness vs inclination might be relatively easy, but such an average fit might have no physical meaning.

It is reasonable to ask at least superficially whether or not such different populations might actually exist. Yes, I think that at this early stage of understanding of the evolution of the dusty ISM in galaxies, it is clear to me that there could be at least a few different classes of spiral galaxy with different large-scale distributions of dust. At least, until it is demonstrated otherwise, is it not safer to *assume* that there are multiple populations of spiral galaxies, each with a different distribution of dust in the disk? We have seen it elsewhere, but at this meeting perhaps the most compelling demonstration of two different general ISM types was the 21 cm HI maps by Bob Braun. He showed two very different distributions of neutral gas: a "grand design" of hydrogen arms and a "flocculent", which might not even be recognized as a spiral galaxy on the basis of the radio line emission alone. If we assume that the dust follows the atomic hydrogen at least approximately, then we might expect the flocculent structure to possess dust more widely distributed through the disk, while the grand design system could have dust more isolated to arm structure. The former might have an "optically thick" disk, while the latter would be "optically thin." This is merely hypothesis, but I consider it to be more plausible than the simple assumption presented at this conference that there is but a single type of dusty disk, viewed at a variety of angles throughout the local Universe.

Acknowledgements

Once again, I would like to thank the organizers of this meeting for their fine support — and patience. Some of this work was supported by NSF Grant AST 91 - 17096.

The best kept till last

LIST OF PARTICIPANTS

R. Arendt — Applied Research Corporation, Code 6853, NASA,/GSFC, Greenbelt, MD 20771, U.S.A

D. L Block — Department of Computational & Applied Mathematics, Witwatersrand University, P O Box 60, Wits 2050, South Africa

A. Boselli — Demirm, Observatoire de Paris, 61 Av.De L'observatoire, 75014 Paris, France

A. Bosma — Observatoire de Marseille, 2 Place Le Verrier, 13248 Marseille Cedex 04, France

L. Bottinelli — Radioastromie/Arpeges, Observatoire de Paris-Meudon, 5 Place Jules Janssen, 91195 Meudon, France

R. Braun — Department of Astronomy, NFRA, Postbus 2, 7990AA Dwingeloo, The Netherlands

A. Broeils — Center For Radiophysics and Space Research, Cornell University, Space Sciences Building, Ithaca NY 14053-6801, U.S.A

G. Bruzual — Landessternwarte - Konigstuhl, Heidelberg, 69117, Germany

D. Burstein — Department of Physics & Astronomy, Box 871504, Arizona State University, Tempe, AZ 85283 - 1504, U.S.A

Y. Byun — Institute for Astronomy, University of Hawaii, 2680 Woodlawn Drive, Honolulu, HI 96822, U.S.A

D. Calzetti — Space Telescope Science Institute, 3700 San Martin drive, Baltimore, MD 21218, U.S.A.

S. Courteau — Kitt Peak National Observatory, NOAO, 950 N. Cherry Avenue, P O Box 26732, Tucson, AZ 85726, U.S.A

B. Cunow — Astronomical Institute, University of Munster, Wilhelm-Klemm-Str. 10, D-48149 Munster, Germany

J. I. Davies — Department of Physics and Astronomy, University of Wales College of Cardiff, PO Box 913, Cardiff, CF2 3YB

G. de Vaucouleurs — Department of Astronomy, University of Texas, Austin. TX 78712, U.S.A.

N. Devereux — Department of Astronomy, New Mexico State University, Dept 4500/Box 30001, NM 88001, U.S.A

378

M. J. Disney — Dept. Physics and Astronomy, University of Wales College of Cardiff, PO Box 913, Cardiff, CF2 3YB

R. Evans — Department of Physics & Astronomy, Swarthmore College, Swarthmore, PA 19081, U.S.A

K. Freeman — Mt. Stromlo Observatory, Private Bag, Weston P.), Act 2611, Australia

R. Giovanelli — Department of Astronomy, Cornell University, Space Sciences Building, Ithaca NY 14853, U.S.A

L. Gouguenheim — Department of Radio Astronomy/Arpeges, Observatoire de Paris-Meudon, 5 Place Jules Janssen, 91195 Meudon, France

J. M. Greenberg — Laboratory of Astrophysics, University of Leiden, Niels Bohr weg 2, 2300 RA Leiden, Postbus 950Y, Netherlands.

P. Grosbøl — European Southern Observatory, Karl-Schwarzschild-Str.2, D-85748 Garching bei Munchen, Germany

R. de Grys — Kapteyn Astronomical Institute, P O Box 800, 9700 AV Groningen, The Netherlands

E. Huizinga — European Southern Observatory, Karl-Schwarzschild-Str.2, D-85748 Garching bei Munchen, Germany

P. James — Department of Astrophysics, Liverpool John Moores University, Byrom Street, Liverpool, L3 3AF

H. Jones — Dept. Physics and Astronomy, University of Wales College of Cardiff, PO Box 913, Cardiff, CF2 3YB

S. Jörsäter — Stockholm Observatory, S-13336 Saltsjobaden, Sweden

W. Keel — Department of Physics and Astronomy, University of Alabama, Box 870324, Tuscaloosa, AL 35487-0324, U.S.A

J. H. Knapen — Institute Astrofisica de Canarias, E-38200 La Laguna, Tenerife, Spain

N. Kylafis — Department of Physics, University of Crete, 71409 Heraklion, Crete, Greece

B. Madore — NASA Extragalactic Database, IPAC 100-22, Caltech, Pasadena CA91125, U.S.A

C. Mc Keith — Department of Physics, Queens University Belfast, Belfast, Northern Ireland, BT7 1NN

M. Näslund — Stockholm Observatory, S-13336 Saltsjobaden, Sweden

N. Neininger — IRAM, 300 Rue de la Piscine, 38406 St Martin d'Heres, France

R. Peletier — Kapteyn Astronomical Institute and La Palma Observatory, Apartado 321, 38700 Santa Cruz de la Palma, Tenerife, Spain

F. Prada — Instituto de Astrofisica de Canarias, E-38200, La Laguna, Tenerife, Spain

H. Rix — School of Natural Science, Institute for Advanced Study, Olden Lane, Princeton, NJ 08540, U.S.A

M. Roberts — National Radio Astronomy Observatory, Edgemont Road, Charlottesville, VA 22903, U.S.A

S. Ryder — Department of Physics and Astronomy, University of Alabama, Box 870324, Tuscaloosa, AL 35487-0324, U.S.A

J. Silk — Department of Astronomy, University of California, Berkeley, CA 94720, U.S.A

H. Thronson — Department of Physics, Wyoming Infared Observatory, University of Wyoming, Laramie WY 82071, U.S.A

M. Trewhella — Department of Physics and Astronomy, University of Wales College of Cardiff, PO Box 913, Cardiff, CF2 3YB

E. A. Valentijn — SRON-Space Research-Groningen, P O Box 800, 9700 AV Netherlands

W. Van Driel — Kiso Observatory, The University of Tokyo, Mitake-Mora, Kiso-Gun, Nagano-Ken, 397-01, Japan

B. Wang — Department of Physics and Astronomy, The Johns Hopkins University, Baltimore, MD21210 (Maryland) U.S.A

R. White — Department of Physics and Astronomy, University of Alabama, Box 870324, Tuscaloosa, AL 35487-0324, U.S.A

A. Witt — Department of Physics and Astronomy, The University of Toledo, 2801 W.Bancroft, Toledo, OH 43606, U.S.A

R. Wolstencroft — Royal Observatory, Blackford Hill, Edinburgh, Scotland, EH9 3HJ

D. Zaritsky — Carnegie Observatories, 813 Sant Barbara St., Pasadena, 91101, U.S.A